<u>Disclaimer</u>

The publisher of this book is by no way associated with the National Institute of Standards and Technology (NIST). The NIST did not publish this book. It was published by 50 page publications under the public domain license.

50 Page Publications.

Book Title: User's Guide for SRM 2494 and 2495: The MEMS 5-in-1, 2011 Edition

Book Author: Janet M. Cassard; Jon C. Geist; Theodore V. Vorburger; David T. Read; David G. Seiler

Book Abstract: The Microelectromechanical Systems (MEMS) 5-in-1 is a standard reference device sold as a NIST Standard Reference Material (SRM) that contains MEMS test structures on a test chip. The two SRM chips (2494 and 2495) provide for both dimensional and material property measurements. SRM 2494 was fabricated on a 1.5 um complementary metal oxide semiconductor (CMOS) process via MOSIS. Material properties of the composite oxide layer are reported on the SRM Certificates and described within this guide. SRM 2495 was fabricated using a polysilicon multi-user MEMS process with a backside etch. The material properties of the first polysilicon layer are reported on the SRM Certificates and described within this guide. The MEMS 5-in-1 contains MEMS test structures for use with five standard test methods on one test chip (from which its name is derived). The five standard test methods are for Young's modulus, step height, residual strain, strain gradient, and in-plane length measurements. The first two of these five standard test methods have been published through the Semiconductor Equipment and Materials International (SEMI) in November 2009. The remaining three standard test methods have been published through the American Society for Testing and Materials (ASTM) in December 2005. All five of these standard test methods include round robin precision and bias data. The Certificates accompanying these SRMs report eight properties. In addition to the five properties mentioned in the previous paragraph, residual stress, stress gradient, and thickness are also reported. The values for the first two of these properties are obtained from equations provided in the Young's modulus standard test method. The value for the third property (thickness) is obtained from step height measurements using the step height standard test method. Therefore, to determine the eight properties reported here, five standard test methods are used.

Citation: NIST SP - 260-179

Keyword: ASTM; cantilevers; fixed-fixed beams; interferometry; length measurements; MEMS; residual strain; residual stress; round robin; SEMI; SRM; step height measurements; strain gradient; stress gradient; test structures; thickness; vibrometry; Young's modulus measurements

NIST Special Publication 260-174, 2011 Ed.

Standard Reference Materials®

User's Guide for SRM 2494 and 2495:
The MEMS 5-in-1,
2011 Edition

Janet M. Cassard, Jon Geist,
Theodore V. Vorburger, David T. Read,
and David G. Seiler

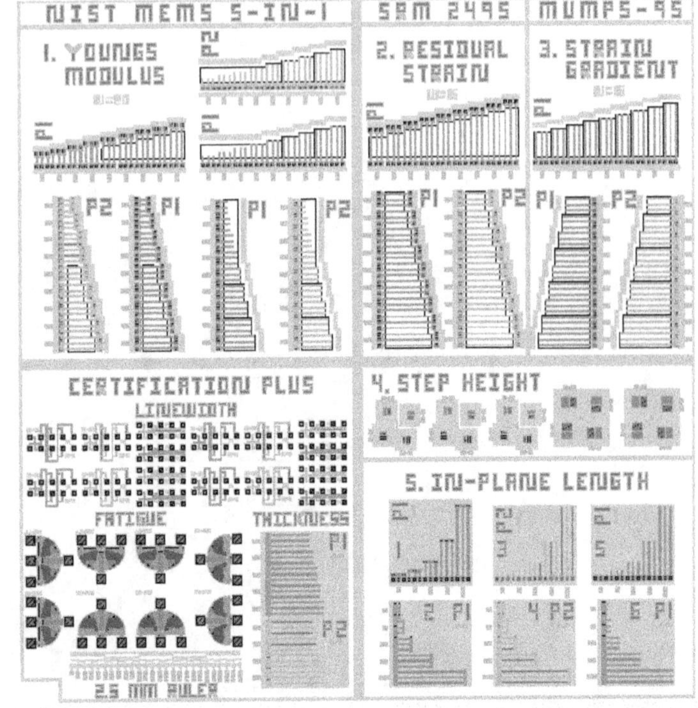

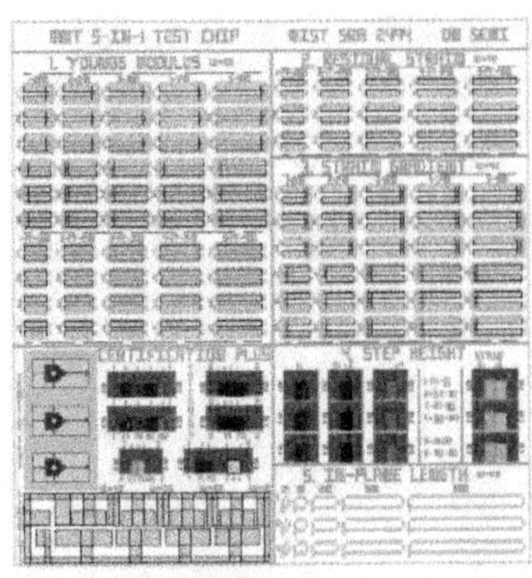

NIST National Institute of Standards and Technology • Technology Administration • U.S. Department of Commerce

NIST Special Publication 260-174, 2011 Ed.

Standard Reference Materials®
User's Guide for SRM 2494 and 2495:
The MEMS 5-in-1, 2011 Edition

Janet M. Cassard
Jon Geist
*Semiconductor Electronics Division
Physical Measurement Laboratory

Theodore V. Vorburger
Mechanical Metrology Division
Physical Measurement Laboratory

David T. Read
Materials Reliability Division
Material Measurement Laboratory

David G. Seiler*

National Institute of Standards and Technology
Gaithersburg, MD 20899

September 2011

U.S. Department of Commerce
Rebecca Blank, Secretary

National Institute of Standards and Technology
Patrick Gallagher, Director

National Institute of Standards and Technology Special Publication 260-174, 2011 Ed.
Natl. Inst. Stand. Technol. Spec. Publ. 260-174, 2011 Ed., 244 pages (September 2011)
CODEN: NSPUE2

Definition of Terms

Terms	**Definitions**
cantilever	a test structure that consists of a freestanding beam that is fixed at one end[1]
fixed-fixed beam	a test structure that consists of a freestanding beam that is fixed at both ends[1]
in-plane length (or deflection) measurement	the experimental determination of the straight-line distance between two transitional edges in a MEMS device[1]
interferometer	a non-contact optical instrument used to obtain topographical 3-D data sets[1]
residual strain	in a MEMS process, the amount of deformation (or displacement) per unit length constrained within the structural layer of interest after fabrication yet before the constraint of the sacrificial layer (or substrate) is removed (in whole or in part)[1]
(residual) strain gradient	a through-thickness variation (of the residual strain) in the structural layer of interest before it is released[1]
residual stress	the remaining forces per unit area within the structural layer of interest after the original cause(s) during fabrication have been removed yet before the constraint of the sacrificial layer (or substrate) is removed (in whole or in part)[2]
(residual) stress gradient	a through-thickness variation (of the residual stress) in the structural layer of interest before it is released[2]
step height	the distance in the z-direction that an initial, flat, processed surface (or platform) is to a final, flat, processed surface (or platform)[2]
stiction	adhesion between the portion of a structural layer that is intended to be freestanding and its underlying layer[1]
test structure	a component (such as, a fixed-fixed beam or cantilever) that is used to extract information (such as, the residual strain or the strain gradient of a layer) about a fabrication process[1]
thickness	the height in the z-direction of one or more designated thin-film layers[2]
vibrometer	an instrument for non-contact measurements of surface motion[2]
Young's modulus	a parameter indicative of material stiffness that is equal to the stress divided by the strain when the material is loaded in uniaxial tension, assuming the strain is small enough such that it does not irreversibly deform the material[2]

---

[1] Reprinted, with permission, from ASTM E 244405 [1] Terminology Relating to Measurements Taken on Thin, Reflecting Films, copyright ASTM International, 100 Barr Harbor Drive, West Conshohocken, PA 19428, January 2006.

[2] Reprinted, with permission, from SEMI MS2, MS3, and MS4 copyright Semiconductor Equipment and Materials International, Inc. (SEMI) © 2010, 3081 Zanker Road, San Jose, CA 95134, www.semi.org.

Definition of Symbols

The definitions of symbols used with the MEMS 5-in-1 are presented in this section, which is divided into eight parts (one part for each of eight parameters) described as follows. The first set of symbols and definitions are associated with Young's modulus measurements using SEMI standard test method MS4 [1]. The second set of symbols and definitions are for residual strain measurements using ASTM standard test method E 2245 [2], the third set for strain gradient measurements using ASTM standard test method E 2246 [3], the fourth set for step height measurements using SEMI standard test method MS2 [4], and the fifth set for in-plane length measurements using ASTM standard test method E 2244 [5]. The above-mentioned test methods are the five standard test methods associated with the MEMS 5-in-1. The sixth and seventh sets of symbols and definitions pertain to residual stress and stress gradient calculations, respectively, as specified in SEMI standard test method MS4 [1] for Young's modulus measurements. The eighth set of symbols and definitions is for thickness measurements, as specified in Sec. 8 of this document. For SRM 2494, the thickness measurements are obtained using the electro-physical technique [6] and for SRM 2495, the thickness measurements are obtained using the optomechanical technique [7]. Both of these techniques utilize SEMI standard test method MS2 [4] for step height measurements.

When cross referencing these symbols and their definitions among documents, the standard test methods, and the data analysis sheets, care should be given with respect to which symbols are considered calibrated and which are not. Although consistent within each document, standard test method, or web page, they may not be consistent between references. The intent of this document is to present definitions of the symbols that are consistent with what the user would view to be the easiest quantities (typically raw, uncalibrated data) to input on the data analysis sheets. If one of the definitions to a symbol presented below is written exactly as it is written in the standard test method's Terminology Section,[3] the applicable standard test method is specified within brackets after the definition.

1. For Young's modulus measurements [1]:

η	= viscosity of the ambient surrounding the cantilever [SEMI MS4]
ρ	= density of the thin film layer [SEMI MS4]
σ_μ	= one sigma uncertainty of the value of μ [SEMI MS4]
σ	= one sigma uncertainty of the value of ■ [SEMI MS4]
$\sigma_{cantilever}$	= uncertainty in the cantilever's resonance frequency due to geometry and/or composition deviations from the ideal [SEMI MS4]
σ_{Einit}	= estimated standard deviation of E_{init} [SEMI MS4]
σ_{fQ}	= the calibrated standard deviation of the frequency measurements (used to obtain f_{can}) that is due to damping
σ_{freq}	= the standard deviation of $f_{undamped1}$, $f_{undamped2}$, and $f_{undamped3}$ (also called $\sigma_{fundamped}$)
$\sigma_{freqcal}$	= the calibrated standard deviation of the frequency measurements (used to obtain f_{can}) that is due to the calibration of the time base for which the uncertainty is assumed to scale linearly [SEMI MS4]
σ_{fresol}	= the calibrated standard deviation of the frequency measurements (used to obtain f_{can}) that is due to the frequency resolution [SEMI MS4]
$\sigma_{fundamped}$	= one sigma uncertainty of the calibrated undamped resonance frequency measurements [SEMI MS4]
σ_L	= one sigma uncertainty of the value of L_{can} [SEMI MS4]
σ_{meter}	= for calibrating the time base of the instrument: the standard deviation of the measurements used to obtain f_{meter} [SEMI MS4]

---

[3] As to be submitted for ballot as of the writing of this document, unless otherwise indicated.

$\blacksquare_{support}$	= the estimated uncertainty in the cantilever's resonance frequency due to a non-ideal support (or attachment conditions) [SEMI MS4]
$\blacksquare_{thick}$	= one sigma uncertainty of the value of t [SEMI MS4]
$\blacksquare_{W}$	= one sigma uncertainty of the value of W_{can} [SEMI MS4]
cal_f	= the calibration factor for a frequency measurement [SEMI MS4]
d	= the gap between the bottom of the suspended cantilever and the top of the underlying layer
E	= calculated Young's modulus value of the thin film layer [SEMI MS4]
E_{init}	= initial estimate for the Young's modulus value of the thin film layer [SEMI MS4]
E_{max}	= maximum Young's modulus value as determined in an uncertainty calculation [SEMI MS4-1109]
E_{min}	= minimum Young's modulus value as determined in an uncertainty calculation [SEMI MS4-1109]
f_{can}	= average calibrated undamped resonance frequency of the cantilever, which includes the frequency correction term [SEMI MS4]
$f_{caninit}$	= estimate for the fundamental resonance frequency of a cantilever
$f_{correction}$	= correction term for the cantilever's resonance frequency [SEMI MS4]
$f_{dampedn}$	= the nth calibrated, damped resonance frequency measurement
$f_{instrument}$	= for calibrating the time base of the instrument: the frequency setting for the calibration measurements (or the manufacturer's specification for the clock frequency) [SEMI MS4]
f_{measn}	= an uncalibrated measurement of the resonance frequency where the trailing subscript n is 1, 2, or 3
f_{meter}	= for calibrating the time base of the instrument: the calibrated average frequency of the calibration measurements (or the calibrated average clock frequency) taken with a frequency meter [SEMI MS4]
f_{resol}	= uncalibrated frequency resolution for the given set of measurement conditions [SEMI MS4]
$f_{undampedn}$	= the nth calibrated undamped resonance frequency calculated from the cantilever's n^{th} damped resonance frequency measurement, if applicable
L_{can}	= suspended cantilever length [SEMI MS4]
p_{diff}	= estimated percent difference between the damped and undamped resonance frequency of the cantilever [SEMI MS4]
Q	= oscillatory quality factor of the cantilever [SEMI MS4]
t	= thickness of the thin film layer [SEMI MS4]
$u^{\blacksquare}$	= component in the combined standard uncertainty calculation for Young's modulus that is due to the uncertainty of $\blacksquare$ [SEMI MS4-1109]
u_{cE}	= combined standard uncertainty of a Young's modulus measurement as obtained from the resonance frequency of a cantilever [SEMI MS4]
u_{certf}	= for calibrating the time base of the instrument: the certified uncertainty of the frequency measurements as specified on the frequency meter's certificate [SEMI MS4]
u_{cmeter}	= for calibrating the time base of the instrument: the uncertainty of the frequency measurements taken with the frequency meter [SEMI MS4]
u_{damp}	= component in the combined standard uncertainty calculation for Young's modulus that is due to damping [SEMI MS4-1109]
U_E	= the expanded uncertainty of a Young's modulus measurement [SEMI MS4]
u_{freq}	= component in the combined standard uncertainty calculation for Young's modulus that is due to the measurement uncertainty of the average resonance frequency
$u_{freqcal}$	= component in the combined standard uncertainty calculation for Young's modulus that is due to the frequency calibration

u_{fresol} = component in the combined standard uncertainty calculation for Young's modulus that is due to f_{resol} [SEMI MS4-1109]

u_L = component in the combined standard uncertainty calc ulation for Young's modulus that is due to the measurement uncertainty of L_{can} [SEMI MS4-1109]

u_{thick} = component in the combined standard uncertainty calculation for Young's modulus that is due to the measurement uncertainty of t [SEMI MS4-1109]

W_{can} = suspended cantilever width [SEMI MS4]

2. For residual strain measurements [2]:

= the misalignment angle [ASTM E 2245]

■ *rcorrection* = the relative residual strain correction term [ASTM E 2245]

■ *r* = the residual strain [ASTM E 2245]

■ *r-high* = the maximum residual strain value as determined in an uncertainty calculation

■ *r-low* = the minimum residual strain value as determined in an uncertainty calculation

■ *6same* = the maximum of two uncalibrated values (σ_{same1} and σ_{same2}) where σ_{same1} is the standard deviation of the six step height measurements taken on the physical step height standard at the same location before the data session and σ_{same2} is the standard deviation of the six measurements taken at this same location after the data session [ASTM E 2245]

■ *cert* = the certified one sigma uncertainty of the physical step height standard used for calibration [ASTM E 2245]

■ *Lrepeat(samp)*′ = the in-plane length repeatability standard deviation (for the given combination of lenses for the given interferometric microscope) as obtained from test structures fabricated in a process similar to that used to fabricate the sample and when the transitional edges face each other [ASTM E 2245]

■ *noise* = the standard deviation of the noise measurement, calculated to be one-sixth the value of R_{tave} minus R_{ave} [ASTM E 2245]

■ *Rave* = the standard deviation of the surface roughness measurement, calculated to be one-sixth the value of R_{ave} [ASTM E 2245]

■ *repeat(samp)* = the relative residual strain repeatability standard deviation as obtained from fixed-fixed beams fabricated in a process similar to that used to fabricate the sample [ASTM E 2245]

■ *samp* = the standard deviation in a height measurement due to the sample's peak-to-valley surface roughness as measured with the interferometer and calculated to be one-sixth the value of R_{tave}

■ *xcal* = the standard deviation in a ruler measurement in the interferomet ric microscope's x-direction for the given combination of lenses [ASTM E 2245]

■ *zcal* = the calibrated standard deviation of the twelve step height measurements taken along the certified portion of the physical step height standard before and after the data session and which is assumed to scale linearly with height

cal_x = the x-calibration factor of the interferometric microscope for the given combination of lenses [ASTM E 2245]

cal_{xmax} = the maximum x-calibration factor

cal_{xmin} = the minimum x-calibration factor

cal_z = the z-calibration factor of the interferometric microscope for the given combination of lenses [ASTM E 2245]

cert = the certified (that is, calibrated) value of the physical step height standard [ASTM E 2245]

L = the in-plane length measurement of the fixed-fixed beam [ASTM E 2245]

L_0 = the calibrated length of the fixed-fixed beam if there are no applied axial-compressive forces [ASTM E 2245]

L_c = the total calibrated length of the curved fixed-fixed beam (as modeled with two cosine functions) with $v1_{end}$ and $v2_{end}$ as the calibrated v values of the endpoints [ASTM E 2245]

L_e = the calibrated effective length of the fixed-fixed beam calculated as the straight-line measurement between v_{eF} and v_{eS} [ASTM E 2245]

L_{offset} = the in-plane length correction term for the given type of in-plane length measurement taken on similar structures when using similar calculations and for the given combination of lenses for a given interferometric microscope [ASTM E 2245]

$n1_t$ = indicative of the data point uncertainty associated with the chosen value for $x1_{uppert}$, with the subscript "t" referring to the data trace. If it is easy to identify one point that accurately locates the upper corner of Edge 1, the maximum uncertainty associated with the identification of this point is $n1_t x_{res} cal_x$, where $n1_t$=1. [ASTM E 2245]

$n2_t$ = indicative of the data point uncertainty associated with the chosen value for $x2_{uppert}$, with the subscript "t" referring to the data trace. If it is easy to identify one point that accurately locates the upper corner of Edge 2, the maximum uncertainty associated with the identification of this point is $n2_t x_{res} cal_x$, where $n2_t$=1. [ASTM E 2245]

R_{ave} = the calibrated surface roughness of a flat and leveled surface of the sample material calculated to be the average of three or more measurements, each measurement taken from a different 2-D data trace [ASTM E 2245]

R_{tave} = the calibrated peak-to-valley roughness of a flat and leveled surface of the sample material calculated to be the average of three or more measurements, each measurement taken from a different 2-D data trace [ASTM E 2245]

$ruler_x$ = the interferometric microscope's maximum field of view in the x-direction for the given combination of lenses as measured with a 10- m grid (or finer grid) ruler [ASTM E 2245]

$scope_x$ = the interferometric microscope's maximum field of view in the x-direction for the given combination of lenses [ASTM E 2245]

U_r = the expanded uncertainty of a residual strain measurement [ASTM E 2245]

u_{cr} = the combined standard uncertainty of a residual strain measurement [ASTM E 2245]

u_{cert} = the component in the combined standard uncertainty calculation for residual strain that is due to the uncertainty of the value of the physical step height standard used for calibration [ASTM E 2245]

$u_{correction}$ = the component in the combined standard uncertainty calculation for residual strain that is due to the uncertainty of the correction term [ASTM E 2245]

u_{drift} = the component in the combined standard uncertainty calculation for residual strain that is due to the amount of drift during the data session [ASTM E 2245]

u_L = the component in the combined standard uncertainty calculation for residual strain that is due to the measurement uncertainty of L [ASTM E 2245]

u_{linear} = the component in the combined standard uncertainty calculation for residual strain that is due to the deviation from linearity of the data scan [ASTM E 2245]

u_{noise} = the component in the combined standard uncertainty calculation for residual strain that is due to interferometric noise [ASTM E 2245]

u_{Rave} = the component in the combined standard uncertainty calculation for residual strain that is due to the sample's surface roughness [ASTM E 2245]

$u_{repeat(samp)}$ = the component in the combined standard uncertainty calculation for residual strain that is due to the repeatability of residual strain measurements taken on fixed-fixed beams processed similarly to the one being measured [ASTM E 2245]

$u_{repeat(shs)}$ = the component in the combined standard uncertainty calculation for residual strain that is due to the repeatability of measurements taken on the physical step height standard [ASTM E 2245]

u_{samp}	= the component in the combined standard uncertainty calculation for residual strain that is due to the sample's peak-to-valley surface roughness as measured with the interferometer [ASTM E 224505]
u_W	= the component in the combined standard uncertainty calculation for residual strain that is due to variations across the width of the fixed-fixed beam [ASTM E 2245]
u_{xcal}	= the component in the combined standard uncertainty calculation for residual strain that is due to the uncertainty of the calibration in the x-direction [ASTM E 2245]
u_{xres}	= the component in the combined standard uncertainty calculation for residual strain that is due to the resolution of the interferometric microscope in the x-direction as pertains to the data points chosen along the fixed-fixed beam [ASTM E 2245]
u_{xresL}	= the component in the combined standard uncertainty calculation for residual strain that is due to the resolution of the interferometric microscope in the x-direction as pertains to the in-plane length measurement
u_{zcal}	= the component in the combined standard uncertainty calculation for residual strain that is due to the uncertainty of the calibration in the z-direction [ASTM E 224505]
u_{zres}	= the component in the combined standard uncertainty calculation for residual strain that is due to the resolution of the interferometric microscope in the z-direction [ASTM E 2245]
$v1_{end}$	= one endpoint of the in-plane length measurement [ASTM E 2245]
$v2_{end}$	= another endpoint of the in-plane length measurement [ASTM E 2245]
v_{eF}	= the calibrated v value of the inflection point of the cosine function modeling the first abbreviated data trace [ASTM E 2245]
v_{eS}	= the calibrated v value of the inflection point of the cosine function modeling the second abbreviated data trace [ASTM E 2245]
$x1_{uppert}$	= the uncalibrated x-value that most appropriately locates the upper corner associated with Edge 1 using Trace t
$x2_{uppert}$	= the uncalibrated x-value that most appropriately locates the upper corner associated with Edge 2 using Trace t
x_{res}	= the uncalibrated resolution of the interferometric microscope in the x-direction
$y_{a'}$	= the uncalibrated y-value associated with Trace a$'$
$y_{e'}$	= the uncalibrated y-value associated with Trace e$'$
$\bar{z}_6$	= the uncalibrated average of the six calibration measurements from which $z_{repeat(shs)}$ is found
$\bar{z}_{6same}$	= the uncalibrated average of the six calibration measurements from which σ_{6same} is found [ASTM E 2245]
$\bar{z}_{ave}$	= the average of the calibration measurements taken along the physical step height standard before and after the data session [ASTM E 2245]
z_{drift}	= the uncalibrated positive difference between the average of the six calibration measurements taken before the data session (at the same location on the physical step height standard used for calibration) and the average of the six calibration measurements taken after the data session (at this same location) [ASTM E 2245]
z_{lin}	= over the instrument's total scan range, the maximum relative deviation from linearity, as quoted by the instrument manufacturer (typically less than 3 %) [ASTM E 2245]
$z_{repeat(shs)}$	= the maximum of two uncalibrated values; one of which is the positive uncalibrated difference between the minimum and maximum values of the six calibration measurements taken before the data session (at the same location on the physical step height standard used for calibration) and the other is the positive uncalibrated difference between the minimum and maximum values of the six calibration measurements taken after the data session (at this same location)
z_{res}	= the calibrated resolution of the interferometric microscope in the z-direction [ASTM E

2245]

3. For strain gradient measurements [3]:

α	= the misalignment angle [ASTM E 2246]
σ_{6same}	= the maximum of two uncalibrated values (σ_{same1} and σ_{same2}) where σ_{same1} is the standard deviation of the six step height measurements taken on the physical step height standard at the same location before the data session and σ_{same2} is the standard deviation of the six measurements taken at this same location after the data session [ASTM E 2246]
σ_{cert}	= the certified one sigma uncertainty of the physical step height standard used for calibration [ASTM E 2246]
$\sigma_{repeat(samp)}$	= the relative strain gradient repeatability standard deviation as obtained from cantilevers fabricated in a process similar to that used to fabricate the sample [ASTM E 2246]
σ_{samp}	= the standard deviation in a height measurement due to the sample's peak-to-valley surface roughness as measured with the interferometer and calculated to be one-sixth the value of R_{tave}
σ_{xcal}	= the standard deviation in a ruler measurement in the interferometric microscope's x-direction for the given combination of lenses [ASTM E 2246]
σ_{zcal}	= the calibrated standard deviation of the twelve step height measurements taken along the certified portion of the physical step height standard before and after the data session and which is assumed to scale linearly with height
cal_x	= the x-calibration factor of the interferometric microscope for the given combination of lenses [ASTM E 2246]
cal_{xmax}	= the maximum x-calibration factor
cal_{xmin}	= the minimum x-calibration factor
cal_z	= the z-calibration factor of the interferometric microscope for the given combination of lenses [ASTM E 2246]
$cert$	= the certified (that is, calibrated) value of the physical step height standard [ASTM E 2246]
$n1_t$	= indicative of the data point uncertainty associated with the chosen value for $x1_{uppert}$, with the subscript "t" referring to the data trace. If it is easy to identify one point that accurately locates the upper corner of Edge 1, the maximum uncertainty associated with the identification of this point is $n1_t x_{res} cal_x$, where $n1_t$=1. [ASTM E 2246]
R_{ave}	= the calibrated surface roughness of a flat and leveled surface of the sample material calculated to be the average of three or more measurements, each measurement taken from a different 2-D data trace [ASTM E 2246]
R_{tave}	= the calibrated peak-to-valley roughness of a flat and leveled surface of the sample material calculated to be the average of three or more measurements, each measurement taken from a different 2-D data trace [ASTM E 2246]
s_g	= the strain gradient as calculated from three data points [ASTM E 2246]
$s_{gcorrection}$	= the strain gradient correction term for the given design length [ASTM E 2246]
s_{g-high}	= the maximum strain gradient value as determined in an uncertainty calculation
s_{g-low}	= the minimum strain gradient value as determined in an uncertainty calculation
u_{cert}	= the component in the combined standard uncertainty calculation for strain gradient that is due to the uncertainty of the value of the physical step height standard used for calibration [ASTM E 2246]
$u_{correction}$	= the component in the combined standard uncertainty calculation for strain gradient that is due to the uncertainty of the correction term [ASTM E 2246]
u_{csg}	= the combined standard uncertainty of a strain gradient measurement [ASTM E 2246]
u_{drift}	= the component in the combined standard uncertainty calculation for strain gradient that is due to the amount of drift during the data session [ASTM E 2246]
u_{linear}	= the component in the combined standard uncertainty calculation for strain gradient that

is due to the deviation from linearity of the data scan [ASTM E 2246]

u_{noise} = the component in the combined standard uncertainty calculation for strain gradient that is due to interferometric noise [ASTM E 2246]

u_{Rave} = the component in the combined standard uncertainty calculation for strain gradient that is due to the sample's surface roughness [ASTM E 2246]

$u_{repeat(samp)}$ = the component in the combined standard uncertainty calculation for strain gradient that is due to the repeatability of measurements taken on cantilevers processed similarly to the one being measured [ASTM E 2246]

$u_{repeat(shs)}$ = the component in the combined standard uncertainty calculation for strain gradient that is due to the repeatability of measurements taken on the physical step height standard [ASTM E 2246]

u_{samp} = the component in the combined standard uncertainty calculation for strain gradient that is due to the sample's peak-to-valley surface roughness as measured with the interferometer [ASTM E 224605]

U_{sg} = the expanded uncertainty of a strain gradient measurement [ASTM E 2246]

u_W = the component in the combined standard uncertainty calculation for strain gradient that is due to the measurement uncertainty across the width of the cantilever [ASTM E 2246]

u_{xcal} = the component in the combined standard uncertainty calculation for strain gradient that is due to the uncertainty of the calibration in the x-direction [ASTM E 2246]

u_{xres} = the component in the combined standard uncertainty calculation for strain gradient that is due to the resolution of the interferometric microscope in the x-direction [ASTM E 2246]

u_{zcal} = the component in the combined standard uncertainty calculation for strain gradient that is due to the uncertainty of the calibration in the z-direction [ASTM E 224605]

u_{zres} = the component in the combined standard uncertainty calculation for strain gradient that is due to the resolution of the interferometric microscope in the z-direction [ASTM E 2246]

$x1_{uppert}$ = the uncalibrated x-value that most appropriately locates the upper corner associated with Edge 1 using Trace t

x_{res} = the uncalibrated resolution of the interferometric microscope in the x-direction (for the given combination of lenses)

y_t = the uncalibrated y-value associated with Trace t

$\bar{z}_6$ = the uncalibrated average of the six calibration measurements from which $z_{repeat(shs)}$ is found

$\bar{z}_{6same}$ = the uncalibrated average of the six calibration measurements from which σ_{6same} is found [ASTM E 2246]

$\bar{z}_{ave}$ = the average of the calibration measurements taken along the physical step height standard before and after the data session [ASTM E 2246]

z_{drift} = the uncalibrated positive difference between the average of the six calibration measurements taken before the data session (at the same location on the physical step height standard used for calibration) and the average of the six calibration measurements taken after the data session (at this same location) [ASTM E 2246]

z_{lin} = over the instrument's total scan range, the maximum relative deviation from linearity, as quoted by the instrument manufacturer (typically less than 3 %) [ASTM E 2246]

$z_{repeat(shs)}$ = the maximum of two uncalibrated values; one of which is the positive uncalibrated difference between the minimum and maximum values of the six calibration measurements taken before the data session (at the same location on the physical step height standard used for calibration) and the other is the positive uncalibrated difference between the minimum and maximum values of the six calibration measurements taken after the data session (at this same location)

z_{res} = the calibrated resolution of the interferometric microscope in the *z*-direction [ASTM E 2246]

4. For step height measurements [4]:

$\blacksquare_{6ave}$ = the maximum of two uncalibrated values (σ_{before} and σ_{after}) where σ_{before} is the standard deviation of the six step height measurements taken along the physical step height standard before the data session and σ_{after} is the standard deviation of the six measurements taken along the physical step height standard after the data session [SEMI MS2]

$\blacksquare_{6same}$ = the maximum of two uncalibrated values (σ_{same1} and σ_{same2}) where σ_{same1} is the standard deviation of the six step height measurements taken at the same location on the physical step height standard before the data session and σ_{same2} is the standard deviation of the six measurements taken at this same location after the data session [SEMI MS2]

$\blacksquare_{cert}$ = the one sigma uncertainty of the physical step height standard used for calibration [SEMI MS2]

$\blacksquare_{repeat(samp)}$ = the relative step height repeatability standard deviation as obtained from step height test structures fabricated in a process similar to that used to fabricate the sample [SEMI MS2]

$\blacksquare_{Wstep}$ = the standard deviation of the calibrated step height measurements taken from the data traces on one step height test structure

cal_z = the *z*-calibration factor of the interferometric microscope or comparable instrument [SEMI MS2]

cert = the certified value of the physical step height standard used for calibration [SEMI MS2]

platNrD = the calibrated average of the reference platform height measurements taken from multiple data traces on one step height test structure, where *N* is the test structure number (*1*, *2*, *3*, etc.), *r* indicates it is from a reference platform, and *D* directionally indicates which reference platform (using the compass indicators *N*, *S*, *E*, or *W* where *N* refers to the reference platform designed closest to the top of the chip) [SEMI MS2]

platNrDt = an uncalibrated reference platform height measurement from one data trace, where *N* is the test structure number (*1*, *2*, *3*, etc.), *r* indicates it is from a reference platform, *D* directionally indicates which reference platform (using the compass indicators *N*, *S*, *E*, or *W* where *N* refers to the reference platform designed closest to the top of the chip), and *t* is the data trace (*a*, *b*, *c*, etc.) being examined [SEMI MS2]

platNXt = an uncalibrated platform height measurement from one data trace, where *N* is the test structure number (*1*, *2*, *3*, etc.), *X* is the capital letter associated with the platform (*A*, *B*, *C*, etc.) as lettered starting with *A* for the platform closest to *platNrW* or *platNrS*, and *t* is the data trace (*a*, *b*, *c*, etc.) being examined [SEMI MS2]

$s_{platNrDt}$ = the uncalibrated standard deviation of the data from Trace t on *platNrD* [SEMI MS2]

$s_{platNXave}$ = the average of the calibrated standard deviation values from the data traces on *platNX* [SEMI MS2]

$s_{platNXt}$ = the uncalibrated standard deviation of the data from Trace t on *platNX* [SEMI MS2]

$s_{platNYt}$ = the uncalibrated standard deviation of the data from Trace t on *platNY* [SEMI MS2]

$s_{roughNX}$ = the uncalibrated surface roughness of *platNX* measured as the smallest of all the values obtained for $s_{platNXt}$; however, if the surfaces of the platforms (including the reference platform) all have identical compositions, then it is measured as the smallest of all the standard deviation values obtained from data traces *a b*, and *c* along these platforms [SEMI MS2]

$s_{roughNY}$ = the uncalibrated surface roughness of *platNY* measured as the smallest of all the values obtained for $s_{platNYt}$; however, if the surfaces of the platforms (including the reference platform) all have identical compositions, then it is measured as the smallest of all the standard deviation values obtained from data traces *a b*, and *c* along these platforms [SEMI MS2]

$stepN_{XY}$	= the average of the calibrated step height measurements taken from multiple data traces on one step height test structure, where N is the number associated with the test structure, X is the capital letter associated with the initial platform (or r is used if it is the reference platform), Y is the capital letter associated with the final platform (or r is used if it is the reference platform), and the step is from the initial platform to the final platform [SEMI MS2]
$stepN_{XYt}$	= a calibrated step height measurement from one data trace on one step height test structure, where N is the number associated with the test structure, X is the capital letter associated with the initial platform (or r is used if it is the reference platform), Y is the capital letter associated with the final platform (or r is used if it is the reference platform), t is the data trace (a, b, c, etc.) being examined, and the step is from the initial platform to the final platform [SEMI MS2]
u_{cal}	= the component in the combined standard uncertainty calculation for step height measurements that is due to the uncertainty of the measurements taken across the physical step height standard [SEMI MS2]
u_{cert}	= the component in the combined standard uncertainty calculation for step height measurements that is due to the uncertainty of the value of the physical step height standard used for calibration [SEMI MS2]
u_{cSH}	= the combined standard uncertainty of a step height measurement [SEMI MS2]
u_{drift}	= the component in the combined standard uncertainty calculation for step height measurements that is due to the amount of drift during the data session [SEMI MS2]
u_{linear}	= the component in the combined standard uncertainty calculation for step height measurements that is due to the deviation from linearity of the data scan [SEMI MS2]
u_{Lstep}	= the component in the combined standard uncertainty calculation for step height measurements that is due to the measurement uncertainty of the step height across the length of the step, where the length is measured perpendicular to the edge of the step [SEMI MS2]
$u_{repeat(samp)}$	= the component in the combined standard uncertainty calculation for step height measurements that is due to the repeatability of measurements taken on step height test structures processed similarly to the one being measured [SEMI MS2]
$u_{repeat(shs)}$	= the component in the combined standard uncertainty calculation for step height measurements that is due to the repeatability of measurements taken on the physical step height standard [SEMI MS2]
U_{SH}	= the expanded uncertainty of a step height measurement [SEMI MS2]
u_{Wstep}	= the component in the combined standard uncertainty calculation for step height measurements that is due to the measurement uncertainty of the step height across the width of the step, where the width is measured parallel to the edge of the step [SEMI MS2]
$\bar{z}_6$	= the uncalibrated average of the six calibration measurements that was used to determine $z_{repeat(shs)}$
$\bar{z}_{6ave}$	= the uncalibrated average of the six calibration measurements from which σ_{6ave} is found [SEMI MS2]
$\bar{z}_{6same}$	= the uncalibrated average of the six calibration measurements used to determine σ_{6same} [SEMI MS2]
$\bar{z}_{ave}$	= the average of the twelve calibration measurements (taken along the physical step height standard before and after the data session) used to calculate cal_z [SEMI MS2]
z_{drift}	= the uncalibrated positive difference between the average of the six calibration measurements taken before the data session (at the same location on the physical step height standard) and the average of the six calibration measurements taken after the data session (at this same location) [SEMI MS2]

z_{lin} = over the instrument's total scan range, the maximum relative deviation from linearity (typically less than 3 %), as quoted by the instrument manufacturer [SEMI MS2]

$z_{repeat(shs)}$ = the maximum of two uncalibrated values; one of which is the positive difference between the minimum and maximum values of the six calibration measurements taken before the data session (at the same location on the physical step height standard) and the other is the positive difference between the minimum and maximum values of the six calibration measurements taken after the data session (at this same location)

$\bar{z}_{same}$ = the uncalibrated average of the twelve calibration measurements that were taken before and after the data session (at the same location on the physical step height standard) and that is used to calculate cal_z [used with SEMI MS2-1109]

5. For in-plane length measurements [5]:

α = the misalignment angle [ASTM E 2244]

■$_{repeat(samp)'}$ = the in-plane length repeatability standard deviation (for the given combination of lenses for the given interferometric microscope) as obtained from test structures fabricated in a process similar to that used to fabricate the sample and for the same or a similar type of measurement [ASTM E 2244]

■$_{xcal}$ = the standard deviation in a ruler measurement in the interferometric microscope's x-direction for the given combination of lenses [ASTM E 2244]

cal_x = the x-calibration factor of the interferometric microscope for the given combination of lenses [ASTM E 2244]

cal_z = the z-calibration factor of the interferometric microscope for the given combination of lenses [ASTM E 2244]

cert = the certified (that is, calibrated) value of the physical step height standard [ASTM E 2244]

L = the in-plane length measurement that accounts for misalignment and includes the in-plane length correction term, L_{offset} [ASTM E 2244]

L_{align} = the in-plane length, after correcting for misalignment, used to calculate L [ASTM E 2244]

L_{meas} = the measured in-plane length used to calculate L_{align} [ASTM E 2244]

L_{offset} = the in-plane length correction term for the given type of in-plane length measurement on similar structures, when using similar calculations, and for a given magnification of a given interferometric microscope [ASTM E 2244]

$n1_t$ = indicative of the data point uncertainty associated with the chosen value for $x1_{uppert}$, with the subscript "t" referring to the data trace. If it is easy to identify one point that accurately locates the upper corner of Edge 1, the maximum uncertainty associated with the identification of this point is $n1_t x_{res} cal_x$, where $n1_t$=1. [ASTM E 2244]

$n2_t$ = indicative of the data point uncertainty associated with the chosen value for $x2_{uppert}$, with the subscript "t" referring to the data trace. If it is easy to identify one point that accurately locates the upper corner of Edge 2, the maximum uncertainty associated with the identification of this point is $n2_t x_{res} cal_x$, where $n2_t$=1. [ASTM E 2244]

$ruler_x$ = the interferometric microscope's maximum field of view in the x-direction for the given combination of lenses as measured with a 10- m grid (or finer grid) ruler [ASTM E 2244]

$scope_x$ = the interferometric microscope's maximum field of view in the x-direction for the given combination of lenses [ASTM E 2244]

u_{align} = the component in the combined standard uncertainty calculation for an in-plane length measurement that is due to alignment uncertainty [ASTM E 2244]

u_{cL} = the combined standard uncertainty for an in-plane length measurement [ASTM E 2244]

U_L = the expanded uncertainty of an in-plane length measurement [ASTM E 2244]

u_L	= the component in the combined standard uncertainty calculation for an in-plane length measurement that is due to the uncertainty in the calculated length [ASTM E 2244]
u_{offset}	= the component in the combined standard uncertainty calculation for an in-plane length measurement that is due to the uncertainty of the value for L_{offset} [ASTM E 2244]
$u_{repeat(L)}$	= the component in the combined standard uncertainty calculation for an in-plane length measurement that is due to the uncertainty of the four measurements taken on the test structure at different locations [ASTM E 2244]
$u_{repeat(samp)}$	= the component in the combined standard uncertainty calculation for an in-plane length measurement that is due to the repeatability of measurements taken on test structures processed similarly to the sample, using the same combination of lenses for the given interferometric microscope for the measurement, and for the same or a similar type of measurement [ASTM E 2244]
u_{xcal}	= the component in the combined standard uncertainty calculation for an in-plane length measurement that is due to the uncertainty of the calibration in the *x*-direction [ASTM E 2244]
u_{xres}	= the component in the combined standard uncertainty calculation for an in-plane length measurement that is due to the resolution of the interferometric microscope in the *x*-direction
$x1_{uppert}$	= the uncalibrated *x*-value that most appropriately locates the upper corner associated with Edge 1 using Trace t [ASTM E 2244]
$x2_{uppert}$	= the uncalibrated *x*-value that most appropriately locates the upper corner associated with Edge 2 using Trace t [ASTM E 2244]
x_{res}	= the uncalibrated resolution of the interferometric microscope in the *x*-direction for the given combination of lenses [ASTM E 2244]
ya'	= the uncalibrated *y*-value associated with Trace a′ [ASTM E 2244]
ye'	= the uncalibrated *y*-value associated with Trace e′ [ASTM E 2244]
$\bar{z}_{ave}$	= the average of the calibration measurements taken along the physical step height standard before and after the data session [ASTM E 2244]

6. For residual stress calculations [1]:

■$_r$	= residual strain of the thin film layer [SEMI MS4]
■$_r$	= residual stress of the thin film layer [SEMI MS4]
E	= calculated Young's modulus value of the thin film layer [SEMI MS4]
$u_{r(}$■$_{r)}$	= component in the combined standard uncertainty calculation for residual stress that is due to the measurement uncertainty of ■$_r$ [SEMI MS4-1109]
U■$_r$	= the expanded uncertainty of a residual stress measurement [SEMI MS4]
u_{cr}	= combined standard uncertainty value for residual strain [SEMI MS4]
u_c■$_r$	= combined standard uncertainty value for residual stress [SEMI MS4]
u_{cE}	= combined standard uncertainty of a Young's modulus measurement as obtained from the resonance frequency of a cantilever [SEMI MS4]
$u_{E(}$■$_{r)}$	= component in the combined standard uncertainty calculation for residual stress that is due to the measurement uncertainty of *E* [SEMI MS4-1109]

7. For (residual) stress gradient calculations [1]:

■$_g$	= stress gradient of the thin film layer [SEMI MS4]
E	= calculated Young's modulus value of the thin film layer [SEMI MS4]
s_g	= strain gradient of the thin film layer [SEMI MS4]
U■$_g$	= the expanded uncertainty of a stress gradient measurement [SEMI MS4]
u_c■$_g$	= combined standard uncertainty value for stress gradient [SEMI MS4]

u_{cE}	= combined standard uncertainty of a Young's modulus measurement as obtained from the resonance frequency of a cantilever [SEMI MS4]
u_{csg}	= combined standard uncertainty value for strain gradient [SEMI MS4]
$u_{E(■g)}$	= component in the combined standard uncertainty calculation for stress gradient that is due to the measurement uncertainty of E [SEMI MS4-1109]
$u_{sg(■g)}$	= component in the combined standard uncertainty calculation for stress gradient that is due to the measurement uncertainty of s_g [SEMI MS4-1109]

8. For thickness measurements:

For SRM 2494 using Data Analysis Sheet T.1:

$■_{SiO2}$	= the permittivity of SiO_2
■	= the resistivity of the thin film
■■	= the estimated standard deviation of $■_{SiO2}$
■■	= the standard deviation of the resistivity
$■_{Ca}$	= the standard deviation of the capacitance value
$■_{resCa}$	= a residual (unclassified) standard deviation of the residual capacitance component
$■_{resRs}$	= a residual (unclassified) standard deviation of the residual sheet resistance or residual resistivity component
$■_{Rs}$	= the standard deviation of the sheet resistance
C_a	= the capacitance per unit area in attofarads per square micrometer, for which the fringing capacitance and stray capacitance have been removed
r_{res}	= the residual (unclassified) capacitance, sheet resistance, or resistivity component, as applicable
R_s	= the interconnect sheet resistance
t	= the thickness
t_{SiO2}	= the thickness of the composite SiO_2 beam
u_{cSiO2}	= the combined standard uncertainty of the composite SiO_2 beam thickness
u_{ctCa}	= the combined standard uncertainty of a thickness value obtained from capacitance measurements
u_{ctRs}	= the combined standard uncertainty of a thickness value obtained from sheet resistance measurements
u_{res}	= a residual (unclassified) uncertainty component for a step height measurement
U_{SiO2}	= the expanded uncertainty of a composite SiO_2 beam thickness measurement

For SRM 2495 using Data Analysis Sheet T.3:

■	= the poly1 or poly2 thickness
$■H$	= range of the anchor etch depth (as provided by the processing facility)
$■_{6aveN}$	= the maximum of two uncalibrated values (σ_{before} and σ_{after}) where σ_{before} is the standard deviation of the six step height measurements taken along the physical step height standard before the data session and σ_{after} is the standard deviation of the six measurements taken along the physical step height standard after the data session and where the subscript N is A for the measurement of A and B for the measurement of B
$■_{6sameN}$	= the maximum of two uncalibrated values (σ_{same1} and σ_{same2}) where σ_{same1} is the standard deviation of the six step height measurements taken on the physical step height standard at the same location before the data session and σ_{same2} is the standard deviation of the six measurements taken at this same location after the data session and where the subscript N is A for the measurement of A and B for the measurement of B
$■_{certN}$	= certified one sigma uncertainty of the physical step height standard used for calibration, where the subscript N is A for the measurement of A and B for the

	measurement of B
$s_{repeat(samp)N}$	= the relative step height repeatability standard deviation as obtained from step height test structures fabricated in a process similar to that used to fabricate the sample, where the subscript N is A for the measurement of A and B for the measurement of B
A	= in a surface micromachining process, the positive vertical distance between the top of the underlying layer to the top of the structural layer in the anchor area
B	= in a surface micromachining process, the vertical distance between the top of the structural layer in the anchor area to the top of a beam composed of that structural layer where adhered to the top of the underlying layer
cal_{zN}	= the z-calibration factor of the interferometer for the given combination of lenses, where the subscript N is A for the measurement of A and B for the measurement of B
$cert_N$	= the certified value of the physical step height standard used for calibration, where the subscript N is A for the measurement of A and B for the measurement of B
H	= the anchor etch depth
J	= the positive vertical distance between the bottom of the suspended structural layer and the top of the underlying layer, which takes into consideration the roughness of each surface, any residue present between the layers, and a tilting component
J_{est}	= the estimated value for the dimension J
$platX$	= the flat, processed poly0 layer (on the opposite side of the anchor than the cantilever) that is used in the measurement of A
$platXt$	= an uncalibrated platform height measurement from one data trace on $platX$ where t is the data trace (a, b, or c) being examined
$platY$	= the flat top surface of the poly1 or poly2 layer (within its anchor to the underlying poly0 layer) that is used in the measurements of both A and B
$platYt1$	= an uncalibrated platform height measurement from one data trace on $platY$ where t is the data trace (a, b, or c) being examined for the measurement of A
$platYt2$	= an uncalibrated platform height measurement from one data trace on $platY$ where t is the data trace (a, b, or c) being examined for the measurement of B
$platZ$	= the top surface of the cantilever beam (where it is adhered to the top of the underlying layer) that is used in the measurement of B
$platZt$	= an uncalibrated platform height measurement from one data trace on $platZ$ where t is the data trace (a, b, or c) being examined
s_{platXt}	= the uncalibrated standard deviation of the data from one data trace on $platX$ where t is the data trace (a, b, or c) being examined
$s_{platYt1}$	= the uncalibrated standard deviation of the data from one data trace on $platY$ where t is the data trace (a, b, or c) being examined for the measurement of A
$s_{platYt2}$	= the uncalibrated standard deviation of the data from one data trace on $platY$ where t is the data trace (a, b, or c) being examined for the measurement of B
s_{platZt}	= the uncalibrated standard deviation of the data from one data trace on $platZ$ where t is the data trace (a, b, or c) being examined
s_{roughX}	= the uncalibrated surface roughness of $platX$ measured as the smallest of all the values obtained for s_{platXt}; however, if the surfaces of $platX$, $platY$, and $platZ$ all have identical compositions, then it is measured as the smallest of all the values obtained for s_{platXt}, $s_{platYt1}$, $s_{platYt2}$, and s_{platZt} in which case $s_{roughX}=s_{roughY}=s_{roughZ}$
s_{roughY}	= the uncalibrated surface roughness of $platY$ measured as the smallest of all the values obtained for $s_{platYt1}$ and $s_{platYt2}$; however, if the surfaces of $platX$, $platY$, and $platZ$ all have identical compositions, then it is measured as the smallest of all the values obtained for s_{platXt}, $s_{platYt1}$, $s_{platYt2}$, and s_{platZt} in which case $s_{roughX}=s_{roughY}=s_{roughZ}$
s_{roughZ}	= the uncalibrated surface roughness of $platZ$ measured as the smallest of all the values obtained for s_{platZt}; however, if the surfaces of $platX$, $platY$, and $platZ$ all have identical compositions, then it is measured as the smallest of all the values obtained for s_{platXt}, $s_{platYt1}$, $s_{platYt2}$, and s_{platZt} in which case $s_{roughX}=s_{roughY}=s_{roughZ}$

$U^{\blacksquare}$	= the expanded uncertainty of a poly1 or poly2 thickness measurement
$u_c^{\blacksquare}$	= the combined standard uncertainty of the poly1 or poly2 thickness
u_{calN}	= the component in the combined standard uncertainty calculation for step height measurements that is due to the uncertainty of the measurements taken across the physical step height standard, where the subscript N is A for the measurement of A and B for the measurement of B
u_{certN}	= the component in the combined standard uncertainty calculation for step height measurements that is due to the uncertainty of the value of the physical step height standard used for calibration, where the subscript N is A for the measurement of A and B for the measurement of B
u_{cJest}	= estimated value for the combined standard uncertainty of J_{est}
u_{driftN}	= the component in the combined standard uncertainty calculation for step height measurements that is due to the amount of drift during the data session, where the subscript N is A for the measurement of A and B for the measurement of B
$u_{linearN}$	= the component in the combined standard uncertainty calculation for step height measurements that is due to the deviation from linearity of the data scan, where the subscript N is A for the measurement of A and B for the measurement of B
u_{LstepN}	= the component in the combined standard uncertainty calculation for step height measurements that is due to the measurement uncertainty of the step height across the length of the step, where the length is measured perpendicular to the edge of the step, and where the subscript N is A for the measurement of A and B for the measurement of B
$u_{repeat(samp)N}$	= the component in the combined standard uncertainty calculation for step height measurements that is due to the repeatability of measurements taken on step height test structures processed similarly to the one being measured, where the subscript N is A for the measurement of A and B for the measurement of B
$u_{repeat(shs)N}$	= the component in the combined standard uncertainty calculation for step height measurements that is due to the repeatability of measurements taken on the physical step height standard, where the subscript N is A for the measurement of A and B for the measurement of B
u_{WstepN}	= the component in the combined standard uncertainty calculation for step height measurements that is due to the measurement uncertainty of the step height across the width of the step, where the width is measured parallel to the edge of the step, and where the subscript N is A for the measurement of A and B for the measurement of B
$\bar{z}_{6aveN}$	= the uncalibrated average of the six calibration measurements from which σ_{6aveN} is found, where the subscript N is A for the measurement of A and B for the measurement of B
$\bar{z}_{aveN}$	= the average of the twelve calibration measurements (taken along the physical step height standard before and after the data session) used to calculate cal_{zN}, where the subscript N is A for the measurement of A and B for the measurement of B
z_{driftN}	= the uncalibrated positive difference between the average of the six calibration measurements taken before the data session (at the same location on the physical step height standard used for calibration) and the average of the six calibration measurements taken after the data session (at this same location) where the subscript N is A for the measurement of A and B for the measurement of B
z_{linN}	= over the instrument's total scan range, the maximum relative deviation from linearity, as quoted by the instrument manufacturer (typically less than 3 %) where the subscript N is A for the measurement of A and B for the measurement of B
$\bar{z}_{6sameN}$	= the uncalibrated average of the six calibration measurements from which σ_{6sameN} is found, where the subscript N is A for the measurement of A and B for the measurement of B

Table of Contents

List of Figures

List of Tables

Standard Reference Materials®
User's Guide for SRM 2494 and 2495:
The MEMS 5-in-1, 2011 Edition

by
Janet M. Cassard and Jon Geist
Semiconductor Electronics Division, PML

Theodore V. Vorburger
Mechanical Metrology Division, PML

David T. Read
Materials Reliability Division, MML

and
David G. Seiler
Semiconductor Electronics Division, PML

The Microelectromechanical Systems (MEMS) 5-in-1 is a standard reference device sold as a NIST Standard Reference Material (SRM) that contains MEMS test structures on a test chip. The two SRM chips (2494 and 2495) provide for both dimensional and material property measurements. SRM 2494 was fabricated on a multi-user 1.5 μm complementary metal oxide semiconductor (CMOS) process followed by a bulk-micromachining etch. Material properties of the composite oxide layer are reported on the SRM Certificates and described within this guide. [4] SRM 2495 was fabricated using a polysilicon multi-user surface-micromachining MEMS process with a backside etch. The material properties of the first or second polysilicon layer are reported on the SRM Certificates and described within this guide.

The MEMS 5-in-1 contains MEMS test structures for use with five standard test methods on one test chip (from which its name is derived). The five standard test methods are for Young's modulus, step height, residual strain, strain gradient, and in-plane length measurements. The first two of these five standard test methods have been published through the Semiconductor Equipment and Materials International (SEMI) in November 2009. The remaining three standard test methods have been published through the American Society for Testing and Materials (ASTM) in December 2005. All five of these standard test methods include round robin precision and bias data.

The Certificates accompanying these SRMs typically report eight properties. In addition to the five properties mentioned in the previous paragraph, residual stress, stress gradient, and thickness are also reported. The values for the first two of these properties are obtained from equations provided in the Young's modulus standard test method. The value for the third property (thickness) is obtained from step height measurements using the step height standard test method. Therefore, to determine the eight properties reported here, five standard test methods are used.

The MEMS 5-in-1 will allow users of the five standard test methods to compare NIST measurements with their own, thereby validating their use of the documentary standard test methods. To perform the calculations, the SRM utilizes the on-line data analysis sheets on the MEMS Calculator Web Site (Standard Reference Database 166) accessible via the National Institute of Standards and Technology (NIST) Data Gateway (http://srdata.nist.gov/gateway/) with the keyword "MEMS Calculator."

Key words: ASTM, cantilevers, fixed-fixed beams, interferometry, length measurements, MEMS, residual strain, residual stress, round robin, SEMI, SRM, step height measurements, strain gradient, stress gradient, test structures, thickness, vibrometry, Young's modulus measurements

---

[4] Contribution of the National Institute of Standards and Technology; not subject to copyright in the U.S.

Introduction

The Microelectromechanical System (MEMS) [5] 5-in-1 is a standard reference device sold as a NIST Standard Reference Material (SRM) in the form of a test chip that contains test structures for five standard test methods. The five standard test methods are for Young's modulus [1], residual strain [2], strain gradient [3], step height [4], and in-plane length [5] measurements as documented in the Semiconductor Equipment and Materials International (SEMI) standard test method MS4, the American Society for Testing and Materials International (ASTM) standard test method E 2245, ASTM standard test method E 2246, SEMI standard test method MS2, and ASTM standard test method E 2244, respectively. SEMI standard test method MS4 also contains equations for the residual stress and stress gradient and SEMI standard test method MS2 can be used to obtain thickness measurements using the electro-physical technique [6] for SRM 2494 and the opto-mechanical technique [7] for SRM 2495.

SRM 2494, as depicted in Fig. 1, was fabricated on a multi-user 1.5 μm complementary metal oxide semiconductor (CMOS) process [8] followed by a bulk-micromachining etch and SRM 2495, as depicted in Fig. 2, was fabricated using a multi-user polysilicon surface-micromachining MEMS process [9] with a backside etch.

The National Institute of Standards and Technology (NIST) organized two round robin experiments: the 2008–2009 SEMI MEMS Young's Modulus and Step Height Round Robin Experiment [10] and the 2002 ASTM MEMS Length and Strain Round Robin Experiment [11,12]. The purpose of these experiments was to obtain round robin precision and bias data for the five standard test methods and to educate the round robin participants concerning these test methods. The incorporation of the round robin data (where each data set passed verification checks) into the test methods had the effect of validating the standard test methods with reproducibility and bias data. Therefore, the MEMS 5-in-1 is associated with five validated standard test methods.

A round robin user's guide, developed at NIST, was written for each round robin experiment. (These can be downloaded from the URL specified in [13].) For the SEMI standard test methods, the technical details and the round robin results are presented in the article entitled, "MEMS Young's Modulus and Step Height Measurements with Round Robin Results" [10]. For the ASTM standard test methods, the technical details can be found in the article entitled, "MEMS Length and Strain Measurements Using an Optical Interferometer" [14] with additional details, parameter variations as a function of time, and some round robin results presented in the article entitled, "Round Robin for Standardization of MEMS Length and Strain Measurements" [12]. A more detailed uncertainty analysis (with round robin data) is presented in the article entitled, "MEMS Length and Strain Round Robin Results with Uncertainty Analysis" [11].

Thicknesses as obtained using the electro-physical technique (which obtains the thicknesses of all the layers in a CMOS process) are presented in the article entitled, "Electro-physical technique for post-fabrication measurements of CMOS process layer thicknesses" [6]. Using this technique (in conjunction with SEMI standard test method MS2 [4]), the beam oxide thickness for SRM 2494 (which is composed of four oxide layers) is obtained, as described in Sec. 8 of this SP 260. Thicknesses obtained using the opto-mechanical technique (which utilizes stiction, the adherence of beams to an underlying layer) are presented in the article entitled, "New Optomechanical Technique for Measuring Layer Thickness in MEMS Processes" [7]. Using this technique (in conjunction with SEMI standard test method MS2 [4]) the first or second polysilicon layer thickness for SRM 2495 is obtained as described in Sec. 8 of this SP 260.

The consolidation of the above-mentioned articles (as pertains to the MEMS 5-in-1) with the addition of most of the material in Sec. 1 (especially Sec. 1.3 through Sec. 1.14 inclusive) forms the bulk of this user's guide for the MEMS 5-in-1. Therefore, this document provides one point of reference (in combination with the standard test methods) for those considering the acquisition of a MEMS 5-in-1 and for those in possession of a MEMS 5-in-1. The purpose of the SRM is to allow users to compare the data and results from their in-house measurements (using the SEMI [1,4] and ASTM [2,3,5] standard test methods) with NIST measurements and results (using the same SEMI and ASTM standard test methods and same test

---

[5] MEMS are also referred to as microsystems technology (MST) and micromachines.

structures), thereby validating their use of the documentary standard test methods for Young's modulus, residual strain, strain gradient, step height, in-plane length, residual stress, stress gradient, and thickness, as described in this SP 260.

The Young's modulus measurements are taken with an optical vibrometer, stroboscopic interferometer, or comparable instrument. The measurements using the other four test methods are taken with an optical interferometer or comparable instrument. However, a stroboscopic interferometer can be used for all five standard test methods. To calculate the MEMS properties, measurements are initially input on the pertinent on-line data analysis sheet on the MEMS Calculator Web Site (Standard Reference Database 166) accessible via the National Institute of Standards and Technology (NIST) Data Gateway (http://srdata.nist.gov/gateway/) with the keyword "MEMS Calculator" [13]. Then, the "Calculate and Verify" button is clicked to obtain the results. The data are verified by checking to see that all the pertinent boxes in the verification section at the bottom of the data analysis sheet say "ok." If one or more of the boxes say "wait," the issue is addressed, if necessary, by modifying the inputs and recalculating.

Each MEMS 5-in-1 is accompanied by a Certificate and completed data analysis sheets using NIST measurements in the calculations. User in-house measurements can then be compared with the NIST measurements supplied on these data analysis sheets to facilitate the validation of the use of the documentary standard test methods.

The SRM Certificate provides a NIST certified value for both Young's modulus and step height and a NIST reference value for the other parameters (residual strain, strain gradient, in-plane length, residual stress, stress gradient, and thickness). A NIST certified value is a value reported on an SRM Certificate/Certificate of Analysis for which NIST has the highest confidence in its accuracy in that all known or suspected sources of bias have been fully investigated or accounted for by NIST [15]. A NIST reference value is a best estimate of the true value provided on a NIST Certificate/Certificate of Analysis/Report of Investigation where all known or suspected sources of bias have not been fully investigated by NIST [15].

The MEMS 5-in-1 is considered an SRM because its Certificate provides certified values for both step height and Young's modulus. For step height, step height test structures were used in a round robin to assess the measurements in the SEMI MS2 standard test method. This measurement procedure establishes traceability to an accurate realization of the meter as defined in the International System of Units [16, 17]. The certified step height value is accompanied by an uncertainty at a stated level of confidence. In like manner, for Young's modulus, cantilevers were used in a round robin to assess the measurements in the SEMI MS4 standard test method and the certified Young's modulus value is accompanied by an uncertainty at a stated level of confidence. However, for both SRMs 2494 and 2495, an effective Young's modulus is reported due to non-idealities associated with the geometry and/or composition of the cantilever and/or beam support (as discussed in Sec. 2). For both SRMs, measurements taken by a user on a cantilever using SEMI standard test method MS4 can still be compared with NIST measurements to validate the user's use of the documentary standard test method. The units for Young's modulus are in pascals (or kilograms per meter second squared) with major contributors to the Young's modulus value [the frequency and the step height measurements (involved in the determination of the thickness)] providing traceability to primary standards at NIST. Contact the NIST SRM Program Office [18] to order a MEMS 5-in-1 which comes with a Certificate and the pertinent data analysis sheets.

Section 1 of this SP 260 provides details associated with the instrumentation used for measurements taken on the MEMS 5-in-1 along with details associated with the design, fabrication, measurement, and certification of the MEMS 5-in-1. Sections 2 through 8 discuss the properties of Young's modulus, residual strain, strain gradient, step height, in-plane length, residual stress and stress gradient, and thickness, respectively. The summary is given in Sec. 9. Reproductions of the MEMS Calculator Web-based data analysis sheets [13] used for recording the MEMS 5-in-1 data and making calculations are given in Appendix 1 through Appendix 7. The propagation of uncertainty technique [19-21] is used throughout this SP 260 to calculate uncertainties. A brief overview of this technique is given in Appendix 8.

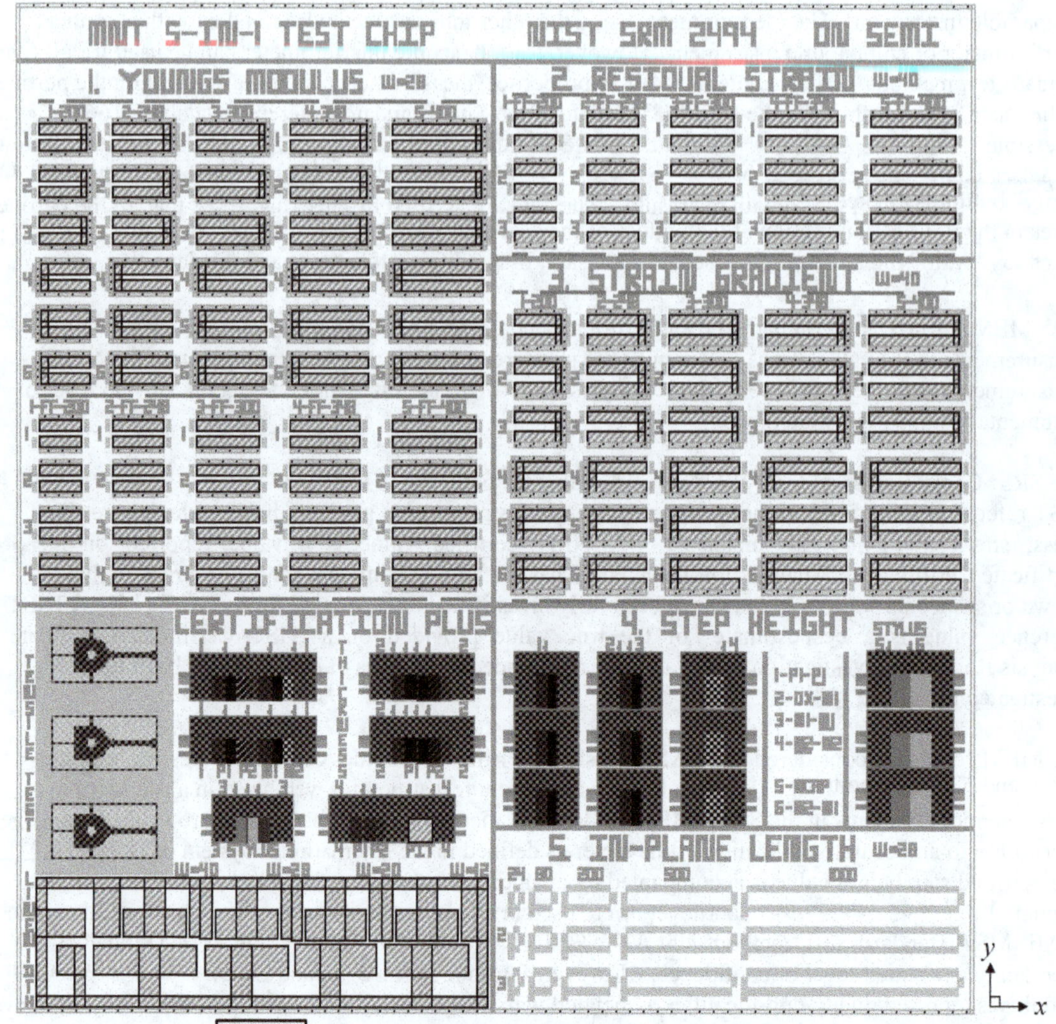

Figure 1 The MEMS 5-in-1 test chip design for SRM 2494,
fabricated on a multi-user 1.5 μm CMOS process [8] followed by a bulk-micromachining etch.
Measurements for the five standard test methods are taken in the applicable group of test structures.
Section 1.2 describes the overall layout of this MEMS 5-in-1 test chip,
with specific test structure design details given in the first subsection of each chapter
(i.e., Sec. 2.1, 3.1, 4.1, 5.1, 6.1, 8.1).

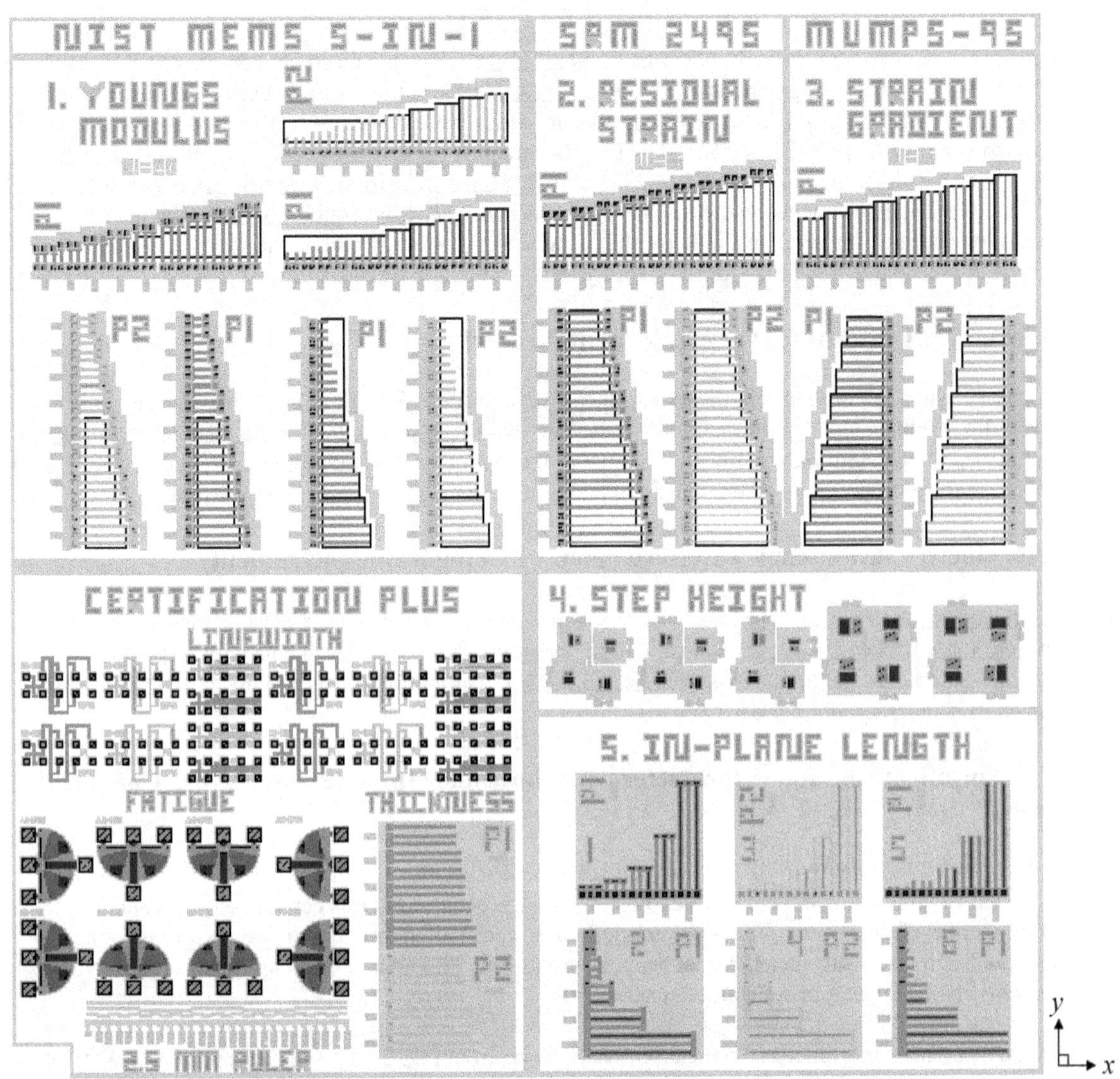

Figure 2. The MEMS 5-in-1 test chip design for SRM 2495, fabricated using a polysilicon multi-user surface-micromachining MEMS process [9] with a backside etch. Measurements for the five standard test methods are taken in the applicable group of test structures. Section 1.2 describes the overall layout of this MEMS 5-in-1 test chip, with specific test structure design details given in the first subsection of each chapter (i.e., Sec. 2.1, 3.1, 4.1, 5.1, 6.1, 8.1).

1 The MEMS 5-in-1

This section provides details concerning the MEMS 5-in-1. It is divided into 14 parts. Section 1.1 provides instrument specifications and validation procedures for the vibrometer, stroboscopic (or optical) interferometer, and comparable instruments. Section 1.2 describes the MEMS 5-in-1 chips shown in Figs. 1 and 2 in the Introduction. The sections (Sec. 1.3 through 1.7) that follow are presented more or less in a time sequence with respect to the tasks performed. Sec. 1.3 describes the classification of the SRM 2494 chips before the post-processing done at NIST and Sec. 1.4 describes the post processing of these chips as well as the post processing of the SRM 2495 chips (before they were delivered to NIST). The SRM 2495 chips were then delivered to NIST and Sec. 1.5 describes the pre-package inspection of both SRM chips (which includes the classification of the SRM 2495 chips), Sec. 1.6 describes the packaging, and Sec. 1.7 describes the NIST measurements on the MEMS 5-in-1. Then, Sec. 1.8 through 1.14 describe the SRM Certificates, traceability, material available for the MEMS 5-in-1, storage and handling, measurement conditions and procedures for the customer, stability tests, and length of certification.

1.1 Overview of Instruments / Equipment Needed

For the MEMS 5-in-1, an optical vibrometer, stroboscopic interferometer, or comparable instrument is required for the Young's modulus measurements, as specified in Sec. 1.1.1. For the residual strain, strain gradient, and in-plane length measurements, an optical interferometer is required, as specified in Sec. 1.1.2. Step height measurements can be taken with an optical interferometer or comparable instrument.

1.1.1 Vibrometer, Stroboscopic Interferometer, or Comparable Instrument

The specifications for an optical vibrometer, stroboscopic interferometer, or comparable instrument are given in Sec. 1.1.1.1. Sec. 1.1.1.2 gives a validation procedure for these instruments with respect to frequency measurements. (If applicable, see Sec. 1.1.2.2 for a validation procedure for these instruments with respect to height measurements.)

1.1.1.1 Specifications for Vibrometer, Stroboscopic Interferometer, or Comparable Instrument

For Young's modulus measurements, a non-contact optical vibrometer, non-contact optical stroboscopic interferometer, or an instrument comparable to one of these is required that is capable of non-contact measurements of surface motion. This section briefly describes the operation and specifications for a typical single beam laser vibrometer, a dual beam laser vibrometer, and a stroboscopic interferometer. The specifications can be applied to comparable instruments.

For a single beam laser vibrometer, a typical schematic is given in Fig. 3. The signal generator shown in this figure excites the sample via a piezoelectric transducer (PZT). The measurement beam is positioned on the sample and is reflected back to the beam splitter where it combines with the reference beam. This interference signal at the beam splitter is comparable to the frequency difference between the beams which is proportional to the instantaneous velocity of the vibration parallel to the measurement beam. The photodetector in this figure records this interference signal as an electrical signal and the velocity decoder provides a voltage proportional to the instantaneous velocity. The Bragg cell is used to determine the sign of the velocity.

For a dual beam laser vibrometer the second beam emanates from the beam splitter in Fig. 3. The measurement beam is positioned on the sample (for example, positioned near the tip of a cantilever). The second beam, the reference beam, is also positioned on the sample (for example, positioned to a different location such as on the support region at the base of the cantilever). The two beams are reflected back to the beam splitter where they optically combine. In this case, the reference beam directly eliminates any movement of the sample also experienced by the measurement beam.

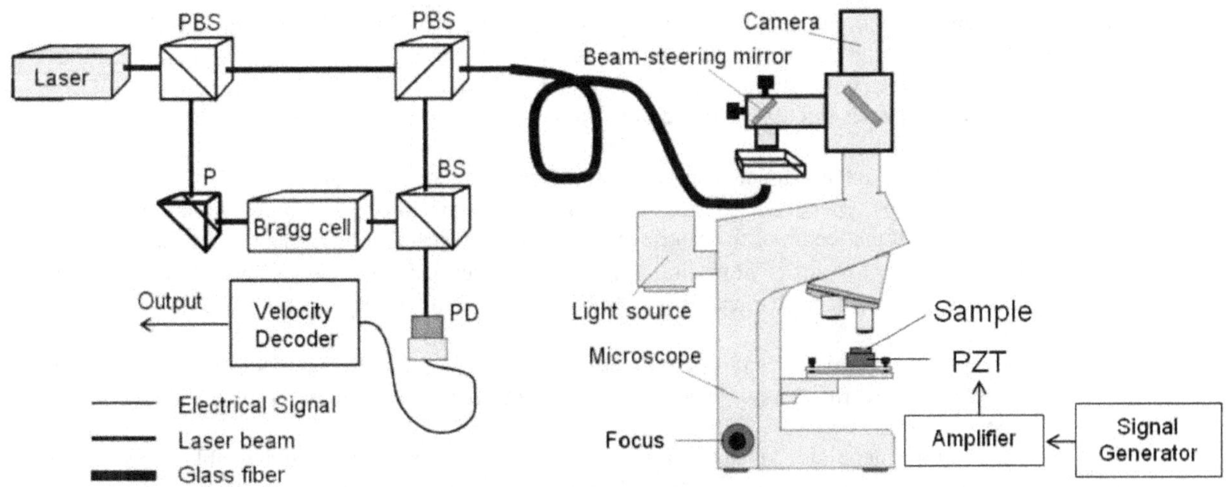

Figure 3. *Schematic of a setup used at NIST for a single beam laser vibrometer.*
(PBS indicates a polarizing beam splitter; BS indicates a beam splitter;
P indicates a prism; and PD indicates a photodetector.)
For a dual beam vibrometer, the reference beam emanates from the beam splitter.

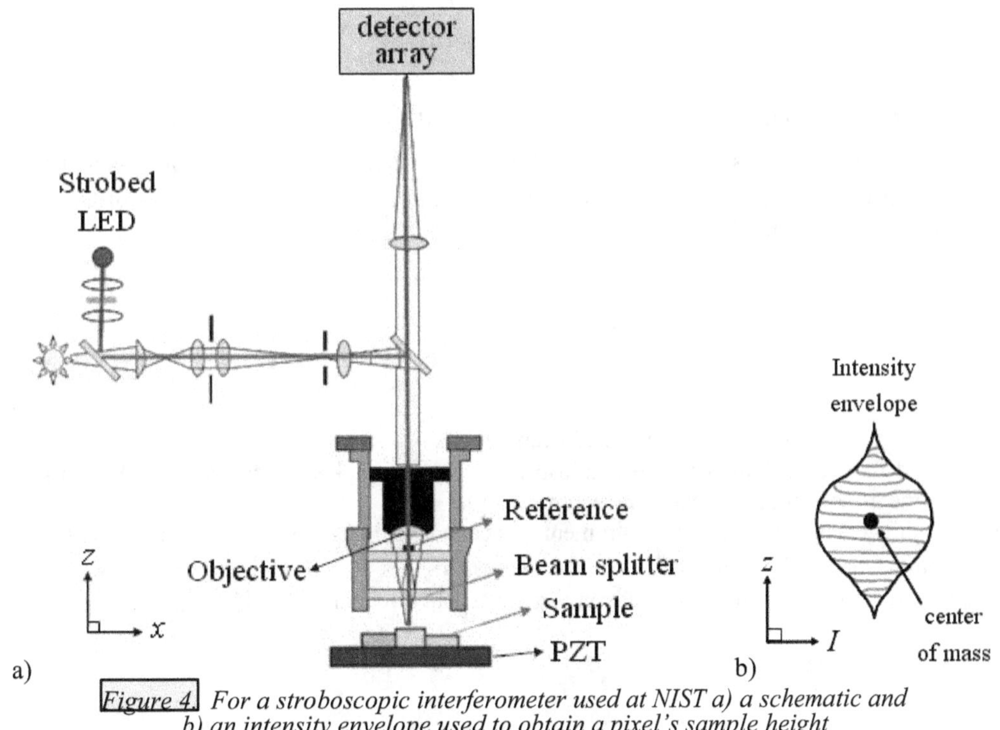

Figure 4. *For a stroboscopic interferometer used at NIST a) a schematic and*
b) an intensity envelope used to obtain a pixel's sample height

For a stroboscopic interferometer, a simplified schematic of a typical setup is shown in Fig. 4. The stroboscopic interferometer can operate in static mode or dynamic mode. In static mode, the topography of the sample can be obtained. The incident light travels to the beam splitter where half of the light travels to the sample then back to the beam splitter and the other half of the light is reflected to a reference surface then back to the beam splitter where the two paths of light form interference light fringes. The software records an intensity envelope incorporating these fringes as seen in Fig. 4(b) as the interferometer scans downward. For each pixel location, the peak contrast of the fringes, phase, or both are used to determine the sample height. This is done for each pixel in the field of view to obtain the topography of the sample surface. In dynamic mode, the sample is typically secured to the top of a PZT then actuated. The incident light is strobed at this same frequency. The interferometer performs a downward scan as done for static measurements for various combinations of phase, frequency, and drive signal to obtain successive 3D images as the sample cycles through its range of motion.

Specifications for the above instruments or a comparable instrument are as follows:
1. The microscope objective used for measurement should be able to encompass in its field of view at least half of the length of the cantilever or fixed-fixed beam being measured. It should also be chosen to allow for sufficient resolution of the cantilever or fixed-fixed beam. Typically, a 4 × and a 20× objective will suffice with the 4 × objective used to initially locate the cantilever or fixed-fixed beam.
2. The resonance frequency is typically determined as the frequency of the tallest peak in a plot of magnitude versus frequency. Therefore, the instrument should be able to produce a magnitude versus frequency plot.
3. To obtain the magnitude versus frequency plot, the signal generator should be able to produce a waveform function (such as a periodic chirp function[6] or a sine wave function[7]).
4. The instrument should be capable of obtaining 3-D images of oscillations in order to identify the oscillation mode shape, in order to confirm that the proper mode has been actuated.
5. A lower bound estimate for the maximum frequency of the instrument needed for a resonating cantilever, $f_{caninit}$, is at least the value calculated using the following equation [10,22]:[8]

$$f_{caninit} = \sqrt{\frac{E_{init}t^2}{38.330\rho L_{can}^4}}, \qquad (1)$$

where E_{init} is an initial estimate for the Young's modulus value of the thin film layer, t is the thickness, ρ is the density, and L_{can} is the suspended cantilever length. A lower bound estimate for the maximum frequency of the instrument needed for a resonating fixed-fixed beam, $f_{ffbinithi}$, is at least the value calculated using the following equation [10,22][9]

$$f_{ffbinithi} = \sqrt{\frac{E_{init}t^2}{0.946\rho L_{ffb}^4}}, \qquad (2)$$

where L_{ffb} is the suspended fixed-fixed beam length.
6. If fixed-fixed beams are measured, an instrument that can make differential measurements (e.g., with the use of two laser beams) is recommended.
7. If cantilevers are measured, an instrument that can make differential measurements is especially recommended a) for estimated resonance frequencies less than 10 kHz and b) if the value for p_{diff} as calculated in the following equation is greater than or equal to 2 % [10,23][10]

---

[6] The periodic chirp function is recommended due to its speed and ability to produce a reproducible resonance frequency without averaging. This function is periodic within the time window with sinusoidal signals (in the selected frequency range and of the same approximate amplitude) emitted at the same time for all fast Fourier transform (FFT) frequencies and with the phases adapted to maximize the energy of the resulting signal.

[7] Although a periodic chirp function is recommended, a sine wave sweep function can produce a resonance frequency however the results can be affected by the direction of the sweep if there is not a sufficient amount of time between measurements.

[8] By inserting the inputs into the correct locations on the appropriate NIST MEMS Calculator Web page [13], the given calculation can be performed on-line in a matter of seconds.

[9] Ibid.

[10] Ibid.

$$p_{diff} = \left(1 - \sqrt{1 - \frac{1}{4Q^2}}\right)100\% .$$ (3)

In the above equation, the Q-factor, Q, for a cantilever can be estimated using the following equation [23].[11]

$$Q = \left[\frac{W_{can}\sqrt{E_{init}}\,\rho}{24\mu}\right]\left(\frac{t}{L_{can}}\right)^2 ,$$ (4)

where μ is the viscosity of the atmosphere surrounding the cantilever (in air, $\mu = 1.84 \times 10^{-5}$ Ns/m^2 at 20° C) and W_{can} is the suspended cantilever width.

1.1.1.2 Validation Procedure for Frequency Measurements

The optical vibrometer, stroboscopic interferometer, or comparable instrument may need to be taken out of service for a variety of reasons. For example, it may need to be sent to the instrument manufacturer (say, every two years) to undergo preventive maintenance. Or, it may undergo a software upgrade. In any event, to accept the instrument back into service to obtain resonance frequency measurements, its performance needs to be validated. The following is done at NIST for a dual beam optical vibrometer (where the steps should be adjusted, as appropriate, for a stroboscopic interferometer or comparable instrument):

1. The instrument is calibrated as specified in Sec. 2.2.
2. While in the acquisition mode, the settings are loaded from a previous file. [This can be done by clicking on "File" then "Load Settings." Then, a file is chosen, for example, a file used for the round robin repeatability data.]
   a. The measurement windows in the vibrometer software are set up, if necessary.
   b. The software settings are checked to ensure they are the same as typically used.
3. The measurement beam and the reference beam are focused.
4. The PZT is connected and checked to ensure that the vibration is audible for a periodic chirp function between 5 kHz and 20 kHz.
5. The resonance frequency is obtained from a previously measured and reliable source. At NIST, a cantilever on the round robin chip from which the repeatability data were extracted is currently used. Therefore, in the Young's modulus section of the round robin chip, resonance frequency measurements are taken on a cantilever with L=200 μm, a cantilever with L=300 μm, or a cantilever with L=400 μm.
6. Data Analysis Sheet YM.3 [13] is filled out using the measured resonance frequencies and the same inputs as used for the round robin (see Table YM4). The resulting Young's modulus value is called M_{200}, M_{300}, or M_{400} for the three different length cantilevers and u_{ML} is the combined standard uncertainty for the given measurement (where the subscript "L" refers to the length of the cantilever that was measured.). The Young's modulus values that were obtained during the round robin are called C_{200}, C_{300}, and C_{400} where u_{CL} is the combined standard uncertainty for the given certified measurement. The difference, D_L, is calculated for the measured cantilever of length L, using the following equation:

$$D_L = |M_L - C_L| ,$$ (5)

where once again the subscript "L" refers to the length of the cantilever that was measured. The uncertainty of the difference, u_{DL}, is calculated using the following equation:

$$u_{DL} = \sqrt{u_{ML}^2 + u_{CL}^2} .$$ (6)

---

[11] Ibid.

The instrument is accepted back into service if, for the measured cantilever, the following equation is satisfied:

$$D_L \leq 2u_{DL}. \tag{7}$$

In addition (or in place of) the above data comparison, the Young's modulus values can be compared as obtained from a cantilever using two different instruments (for example, an optical vibrometer and a stroboscopic interferometer). The above equations [Eq. (5), Eq. (6), and Eq. (7)] can be used where M_L (and u_{ML}) refer to the Young modulus value (and its uncertainty) obtained from Data Analysis Sheet YM.3 [13] using measurements taken with one of the instruments and C_L (and u_{CL}) refer to the Young's modulus value (and its uncertainty) obtained from Data Analysis Sheet YM.3 using measurements taken with the other instrument.

1.1.2 Interferometer or Comparable Instrument

The specifications for an optical interferometer or comparable instrument are given in Sec. 1.1.2.1. Sec. 1.1.2.2 gives a validation procedure for these instruments with respect to height measurements. (If applicable, see Sec. 1.1.1.2 for a validation procedure for these instruments with respect to frequency measurements.)

1.1.2.1 Specifications for Interferometer or Comparable Instrument

For residual strain, strain gradient, and in-plane length measurements, an optical interferometric microscope is used which is capable of obtaining topographical 2-D data traces. (The stroboscopic interferometer operated in the static mode, as described in Sec. 1.1.1, can be used for these measurements.) For step height measurements, an optical interferometric microscope or comparable instrument is used.

Figure 5 is a schematic of a typical optical interferometric microscope that uses the method of coherence scanning interferometry [24], also called vertical scanning interferometry or scanning white light interferometry, for these measurements. However, any calibrated topography measuring instrument that has pixel-to-pixel spacings or sampling intervals as specified in Table 1 and that is capable of performing the test procedure with a vertical resolution finer than 1 nm is permitted. The interferometric microscope or comparable instrument must be capable of measuring step heights to at least 5 μm higher than the step heights to be measured and must be capable of extracting standard deviation and surface roughness values.

Table 1. Interferometer Pixel-to-Pixel Spacing Requirements[a]

Magnification, $\times$	Pixel-to-Pixel Spacing, μm
5	< 2.0
10	< 1.0
20	< 0.50
40	< 0.40
80	< 0.20

[a] This table does not include magnifications at or less than 2.5$\times$ for optical interferometry because the pixel-to-pixel spacings will be too large for this work and the possible introduction of a second set of interferometric fringes in the data set at these magnifications can adversely affect the data. Therefore, magnifications at or less than 2.5$\times$ are not used for measurements taken on SRMs 2494 and 2495.

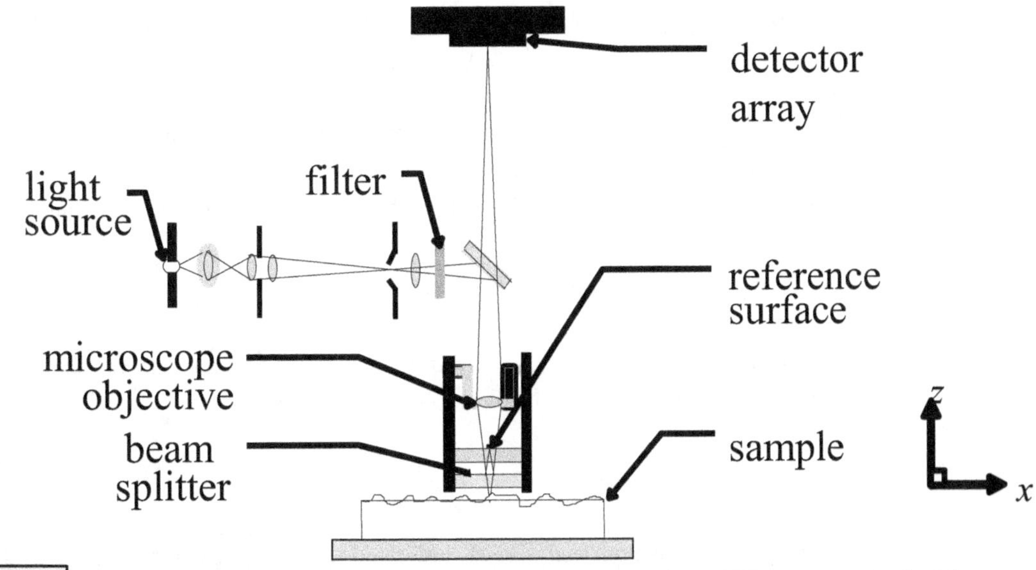

Figure 5. *Schematic of an optical interferometric microscope used at NIST operating in the Mirau configuration where the beam splitter and the reference surface are between the microscope objective and the sample.*

1.1.2.2 Validation Procedure for Height Measurements

The optical interferometer or comparable instrument may need to be taken out of service for a variety of reasons. For example, it may undergo preventive maintenance (say, every two years) or a software upgrade. To accept the instrument back into service, its performance needs to be validated. This can be done with two double-sided physical step height standards.

At NIST, two double-sided, commercial, physical step height standards (for example, a 1.0 µm physical step height standard and a 4.5 µm physical step height standard) are used as a double check to verify static interferometric measurements. These physical step height standards have certified values that are traceable to NIST measurements. The certified value of the physical step height standard is called $cert_{xx}$ where the subscript "xx" denotes the approximate step height value of the physical step height standard (in micrometers) being discussed. This subscript (as well as "yy") will be added to other parameters in this section to faciliate the discussion. (Therefore, $cert_{4.5}$ would refer to the certified value of a 4.5 µm physical step height standard and $cert_{1.0}$ would refer to the certifed value of a 1.0 µm physical step height standard.)

The following six steps (requiring familiarity with the step height measurements in Sec. 5 [4]) should be taken to ensure the calibration of an optical interferometer (where the steps can be adjusted, as appropriate, for a comparable instrument):

1. Initially calibrate the instrument using the instruments prescribed calibration procedure, in order to obtain a reasonable "slope" value, if applicable.

2. The measurements and calculations specified in Sec. 5.2 should be taken and performed on what we will call the "second" physical step height standard.

3. Then, the data for a calibrated step height measurement, M_{xx} (as specifed in Sec 5.3 where M_{xx} would equate with $stepN_{XY}$) should be taken on the left (or right) hand side of what we will call the "first" physical step height standard using three 2-D data traces somewhat evenly spaced across the width of the certified portion of the physical step height standard. Therefore, from the first platform, called $platNX$, three platform measurements ($platNXa$, $platNXb$, and $platNXc$) are recorded (one for each data trace) along with the corresponding standard deviation values ($s_{platNXa}$, $s_{platNXb}$, and $s_{platNXc}$), being careful to extract the measurements within the

certified portion along the length of the step. Similarly, from the second platform, called *platNY*, three platform measurements (*platNYa*, *platNYb*, and *platNYc*) are recorded (one from each of the data traces as was used for the *platNX* meaurements) along with the standard deviation values ($s_{platNYa}$, $s_{platNYb}$, and $s_{platNYc}$). Refer to Sec. 5.3 [4] for measurement and calculation details. The calibration factor, cal_{yy}, from the "second" physical step height standard is used to obtain the calibrated step height measurement using the following equation:

$$M_{xx} = m_{xx} cal_{yy} \ .$$
(8)

where m_{xx} is the uncalibrated step height measurement. For the measurements taken on the 1.0 μm step height standard, the equation would be as follows:

$$M_{1.0} = m_{1.0} cal_{4.5} \ .$$
(9)

4. With a similar uncertainty analysis as presented in Sec. 5.4, the combined standard uncertainty, u_{Mxx}, for the measurement of M_{xx} is determined using the following equation:

$$u_{Mxx} = \sqrt{u_{Lstepxx}^2 + u_{Wstepxx}^2 + u_{certxx}^2 + u_{calxx}^2 + u_{repeat(shs)xx}^2 + u_{driftxx}^2 + u_{linearxx}^2 + u_{repeat(samp)xx}^2} \ ,$$
(10)

where

$$u_{Lstepxx} = \sqrt{\left[\left(\frac{s_{platNXa} + s_{platNXb} + s_{platNXc}}{3}\right)cal_{zyy}\right]^2 - (s_{roughNX}cal_{zyy})^2 + \left[\left(\frac{s_{platNYa} + s_{platNYb} + s_{platNYc}}{3}\right)cal_{zyy}\right]^2 - (s_{roughNY}cal_{zyy})^2} \ ,$$
(11)

$$u_{Wstepxx} = STDEV(M_{xxa}, M_{xxb}, M_{xxc}) \ ,$$
(12)

$$u_{certxx} = u_{certyy}\left|\frac{M_{xx}}{M_{yy}}\right| = \frac{\sigma_{certyy}}{cert_{yy}}\left|M_{xx}\right| \ ,$$
(13)

$$u_{calxx} = u_{calyy}\left|\frac{M_{xx}}{M_{yy}}\right| = \frac{\sigma_{6aveyy}}{\bar{z}_{6aveyy}}\left|M_{xx}\right| \ ,$$
(14)

$$u_{repeat(shs)xx} = u_{repeat(shs)yy}\left|\frac{M_{xx}}{M_{yy}}\right| = \frac{\sigma_{6sameyy}}{\bar{z}_{6sameyy}}\left|M_{xx}\right| \ ,$$
(15)

$$u_{driftxx} = u_{driftyy}\left|\frac{M_{xx}}{M_{yy}}\right| = \frac{z_{driftyy}cal_{zyy}}{2\sqrt{3}cert_{yy}}\left|M_{xx}\right| \ ,$$
(16)

$$u_{linearxx} = u_{linearyy} \left| \frac{M_{xx}}{M_{yy}} \right| = \frac{z_{linyy}}{\sqrt{3}} \left| M_{xx} \right| , \quad \text{and} \tag{17}$$

$$u_{repeat(samp)xx} = \sigma_{repeat(samp)xx} \left| M_{xx} \right| , \tag{18}$$

where the equations for u_{Lstep}, u_{Wstep}, u_{cert}, u_{cal}, $u_{repeat(shs)}$, u_{drift}, u_{linear}, and $u_{repeat(samp)}$ are given in Table SH3 (in Sec. 5.4) replacing occurances of "$stepN_{XY}$" with "M_{xx}" or "M_{yy}," as appropriate. Consult Sec. 5.4 for details including the identification of each uncertainty component and the identification of the other parameters specified in the table. For the measurements taken on the 1.0 μm step height standard, Eq. (10) becomes:

$$u_{M1.0} = \sqrt{ \begin{array}{l} u^2_{Lstep1.0} + u^2_{Wstep1.0} + u^2_{cert1.0} + u^2_{cal1.0} + \\ u^2_{repeat(shs)1.0} + u^2_{drift1.0} + u^2_{linear1.0} + u^2_{repeat(samp)1.0} \end{array} } , \tag{19}$$

where the calibration factor, $cal_{4.5}$, is used to obtain calibrated values, as appropriate.

5. The calibrated step height measurement, M_{xx}, is compared with the certified value, $cert_{xx}$, for that physical step height standard. To do this, the difference, D_{xx}, between the values is examined using the following equation:

$$D_{xx} = \left| M_{xx} - cert_{xx} \right| . \tag{20}$$

For the measurements taken on the 1.0 μm step height standard, the equation would be as follows:

$$D_{1.0} = \left| M_{1.0} - cert_{1.0} \right| . \tag{21}$$

6. The uncertainty of the difference, u_{Dxx}, is calculated using the following equation:

$$u_{Dxx} = \sqrt{ u^2_{Mxx} + \sigma^2_{certxx} } . \tag{22}$$

For the measurements taken on the 1.0 μm step height standard, the equation would be as follows:

$$u_{D1.0} = \sqrt{ u^2_{M1.0} + \sigma^2_{cert1.0} } . \tag{23}$$

7. Repeat the above five steps by calling the "second" physical step height standard the "first" and the previous "first" physical step height standard the "second."

8. The instrument is accepted back into service if the following equations are satisfied:

$$D_{1.0} \leq 2u_{D1.0} \quad \text{and} \tag{24}$$

$$D_{4.5} \leq 2u_{D4.5} . \tag{25}$$

If measurements taken on the two physical step height standards are in agreement, according to Eq. (24) and Eq. (25), then measurements on the MEMS 5-in-1 can be taken using either of the physical step height standards, preferrably the one closest in size to the step to be measured. If the measurements are not in agreement, consider increasing the value of z_{lin} until the equations are in agreement. This may be done if the value for z_{lin} is less than 5 %. Otherwise, repeat the above procedure or contact the instrument manufacturer for advice.

If the instrument is such that the calibration can be changed by another user of the instrument and it is not possible to retrieve the precise state of the calibration that was used above (for example, if it is not possible to use the same "slope" value as obtained in the first step) , the measurements on the MEMS 5-in-1 can still be taken using either of the physical step height standards; however, the instrument must first be recalibrated with respect to the chosen physical step height standard and the value of z_{lin} (as obtained or verified in the above calculation) is used in subsequent uncertainty calculations.

1.2 MEMS 5-in-1 Chips

There are currently two types of MEMS 5-in-1 chips; SRM 2494 is fabricated on a multi-user 1.5 μm CMOS process [8] followed by a bulk-micromachining etch and SRM 2495 is fabricated using a polysilicon multi-user surface-micromachining MEMS process [9] with a backside etch. A design rendition of these chips is given in Fig. 1 and Fig. 2, respectively, in the Introduction. As can be seen in these figures, the fabrication process designation is specified in the upper right hand corner. Participants can obtain the design file (in GDS-II format) for each MEMS 5-in-1 from the NIST MEMS Calculator Website [13].

The MEMS 5-in-1 chip for SRM 2494, as shown in Fig. 1, has a maximum designed x dimension of 4600 μm and a maximum designed y dimension of 4700 μm. The mechanical layer used as the suspended portion of the applicable test structures consists of all oxide: namely, the field oxide, the deposited oxide before and after the metal1 deposition, and the glass layer. (The nitride cap, present atop the glass layer when the chips are received from the semiconductor fabrication service, was removed after fabrication using a CF_4+O_2 etch before a post-processing XeF_2 etch that released the beams, as discussed in Sec. 1.4.1.)

The MEMS 5-in-1 chip for SRM 2495, as shown in Fig. 2, has a maximum designed x dimension of 1 cm and a maximum designed y dimension of 1 cm. The mechanical layer of the suspended portion of the applicable test structures is composed of either poly1 (or P1) or poly2 (or P2). These test structures have a "P1" or "P2" label designed in close proximity to them, as can be seen in Fig. 2.

As seen in both Fig. 1 and Fig. 2, each test chip contains six groupings of test structures with the following headings:
1. Young's Modulus,
2. Residual Strain,
3. Strain Gradient,
4. Step Height,
5. In-Plane Length, and
6. Certification Plus.

For the MEMS 5-in-1, we will mainly be concerned with the first through fifth groupings of test structures. Grouping 1 contains the test structures (namely, cantilevers and fixed-fixed beams) for Young's modulus measurements. Grouping 2 contains fixed-fixed beams for residual strain measurements. Grouping 3 contains cantilevers for strain gradient measurements. Grouping 4 contains step height test structures for step height measurements. And, Grouping 5 contains features for in-plane length measurements.

The Certification Plus section contains additional test structures that may complement the existing set of geometrical and material properties. On SRM 2494 depicted in Fig. 1, these additional test structures include tensile test structures, thickness test structures, and a linewidth test structure that can be used to obtain the Young's modulus of the metal2 layer, the thicknesses of all the layers in the process, and the linewidth of select oxide beam widths, respectively. On SRM 2495 depicted in Fig. 2, linewidth, thickness, and fatigue test structures can provide a) the linewidth of either poly1 or poly2 for select beam widths, b) the thickness of the poly1 or poly2 layer along with data for stiction [12] studies, and c) Young's modulus, ultimate strength, and fatigue for the poly1 layer, respectively. A 2.5 mm ruler may also be present.

1.3 Classification of the SRM 2494 Chips

After the SRM 2494 chips are received from the semiconductor fabrication service, they are stored in a plastic storage container to await an initial inspection for classification purposes. The goal of the inspection for the first, second, third, and fifth groupings of test structures mentioned in Sec. 1.2, is to ensure the existence of at least one suitable test structure in each grouping. This implies at least one cantilever (out of 30) within the Young's modulus grouping of test structures , one fixed-fixed beam (out of

---

[12] Refer to the List of Terms.

15) within the residual strain grouping, one cantilever (out of 30) within the strain gradient grouping, and one in-plane length test structure (out of 15) within the in-plane length grouping. For the fourth grouping of test structures, the goal of the inspection is to ensure the existence of at least one step height test structure (out of three) for each of the four uniquely designed step height test structures.

Those beams (both cantilevers and fixed-fixed beams) in the first three groupings of test stuctures are sought with the best looking attachment points of the beam to the beam support, in other words, beams with no "gunk" in these corners or a minimum amount of gunk that is symmetrically located with respect to a line drawn along the length of the beam. The gunk is believed to be metal that was not completely removed during the fabrication process. Typically, good cantilevers can be found because they are designed with both a 0° orientation and a 180° orientation and the gunk tends to be less present or non-existent for one of these orientations. Therefore, with the existence of suitable cantilevers along with a suitable step height test structure and a suitable in-plane length test structure, the inspection criteria tend to be focused on finding a suitable fixed-fixed beam.

The chips we re classified as "acceptable," or "unacceptable" using the following subjective criteria (as given below for the chosen fixed-fixed beam):

1. Chips with no gunk or a minimum amount of gunk in the corners of the chosen fixed-fixed beam would be given an excellent or "acceptable" classification.
2. If the gunk makes the beam appear unsymmetrical with respect to a line drawn along the length of the beam (for example, if there is gunk in one of the three attachment points of the chosen fixed-fixed beam to the beam support), the classification could be either "acceptable" or "unacceptable."
3. As a rule of thumb, as the amount of gunk in the key attachment points increases and as the chosen beams become less symmetrical with respect to a line drawn along the length of the beam, the classification tends to degrade from "acceptable" to "unacceptable."
4. Sample interferometric data is typically obtained on a chosen beam (usually with a $50\times$ objective and a $0.5\times$ field of view lens) as a guarantee that reasonable data can be extracted.

The dividing line between the classification categories is somewhat subjective, so to help ensure consistency with respect to the same criteria, all of the chips received are classified during the same data session and by the same person.

After the chips are assigned a classification, they are stored in a plastic storage container to await post processing.

1.4 Post Processing

The SRM 2494 and SRM 2495 chips both underwent post processing. The post processing of the SRM 2494 chips was performed at NIST and the post processing of the SRM 2495 chips was performed at the fabrication facility before the chips were delivered [9].

1.4.1 Post Processing of the SRM 2494 Chips

The SRM 2494 chips were transported in a plastic container within a zippered bag to a class 100 clean room at NIST for post fabrication. To remove the nitride cap, the chips were etched one at a time with CF_4+O_2 (with an etch rate of approximately 220.0 nm/min) for about two minutes. A slight overetch helps to eliminate stringers.

To ascertain whether or not the nitride cap was removed, the fourth step height test structure, as depicted in Fig. 6(a) in the step height grouping of test structures is used. A stylus instrument is required for this measurement since the top layer is not reflective (a requirement for optical inteferometry). Therefore, in a laboratory environment, stylus step height measurements associated with the first arrow in Fig. 6(a) that is labeled 5 were taken, with the resulting measurement called $step5_{rA+}$ if it was taken before the CF_4+O_2 etch and called *step5 $_{rA}$−* if it was taken after the XeF $_2$ etch,[13] which comes next. Cross-sectional sideviews of the applicable portion of the test structure given in Fig. 6(a) are given in Figs. 6(b and c) for the measurements of *step5 $_{rA}$+* and *step5 $_{rA}$−*, respectively. (See Sec. 5 [4] for guidance in taking this measurement.) Before the etch, as shown in Fig. 6(b), the combined oxide (or ox) atop metal2 (or m2) and nitride (or ni) thickness (or *step5 $_{rA}$+*) is approximately 1.0 μm. For this step [i.e., *step5 $_{rA}$+* shown in Fig. 6(b)], the thickness of the oxide atop m2 is approximately equal to the thickness of the nitride cap (ni), according to the semiconductor fabrication service [8] implying that each of these layers is approximately 0.5 μm, in this case. Therefore, it is assumed that the nitride cap (on this test structure and also on the entire chip) is removed during the etch if the step height measurement on this test structure after the etch [namely, *step5$_{rA}$−* as shown in Fig. 6(c) without the nitride cap] is less than or equal to *step5 $_{rA}$+* divided by 2, or in this case less than or equal to 0.5 μm.

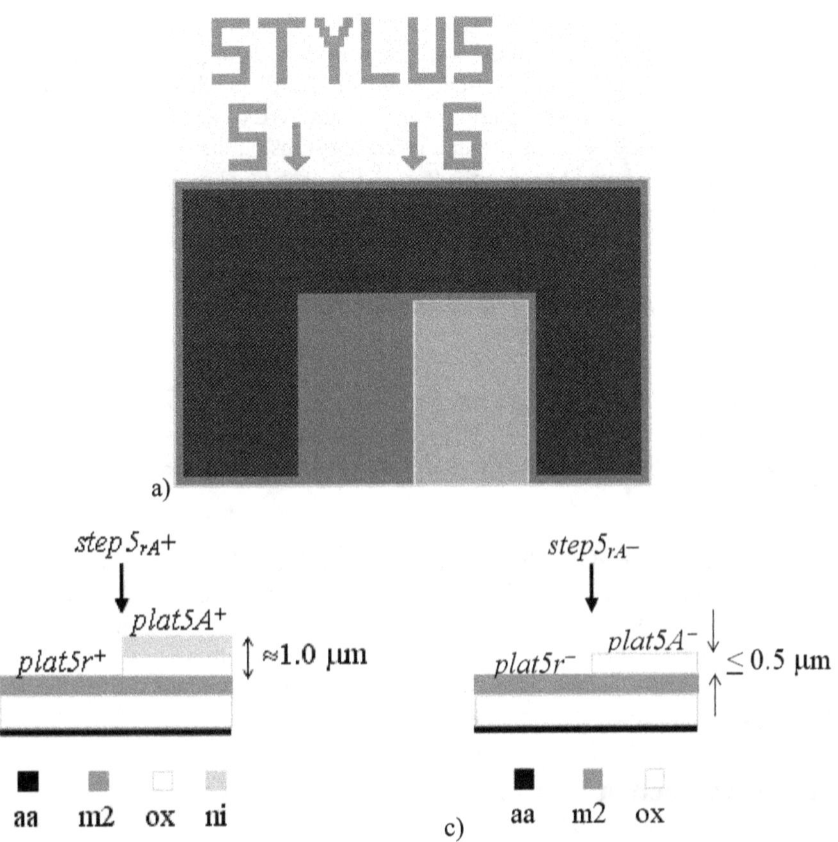

Figure 6. *For the step height test structure used to determine if the nitride cap has been removed, (a) a design rendition, (b) a cross-sectional sideview before the CF$_4$+O$_2$ etch for the step associated with the first arrow in (a) labeled "5," and (c) a cross-sectional sideview of this step after the XeF$_2$ etch. (The active area is labeled "aa.")*

After the removal of the nitride cap, the SRM 2494 chips are isotropically etched one at a time with XeF $_2$ [25] to release the cantilevers and fixed-fixed beams by removing the silicon around and beneath each beam. Fifteen cycles of a XeF$_2$ etch are used where one cycle is as follows:

    1. Starting with a pressure inside the etch chamber of 133.32 Pa (1.0 Torr), XeF$_2$ is released into the chamber until the pressure rises to 399.97 Pa (3.0 Torr).

    2. After 10 s, the XeF$_2$ gas is pumped out.

---

[13] Stylus step height measurements can also be taken in the class 100 clean room after the CF$_4$+O$_2$ etch.

After a cursory microscopic inspection of the chips to ensure that the widest beams (present in the residual strain and strain gradient groupings) are released, the chips are transported in a plastic container within a zippered bag to a laboratory across campus where they are stored in a N_2-filled dry box.

1.4.2 Post Processing of the SRM 2495 Chips

For the SRM 2495 chips processed in a surface-micromachined polysilicon multi-user MEMS process [9], an additional backside etch is required to eliminate any stiction or squeeze film damping phenomena (see Sec. 2.1 for details) associated with cantilever and fixed-fixed beam resonance frequency measurements. An additional layer was added to the designs [as shown in Figs. YM3(a) and YM4(a) in Sec. 2.1] with dimensions that represent the dimensions of the requested opening in the backside of the wafer. The processing facility should be consulted for the appropriate design rules for this layer.

During the post processing, the front side of the wafer is protected, then the wafer thinned (removing the backside films in the process) to a thickness similar to what is used in a silicon-on-insulator (SOI) multi-user MEMS process [9]. Then, a backside photo patterning and RIE (reactive ion etch) are performed, removing the silicon and stopping on the first oxide layer. [It stops on the first oxide layer because the nitride layer, which is normally found between the 2 μm sacrificial oxide layer and the Si wafer, is patterned earlier in the process using a mask derived from the mask used to define the opening requested on the backside of the wafer. Due to the patterning of this nitride layer, the edge of the nitride layer is responsible for an approximate 600 nm vertical transition seen in the structural layer of suspended cantilevers and fixed-fixed beams (as seen in Figs. YM3, YM4, and YM5 for a cantilever). There are two such vertical transitions for fixed-fixed beams.]

An RIE etch is used due to its ability to create more vertical sidewalls (with a 10 $^\circ$ angle) as opposed to the (55° angle) sidewalls from a plasma etch. The RIE-etched sidewalls are sloped so that the digitized areas representing the openings on the backside of the wafer are smaller than the desired openings on the front side of the wafer.

After the backside etch, the exposed sacrificial oxides are etched thereby releasing the beams. A super critical CO_2 dry is performed to minimize stiction for any beams that are designed with an underlying layer. The chips are delivered to NIST with no protective coating, as requested. For delivery, they are typically placed within a sealed clamshell package. Once the chips get to NIST, they are stored in a N_2-filled dry box. They are removed from the clamshell package before the expiration date of the adhesive, if applicable, then inspected, as given in the following section.

1.5 Pre-Package Inspection

Before the SRM 2494 and SRM 2495 chips are packaged, they are inspected in a laboratory environment to determine if they are suitable SRM candidates. The SRM 2495 chips are also inspected for classification purposes. After the pre-package inspection and/or classification, the SRM chips are returned to the N_2-filled dry box to await packaging.

1.5.1 Pre-Package Inspection of the SRM 2494 Chips

The SRM 2494 chips are inspected interferometrically (and/or microscopically) to ensure the following:

1. That there is at least one suitable test structure in the first, second, third, and fifth groupings of test structures (namely, one cantilever in the Young's modulus grouping, one fixed-fixed beam in the residual strain grouping, one cantilever in the strain gradient grouping, and one in-plane length test structure in the in-plane length grouping).
2. That there is at least one suitable step height test structure in the fourth grouping of test structures for each of the four uniquely designed step height test structures. The resulting data from these test structures in combination with data supplied by the semiconductor fabrication

service [8] can be used to obtain the thickness of the composite oxide beams, as given in Sec. 8, for the Young's modulus measurements in Sec. 2.

3. That the chip is correctly classified, as specified in Sec. 1.3, as "acceptable." Only "acceptable" chips are SRM 2494 candidates.

4. That the etched cavities have not merged. Although differences in color between a region that is suspended and a region that is not suspended can be an indicator of non-merged cavities, the existence of a flat region between the cavities (as verified upon examination of an interferometric 2-D data trace between the etched cavities) can be used as additional proof.

5. That the widest beams (located in the residual strain and strain gradient groupings) are released. An interferometric 2-D data trace (taken along the length of one of the cantilevers) with data that curve out-of-plane can be used to verify that the widest beams are released.

6. That the etched cavity is deep enough so that squeeze film damping will not be an issue (see Sec. 2.1 for details). This implies that the gap, d, between the bottom of the suspended cantilever (for Young's modulus measurements) and the top of the underlying layer is greater than or equal to one-third the cantilever width [26].

   The cantilevers for Young's modulus measurements are all designed to be 28 μm in width. Therefore, d should be at least one-third that or 9.3 μm. Since an isotropic XeF$_2$ etch is used to release the beams and since the widest beam on the chip is 40 μm, this implies that the lateral etch must traverse at least half this width (or 20 μm) in order for the beams to release. Therefore, if the vertical etch rate is comparable to the lateral etch rate, then the etched cavity, or d_{cav} as shown in Fig. 7(c), should be at least 20 μm. If we assume here that $d = d_{cav}$, which may be off by up to 0.5 μm for flat beams,[14] then 20 μm is much more than the minimum value of 9.3 μm that is needed to ensure that squeeze film damping will not be an issue.

   As a double-check or if there is uncertainty about the vertical and lateral etch rates of XeF$_2$, interferometric measurements can be taken (both before the CF$_4$+O$_2$ etch and after the XeF$_2$ etch) from the fourth thickness test structure, as shown in Fig. 7(a), in the Certification Plus grouping. In particular, an approximate step height measurement can be taken for the step corresponding to either the 3rd or 4th arrow in this test structure. The absolute value of the step height measurement taken before the CF$_4$+O$_2$ etch, as given by the absolute value of *step4*$_{DE}$+ or *step4*$_{EF}$+ in Fig 7(b), is subtracted from the absolute value of the step height measurement taken after the XeF$_2$ etch, as given by the absolute value of *step4*$_{DE}$− or *step4*$_{EF}$− in Fig. 7(c), to obtain an approximate measurement of the depth of the cavity, d_{cav}, as shown in Fig. 7(c).

7. That there are no stringers or debris present that would adversely affect the data taken on the chosen test structures.

8. That sample static interferometric cantilever and fixed-fixed beam data can be taken on the chosen beams used to obtain strain gradient and residual strain, respectively.

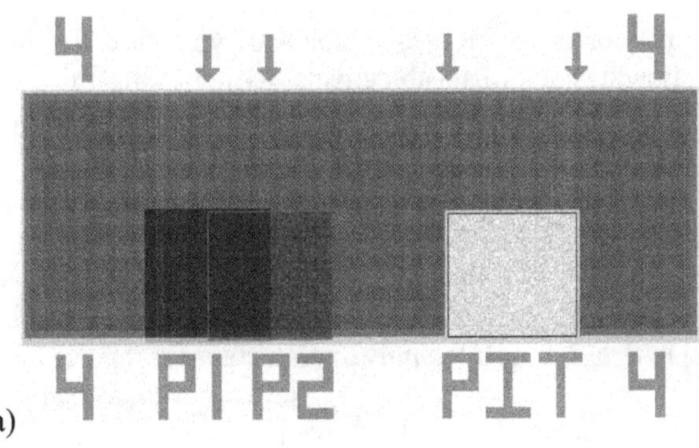

(a)

---

[14] It should be noted that most, if not all, of the beams bend up.

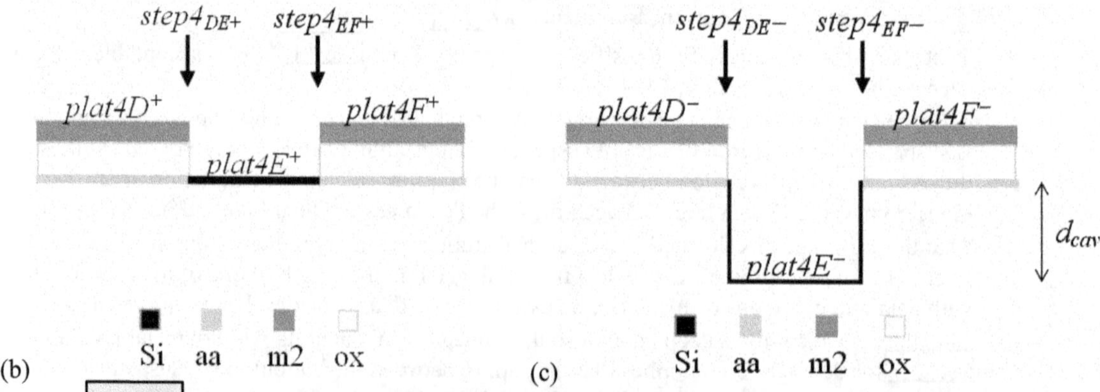

Figure 7. *For the thickness test structure used to determine the depth of the etched cavity (a) a design rendition, (b) a cross-sectional sideview before the CF_4+O_2 etch for the steps associated with the third and fourth arrows in (a), and (c) a cross-sectional sideview after the XeF_2 etch for the steps associated with the third and fourth arrows in (a).*

1.5.2 Pre-Package Inspection of the SRM 2495 Chips

The SRM 2495 chips are inspected interferometrically (and/or microscopically) to ensure the following:

1. That there is at least one suitable test structure in each of the five groupings of test structures for the layer of interest (namely, if poly1 is the layer of interest, then one poly1 cantilever in the Young's modulus grouping, one poly1 fixed-fixed beam in the residual strain grouping, one poly1 cantilever in the strain gradient grouping, one step height test structure in the step height grouping, and one poly1 in-plane length test structure in the in-plane length grouping).

2. The existence of a pegged cantilever (i.e., a cantilever exhibiting stiction) for the layer of interest in the Certification Plus grouping of test structures under the label "Thickness," as seen in Fig. 2. (The resulting data from this test structure can be used to obtain the poly1 or poly2 thickness, as given in Sec. 8 for Young's modulus measurements in Sec. 2.)

3. That useful static interferometric cantilever and fixed-fixed beam data can be taken on the chosen poly1 or poly2 beams. For example, can residual strain and strain gradient data still be taken on the beams if there is no curvature of the beams?

The layer of interest is determined for the given chip. Typically if the above criteria are satisfied for poly1 structures, the poly1 layer is considered the layer of interest unless, for example, the longest poly1 cantilevers do not bend out-of-plane, in which case the poly2 layer is considered the layer of interest assuming the above criteria are met for poly2 structures.

1.5.3 Classification of the SRM 2495 Chips

During the pre-package inspection of the SRM 2495 chips, they are also classified as "acceptable" or "unacceptable" using the following somewhat subjective criteria (as given below):

1. If one of the criteria in Sec. 1.5.2 is not satisfied for the layer of interest, the chip is classified as "unacceptable," and the following criteria can be overlooked.

2. If the anchor attachment point of most of the cantilevers and fixed-fixed beams in at least one key array per grouping of test structures is intact and all or most of the beams are still present in this key array, the chip is given an excellent or "acceptable" classification. [The poly1 arrays are considered key arrays if the poly1 parameters are reported on the SRM Certificates as determined in Sec. 1.5.2. If the poly2 parameters are reported on the SRM Certificates then the poly2 arrays are considered the key arrays. And, the same applies to the arrays in the Young's modulus grouping of test structures.]

3. If a portion of the anchor attachment point of most of the cantilevers and fixed-fixed beams in
    at least one key array per grouping of test structures is missing due to an overetch from the
    backside of the wafer, and all or most of the beams are still present in this key array, the
    classification may be either "acceptable" or "unacceptable."

4. As a rule of thumb, as less of the anchor attachment point remains intact and as the number of
    beams in the key arrays decreases, the classification degrades from "acceptable" to
    "unacceptable."

The dividing line between the classification categories can be somewhat subjective, so to help ensure
consistency with respect to the same criteria, all of the chips received are inspected and classified during
the same data session and by the same person.

1.6 Packaging

After an SRM 2494 or SRM 2495 chip passes the post processing inspection, it is packaged as shown in
Fig. 8; however, the PZT wires may be connected to different pins. The chips are packaged in a laboratory
environment in the following way:

1. The hybrid package has a pin arrangement similar to that shown in Fig. 8(a) with a 1 mm pin-
    to-pin separation. The package is 20.32 mm wide, 27.94 mm long, and 19.05 mm tall (with
    the lid on and including the height of the exposed pins on the bottom of the package). The
    height of the exposed pins on the bottom of the package is 5.08 mm.

2. The PZT is secured to the top of the chip cavity using two thin layers of low stress, non-
    conducting epoxy (Ellsworth Adhesives Resinlab UR 3010 Clear Urethane Encapsulant).
    (The first layer of epoxy ensures that there will not be a conducting path between the package
    and the PZT.)

3. The PZT has the following properties:
    a. The dimensions of the PZT are approximately 5 mm by 5 mm and 2 mm in height.
    b. It is provided with a red and a black wire.
    c. It can achieve a 2.2 μm ($\pm$20 %) displacement at 100 V from DC to 100 kHz.
    d. It has an electrical capacitance of 250 nF ($\pm$20 %).
    e. It has a resonance frequency greater than 300 kHz, at which or above which it shall
       not be operated because that could damage the PZT.

4. Each PZT wire is secured to a package pin using a wire wrap tool. A small amount of solder
    may also be used.

5. The PZT is activated at 10 V and 7000 Hz to ensure that the resulting PZT vibration is barely
    audible and properly connected. (Alternatively, the PZT can be checked with a vibrometer to
    ensure that the vibration is audible for a periodic chirp function between 5 kHz and 20 kHz.)

6. The SRM chips are secured to the top of the PZT.
    a. The SRM 2494 chips are secured using two thin layers of a low stress non-
       conducting epoxy. (The first layer of epoxy ensures that there will not be a
       conducting path between the PZT and the SRM.)
    b. For the SRM 2495 chips, one layer of the low stress non-conducting epoxy is allowed
       to dry on top of the PZT. Then, a thin layer of this epoxy is spread on top of the
       bottom half of the existing epoxy so that only the bottom half of the SRM 2495 chip
       when placed on top of it will be secured to the PZT. Then, the bottom half of the
       SRM 2495 chip is placed atop this thin layer of epoxy so that none of it seeps through
       the portions of the chip that were etched from the backside.

7. The lid (or can) is placed on top of the package to protect the chip and the lid is secured to
    the package with tape before shipment.

The packaged chips are stored in a N$_2$-filled dry box.

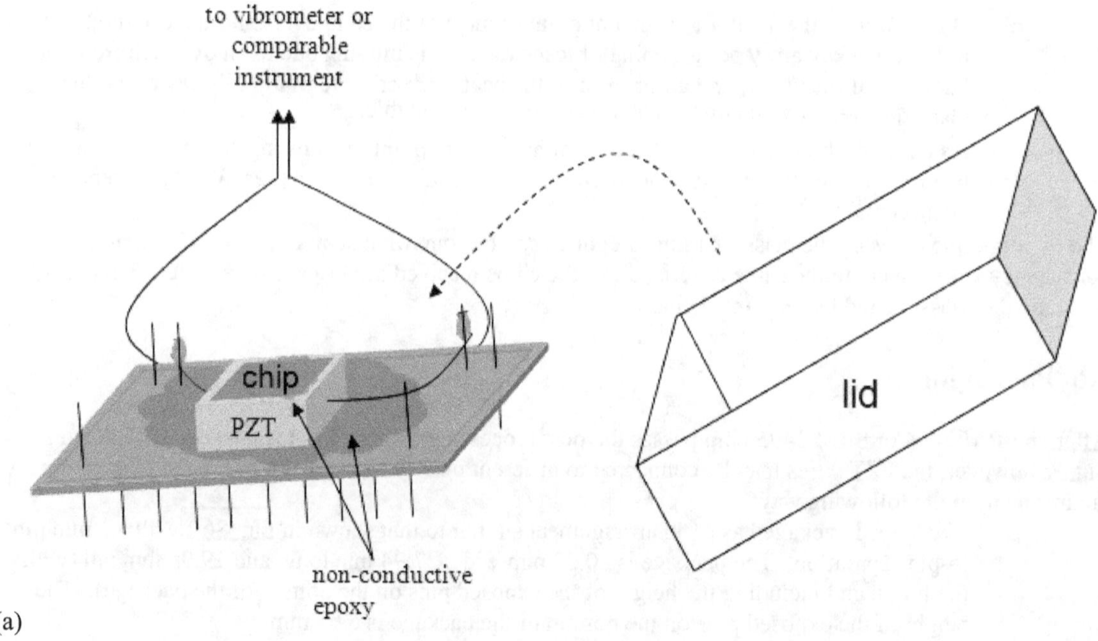

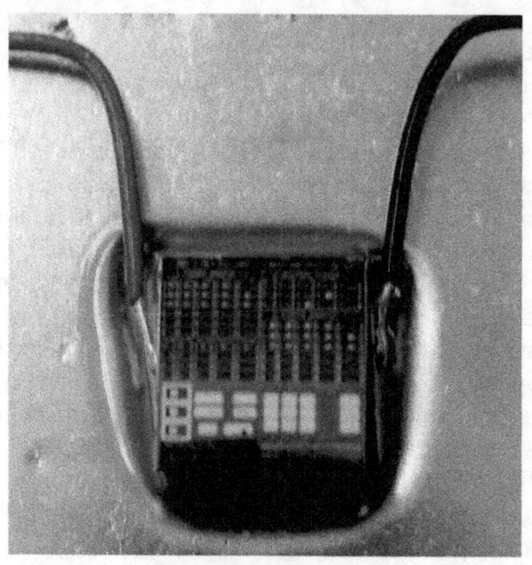

Figure 8. For the MEMS 5-in-1 (a) a drawing of the packaged chip and (b) a photograph of one of the chips inside the package cavity

1.7 NIST Measurements on the MEMS 5-in-1

NIST measurements are taken on the packaged SRMs in a laboratory environment. The chosen test structures are identified and the measurements are taken following the procedures in the applicable standard test method [1-5]. (See also Sec. 2 through Sec. 8, inclusive, for overall guidance.) The data are entered into the pertinent data analysis sheet [3] in order to perform the calculations and verify the data. After verification of the data, the resulting calculated values are entered on the appropriate SRM 2494 or SRM 2495 Certificate.

1.8 The SRM Certificates

SRM Certificates tend to evolve over time. For a current example of the SRM 2494 and 2495 Certificates, see the **Data and Information Files** link on https://www-s.nist.gov/srmors/view_detail.cfm?srm=8096 and https://www-s.nist.gov/srmors/view_detail.cfm?srm=8097, respectively. Both of these certificates as of the writing of this SP 260 include the following:

1. The DOC logo,
2. A serial number for the individually tested SRM,
3. NIST certified values and uncertainties for Young's modulus and step height,
4. NIST reference values and uncertainties for residual strain, residual stress, strain gradient, stress gradient, in-plane length, and thickness,
5. The issue date,
6. The expiration date,
7. An overview,
8. A brief description of the MEMS 5-in-1,
9. Instructions for use,
10. Material available for the MEMS 5-in-1, and
11. References.

1.9 Traceability

Measurement data from the optical interferometer and the stylus profilometer are calibrated from data taken on a certified physical step height standard as specified in Sec. 5.2. The instrument calibration is also checked with the certified value of a second physical step height standard (as detailed in Sec. 1.1.2.2). These commercial physical step height standards are calibrated directly at NIST. Therefore, the certified step height values for the MEMS 5-in-1 are considered NIST-traceable measurements.

Before each data session, the maximum frequency of the optical vibrometer and the stroboscopic interferometer is measured with a 10-digit/s frequency counter that is calibrated by the vendor to provide NIST-traceable measurements. A calibration factor, cal_f, is determined as specified in Sec. 2.2. From this maximum frequency, all other signals are derived and the frequency measurements are multiplied by cal_f to obtain calibrated values. As specified in Sec. 1.1.1.2, the instrument calibration is also checked by comparing one or more resonance frequency measurements   (via Young's modulus calculations) a) with previous resonance frequency measurements (or Young's modulus calculations) us ing the same cantilevers and/or b) with resonance frequency measurements (or Young's modulus   calculations) obtained from another calibrated instrument using the same resonating cantilevers. Therefore, the measured resonance frequency for the MEMS 5-in-1 is considered a NIST-traceable measurement.

The traceability chain for the physical step heights as well as the traceability chain for the certification of frequency can be obtained by sending an email request to mems-support@nist.gov.

1.10 Material Available for the MEMS 5-in-1

After you order a MEMS 5-in-1, you will receive a packaged SRM chip, a Certificate, completed data analysis sheets, and this SP 260. Besides this SP 260, one of the best places from which to obtain

information associated with the MEMS 5-in-1 is the MEMS Calculator Website [13]. The symbol is used on this website to help you quickly find material associated with the MEMS 5-in-1. From this website you can obtain the following:

1. This SP 260,
2. The data analysis sheets (e.g., YM.3, RS.3, SG.3, SH.1.a, L.0, T.1, and T.3),
3. Ordering information for a MEMS 5-in-1,
4. The design file and an accompanying tiff file of each MEMS 5-in-1 chip,
5. The list of the following two SEMI standard test methods [1,4] and three ASTM standard test methods [2-3,5] along with links for ordering information:

- SEMI standard test method MS4, Test Method for Young's Modulus Measurements of Thin, Reflecting Films Based on the Frequency of Beams in Resonance,
- ASTM standard test method E 2245, Test Method for Residual Strain Measurements of Thin, Reflecting Films Using an Optical Interferometer,
- ASTM standard test method E 2246, Test Method for Strain Gradient Measurements of Thin, Reflecting Films Using an Optical Interferometer,
- SEMI standard test method MS2, Test Method for Step Height Measurements of Thin Films, and
- ASTM standard test method E 2244, Test Method for In-Plane Length Measurements of Thin, Reflecting Films Using an Optical Interferometer, and

6. Pertinent references (for downloading from the website) grouped below by topic area:
   - Young's modulus, step height, and thickness [10,27],
   - Step height and thickness [6,7], and
   - Residual strain, strain gradient, and in-plane length [11,14].

This SP 260 focuses on how to use the MEMS 5-in-1 to successfully take measurements with the standard test methods. The above-mentioned references were stepping stones in creating this guide. The references are more general in nature, contain more background information, and provide for more applications of the various test methods.

In addition, during the course of this work, an assessment was performed, as listed below, which specifies the importance of standardization efforts and can be used to guide future standardization efforts:
   - An assessment of the US Measurement System [28]

The appropriate standard test method is used to guide you through the measurements. If you have difficulties understanding the technical basis for the steps in the standard test methods, you can consult this SP 260 or one of the pertinent references [6-7,10-11,14,27-28], which can be downloaded from the website.

1.11 Storage and Handling

The packaged SRM is placed in a small foam-padded wooden box for protection and delivered along with the Certificates, the data analysis sheets, and this SP 260 to NIST's Building 301 where it is labeled, sealed in a plastic bag, and stored in a N_2-filled dry box (or an acceptable alternative) to await shipment to a customer.

The semiconductor test chip is subject to surface contamination and oxidation during storage and handling. The SRM should be handled using the metal package, without contacting the semiconductor test chip. The lid (also called the can) provided with the SRM should be carefully placed atop the package and secured to the package when the SRM is not in use. The SRM should be stored in a dust-free N_2 atmosphere or under vacuum at a temperature of $20.5\ ^{\circ}C \pm 1.1\ ^{\circ}C$. Incidental exposure to air for transport to or use in an analysis system should not produce significant contamination. The customer should avoid exposing the units to large temperature variations, temperature cycling, large humidity variations, or mechanical shock. Particulate contamination of the semiconductor surface may be removed with a low velocity dry N_2 flow. Too high or turbulent flow can break the cantilevers.

1.12 Measurement Conditions and Procedures for the Customer

The SRM is intended to be used in a laboratory environment and stored in a N_2-filled dry box. To take measurements on the MEMS 5-in-1 SRM for comparison with the NIST measurements, the lid is carefully removed. The step height, residual strain, strain gradient, and in-plane length measurements can now be taken using the proper equipment and appropriate standard test method [2-5] to guide you through the measurements.

For Young's modulus measurements, to operate the PZT, the red wire should be driven with a voltage that is positive relative to the black wire. To ensure that you have successfully connected to the PZT, when activated at 10 V and 7000 Hz (or when activated with a periodic chirp function between 5 kHz and 20 kHz), the resulting PZT vibration should be barely audible. The PZT has a resonance frequency greater than 300 kHz, at which or above which it shall not be operated because that could damage the PZT. The Young's modulus measurements are taken using the appropriate standard test method [11] to guide you through the measurements.

The standard test methods [1-5] in conjunction with Sec. 2 through Sec. 8, inclusive, of this SP 260 can be consulted to provide details concerning the measurements taken on the MEMS 5-in-1. Table 2 can be used to navigate through this SP 260. It lists the grouping in which the chosen test structure can be found, the parameter associated with that grouping, the applicable section in this SP 260, the data analysis sheet to use for that parameter, and the applicable appendix in this SP 260, which provides a reproduction of the pertinent data analysis sheet. As an example, details concerning the Young's modulus measurements in the first grouping of test structures are discussed in Sec. 2 of this SP 260 and are recorded in Data Analysis Sheet YM.3 [13] (a reproduction of which is given in Appendix 1).

The calculations are performed on-line by pressing the "Calculate and Verify" button located near the top and/or middle of the applicable data analysis sheet. (These calculations have been checked with similar calculations performed in Excel.) Any pertinent warnings flagged at the bottom of the data analysis sheet should be addressed before comparing your in-house measurements with the NIST measurements (as supplied on the applicable data analysis sheet that accompanies each unit of the MEMS 5-in-1). Consult Sec. 2 through Sec. 8 for specifics associated with the data comparison. Any questions concerning these measurements or comparisons can be directed to mems-support@nist.gov.

Table 2. Grouping, Parameter, Section, Associated Data Sheet, and Appendix

Grouping on the MEMS 5-in-1 (see Sec. 1.2)	Parameter	Section in this SP 260	Data Analysis Sheet	Appendix
1	Young's modulus	2	YM.3	1
	Residual stress	7	YM.3	1
	Stress gradient	7	YM.3	1
2	Residual strain	3	RS.3	2
3	Strain gradient	4	SG.3	3
4	Step height	5	SH.1.a	4
	Thickness	8		
4	for SRM 2494		T.1	6
6	for SRM 2495		T.3	7
5	In-plane length	6	L.0	5

1.13 Stability Tests

For stability tests, one packaged part is stored in a N_2-filled dry box and one packaged part is stored in a plastic storage container. These two packaged parts are called SRM monitors. When the MEMS 5-in-1 chips are measured, so are the SRM monitors from the same processing run, and the same measurements are taken. A parametric value resulting from the first measurements taken on the SRM monitors is considered a certified value, C, with its combined standard uncertainty value, u_C.

For certain parameters on SRM 2494 as listed in Table 3, correction terms and specific standard deviations are given for use in the applicable data analysis sheets. Similar quantities are not used in all the data sheets.

For example, a relative repeatability standard deviation, $\sigma_{repeat(samp)}$ for Young's modulus is not provided in Table 3 because a repeatability component is not needed in its uncertainty equation because the propagation of uncertainty technique [19-21] (a brief overview of which is given in Appendix 8) is used to obtain this uncertainty equation and including this repeatability component would be double counting uncertainties. This is the case for residual stress and stress gradient as well. In addition, thickness values are obtained from step height measurements, and since the step height uncertainty results (which already include a repeatability component) will be utilized in the thickness uncertainty calculations, a repeatability standard deviation is not needed for thickness. It should be pointed out that $\sigma_{repeat(samp)}$ is a relative repeatability standard deviation and $\sigma_{repeat(samp)'}$, which has a trailing "'" in the subscript, is an absolute repeatability standard deviation which is used for the in-plane length measurements.

Table 3. *Correction Terms and Specific Standard Deviations For Use with SRM 2494*

Parameter		Correction Term[a]	Standard Deviation[a]	Values used with SRM 2494
1. Young's modulus	$L = 200$ μm	$f_{correction}$		2.70 kHz
			$\sigma_{support}$	0.45 kHz
			$\sigma_{cantilever}$	0.45 kHz
	$L = 300$ μm	$f_{correction}$		0 kHz
			$\sigma_{support}$	0 kHz
			$\sigma_{cantilever}$	0 kHz
	$L = 400$ μm	$f_{correction}$		−0.20 kHz
			$\sigma_{support}$	0.07 kHz
			$\sigma_{cantilever}$	0.07 kHz
2. Residual strain		$\delta_{\varepsilon r correction}$		0
			$\sigma_{repeat(samp)}$	2.49 %
3. Strain gradient	$L = 200$ μm	$s_{g correction}$		0 m^{-1} [b]
	$L = 248$ μm	$s_{g correction}$		0 m^{-1} [b]
			$\sigma_{repeat(samp)}$	3.02 %
4. Step height			$\sigma_{repeat(samp)}$	3.95 % (from SEMI MS2)
5. In-plane length (for edges that face each other)	$L = 200$ μm (at 25×)	L_{offset}		2.632 μm (at NIST on SRM 2494)
	$L = 200$ μm (at 25×)		$\sigma_{repeat(samp)'}$ [c]	1.1565 μm

[a] Consult the pertinent section in this SP 260 for details associated with these correction terms and specific standard deviations.

[b] This is an assumption since it is difficult to obtain data points beyond 250 μm along the length of the cantilever due to excessive curvature of the cantilever.

[c] The trailing prime in this subscript implies an absolute measurement.

Stability tests can be used to track parametric changes as a function of, for example, time and temperature. For the MEMS 5-in-1, the Certificates state rigid temperature requirements (i.e., 20.5 °C ± 1.1 °C) within which the parameters are not expected to significantly change. Therefore, for the MEMS 5-in-1, only variations as a function of time are tracked for the SRM monitor stored in a plastic storage container and the SRM monitor stored in a N_2-filled dry box. And, given the domino effect discussed in the next section, only the residual strain needs to be monitored quarterly. This measurement is performed on the same SRM monitor test structures that are used to obtain the certified values for that chip. A parametric value resulting from a measurement taken on a quarterly basis is considered a measured value, M, with its combined standard uncertainty value, u_M.

Each quarter, the positive difference, D, between the certified and measured values of residual strain is calculated along with the uncertainty of the difference, u_D, using the following equations:

$$D = |M - C|, \quad \text{and} \tag{26}$$

$$u_D = \sqrt{u_M^2 + u_C^2}. \tag{27}$$

For the SRM monitor stored in a plastic storage container, if the following equation is not satisfied:

$$D \geq u_D, \tag{28}$$

the residual strain begins to be monitored monthly. Once Eq. (28) is not satisfied for the SRM monitor stored in a N_2-filled dry box, the material parameters can be considered out of calibration and the customers in possession of those SRMs from the same processing run are notified, assuming their SRM Certificate has not already expired (as discussed in the next section).

1.14 Length of Certification

It is recommended that the customer purchase a MEMS 5-in-1 SRM every two years. For a chip that was stored in a plastic storage container (not a N_2-filled dry box), residual strain data taken on this chip between May 2002 and August 2005, as shown in Fig. 9, indicate that the residual strain increases as a function of time.[15] Therefore, at the very least for the MEMS 5-in-1, this parameter should be periodically checked at NIST.

In Fig. 9 the data are smoothed out by connecting the second and fourth data points. Let us assume that the first data point is the certified value C, with combined standard uncertainty u_C, and expanded uncertainty U_C, where $U_C = ku_C = 2u_C$ (with $k=2$ to approximate a 95 % level of confidence). As an approximation for the maximum separation between parametric values, let us assume that the third data point corresponds to M, as defined in Sec. 1.13, with combined standard uncertainty u_M, and expanded uncertainty U_M, where $U_M = ku_M = 2u_M$ (with $k=2$ to approximate a 95 % level of confidence). The value for D is then calculated using Eq. (26) and U_D is calculated using the following equation:

$$U_D = 2\sqrt{u_M^2 + u_C^2}. \tag{29}$$

Then, the horizontal dotted line is plotted in Fig. 9, which corresponds to C plus U_D. This dotted line intersects the dotted line corresponding to the smoothed out data. The time between this intersection point and the measurement of C is slightly more than two years. Therefore, purchasing a MEMS 5-in-1 SRM every two years is recommended to ensure that $D \leq U_D$ for residual strain.

In addition, if the residual strain is out of calibration, it implies that the residual stress may also be out of calibration (since the residual strain is used to calculate residual stress). This domino effect continues for the remaining material parameters, as indicated in Table 4. Therefore, to ensure the calibration of the five material parameters (Young's modulus, residual strain, strain gradient, residual stress, and stress gradient) it is recommended that a MEMS 5-in-1 SRM be purchased every two years. Since only one date can be supplied on the SRM Certificate indicating the length of certification, for all the 5-in-1 parameters, it is recommended that a MEMS 5-in-1 SRM be purchased every two years. This will allow for improvements

---

[15] The reason for this increase is not known.

in equipment and procedures for the three dimensional parameters (step height, in-plane length, and thickness) which are not expected to change as a function of time.

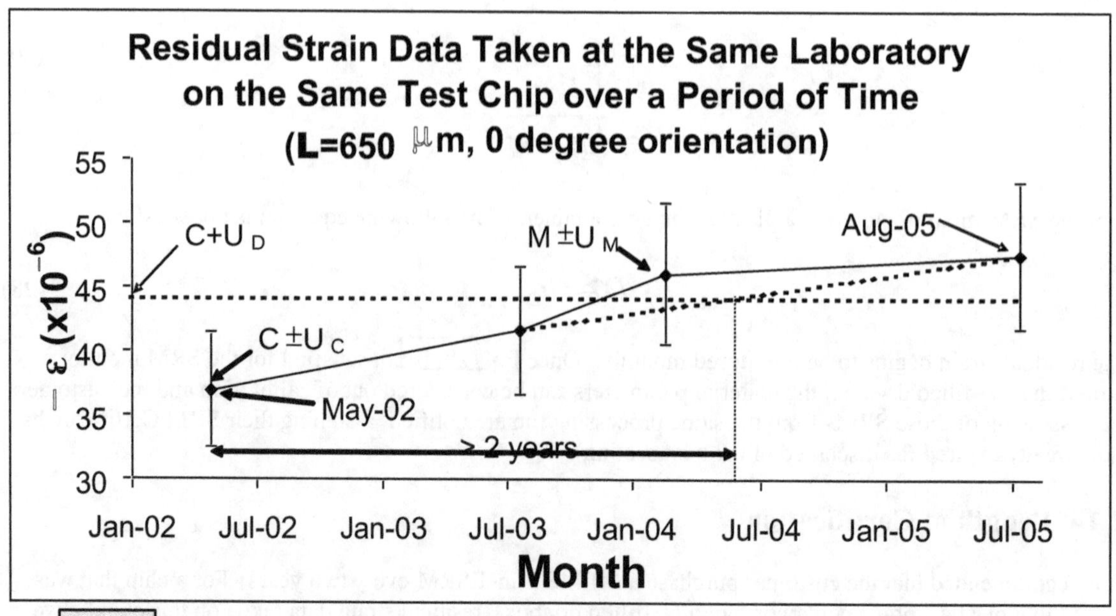

Figure 9. Residual strain data as a function of time for a surface micromachined chip where the uncertainty bars correspond to ±12.0 % to represent the estimated expanded uncertainty values. The chip was stored in a plastic storage container.

Note: In this figure, the uncertainty bars are calculated using $U = 2[u_{cave}^2 + \sigma_{repeat(samp)}^2]^{1/2}$ with u_{cave}=1.8 % as given in [11] and $\sigma_{repeat(samp)}$=5.70% as derived from data given in [11].

Table 4. Length of Certification for the MEMS 5-in-1 Parameters

Parameter	Length of Certification	Reason for Expiration of Calibration
1. residual strain	two years	in case the residual strain varies as a function of time (see Fig. 9)
2. residual stress	two years	residual strain is used to calculate residual stress
3. Young's modulus	two years	Young's modulus is used to calculate residual stress (even though this parameter is not expected to vary as a function of time, its value may be questioned if the chip experienced unexpected environmental variations)
4. stress gradient	two years	Young's modulus is used to calculate stress gradient
5. strain gradient	two years	strain gradient is used to calculate stress gradient (even though Fig. 10 indicates that this parameter is not expected to vary as a function of time, since the value of the other material parameters may be in question, this one may be as well if the chip experienced environmental variations not experienced by the chips from which the data were taken in Fig. 10)
6. step height	two years	to allow for improvements in equipment and procedures
7. thickness	two years	to allow for improvements in equipment and procedures
8. in-plane length	two years	to allow for improvements in equipment and procedures

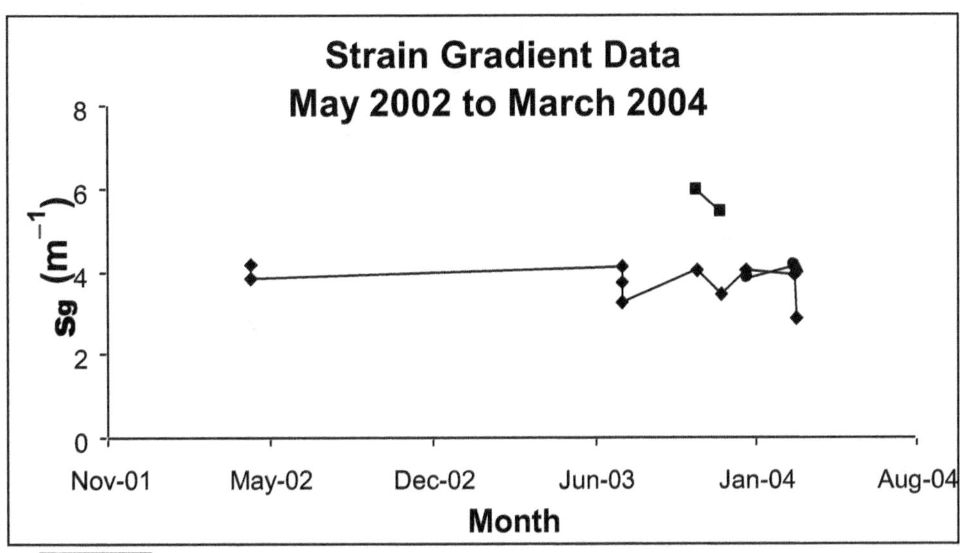

Figure 10. *Strain gradient round robin data as a function of time for lengths ranging from 500 μm to 650 μm. The chip was stored in a plastic storage container.*

Recalibration of the MEMS 5-in-1 SRM is not recommended (and also currently not offered at NIST) for the following reasons:

1. *Improvements inevitable*: Improvements in device design, processing, equipment and procedures are inevitable. Maintaining an "out -of-date" SRM would not be keeping up with the current state of technology.
2. *Price*: Since the fabrication costs of the MEMS 5-in-1 chips are supplied free of charge by both the fabrication service for SRM 2494 [8] and the fabrication facility for SRM 2495 [9], the cost to recalibrate a MEMS 5-in-1 SRM is almost comparable to the purchase price of a new MEMS 5-in-1 SRM. In other words, the cost for both mainly revolves around the cost to measure the parameters on the chips.
3. *Degrading quality*: With continued use is air, the pertinent test structures on the SRMs are bound to get dirty.
4. *Turn-around time*: It is more desirable to be in a position of offering SRMs when they are available as opposed to the challenge of getting an SRM recalibrated within a reasonable amount of time which would mean in the least that:
   a. The step height standards and frequency meter are in calibration,
   b. The instrumentation has been maintained, is operational, and in calibration,
   c. The personnel is available to perform the recalibrations,
   d. The SRM is not inadvertently mishandled, and
   e. Nothing goes wrong.
   In other words, it is easier to guarantee a more reasonable turn-around time by not offering recalibrations. This assumes of course that the SRM Program Office has samples in its possession to sell.
5. *Lack of efficiency and cost-effectiveness*: Providing calibration services requires equipment devoted solely to that service, which would be an added expense, not only for the equipment and space, but also in terms of personnel, overhead, and time. The alternative of providing the SRMs (without recalibrations) enables the equipment and space to be shared. This is a more efficient, cost-effective approach.

As specified previously, it is recommended that a MEMS 5-in-1 SRM be purchased every two years. The customer should contact the SRM Program Office for specifics.

2   Grouping 1: Young's Modulus

Young's modulus is a parameter indicative of material stiffness that is equal to the stress divided by the strain when the ma terial is loaded in uniaxial tension, assuming the strain is small enough such that it does not irreversibly deform the material [1]. The Young's modulus measurement obtained using SEMI standard test method MS4 [1] is based on the average resonance frequency of a single-layered cantilever. These measurements are an aid in the design and fabrication of MEMS devices [28-29] and ICs. Failure mechanisms in ICs such as electromigration, stress migration, and delamination can result due to high values of residual stress (as calculated given the Young's modulus value). Therefore, methods for its characterization are of interest for IC process development and monitoring to improve the yield in CMOS fabrication processes [27].

This section on Young's modulus is not meant to replace but to supplement the SEMI standard test method MS4 [1], which more completely presents the purpose, scope, limitations, terminology, apparatus, and test structure design as well as the calibration procedure, measurement procedure, calculations, precision and accuracy, etc. In this section, the NIST-developed Young's modulus test structures on SRM 2494 and SRM 2495, as shown in Fig. 1 and Fig. 2, respectively, in the Introduction, are given in Sec. 2.1. Sec. 2.2 discusses the calibration procedure for Young's modulus measurements, and Sec. 2.3 discusses the Young's modulus measurement procedure. Following this, the uncertainty analysis is presented in Sec. 2.4, the round robin results are presented in Sec. 2.5, and Sec. 2.6 describes how to use the MEMS 5-in-1 to verify Young's modulus measurements.

2.1 Young's Modulus Test Structures

Young's modulus measurements are taken in the first grouping of test structures, as shown in Fig. YM1(a) for SRM 2494 depicted in Fig. 1 and as shown in Fig. YM1(b) for SRM 2495 depicted in Fig. 2.

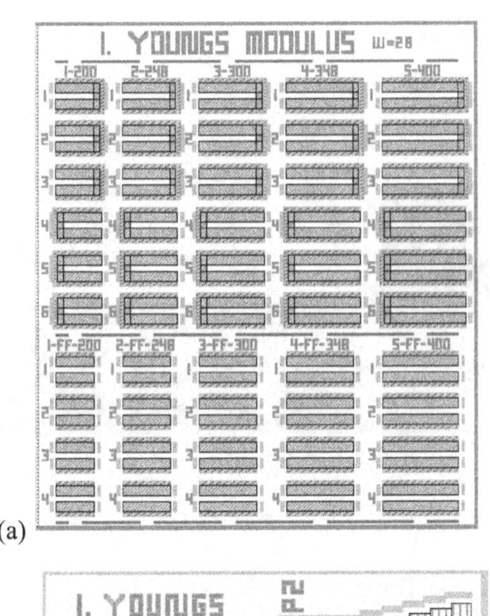

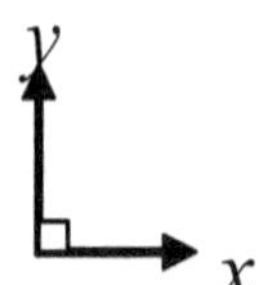

(a)

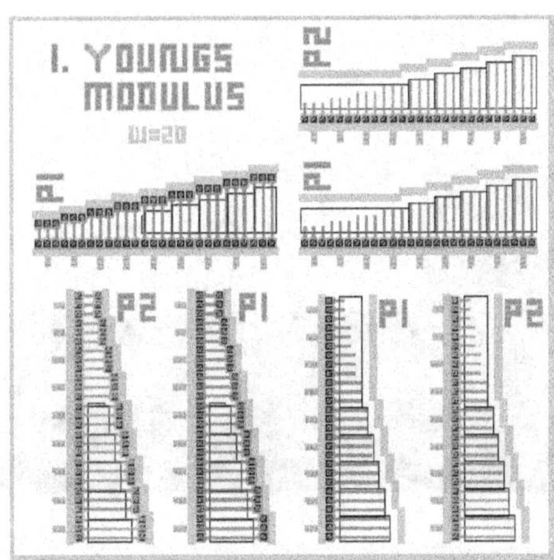

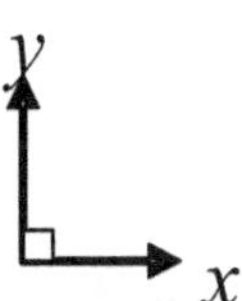

(b)

Figure YM1. The Young's modulus grouping of test structures on (a) SRM 2494, fabricated on a multi-user 1.5 μm CMOS process [8] followed by a bulk-micromachining etch, as depicted in Fig. 1 and (b) SRM 2495, fabricated using a polysilicon multi-user surface-micromachining MEMS process [9] with a backside etch, as depicted in Fig. 2.

The Young's modulus is obtained from resonance frequency measurements of a cantilever. A design rendition of a cantilever test structure in the Young's modulus grouping of test structures, as shown in Figs. YM1(a and b), can be seen respectively in Fig. YM2(a) for the bulk-micromachined SRM 2494 chip and in Figs. YM3(a) and YM4(a) for the surface-micromachined SRM 2495 chip with a backside etch for a poly1 (or p1) cantilever and a poly2 (or p2) cantilever, respectively. Cross sections of these test structures can be seen in Figs. YM2(b and c), Figs. YM3(b and c), and Figs. YM4(b and c) respectviely.

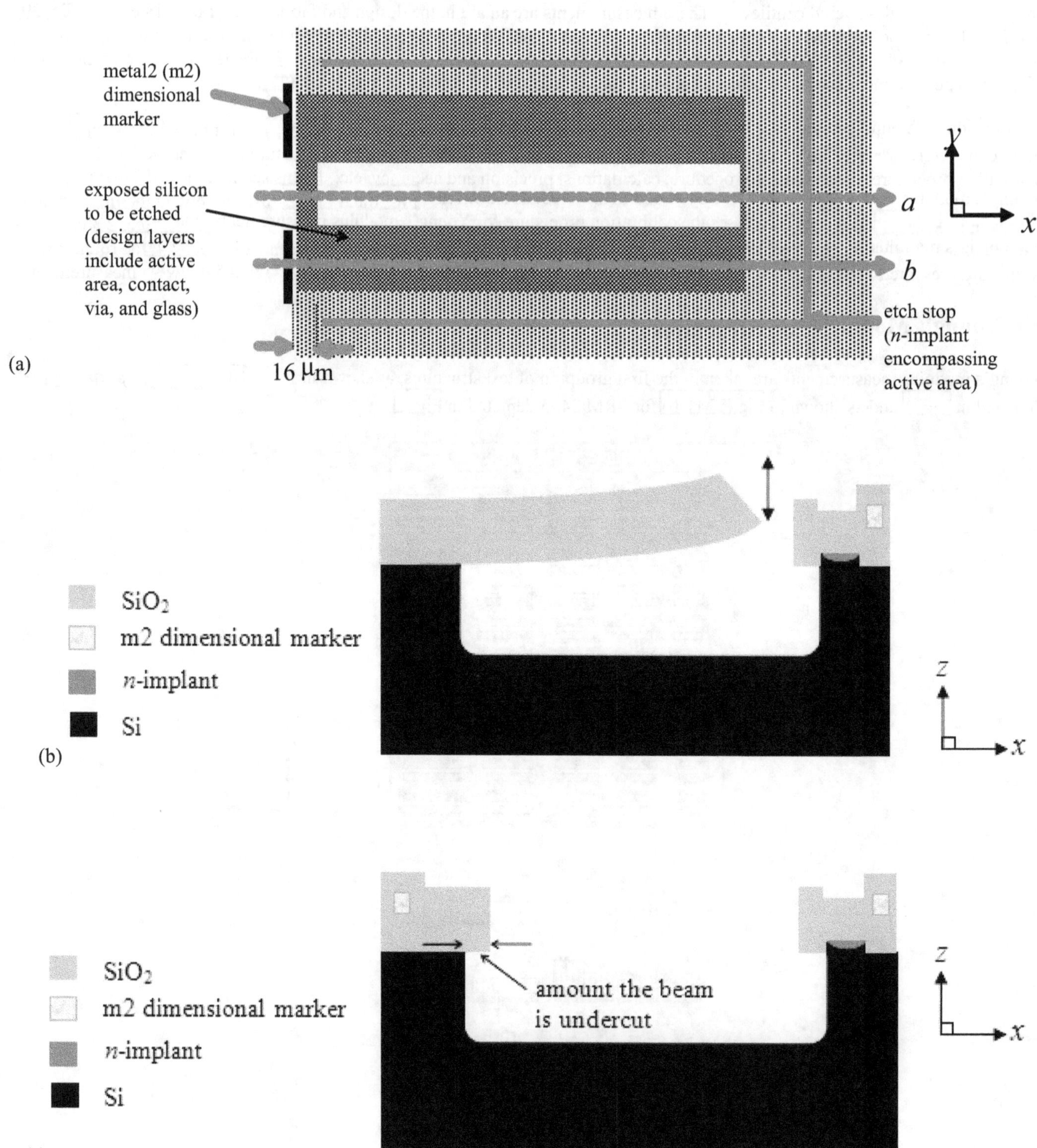

Figure YM2. *For a cantilever test structure on a bulk-micromachined SRM 2494 chip shown in Fig. 1 (a) a design rendition, (b) a cross section along Trace a in (a), and (c) a cross section along Trace b in (a).*

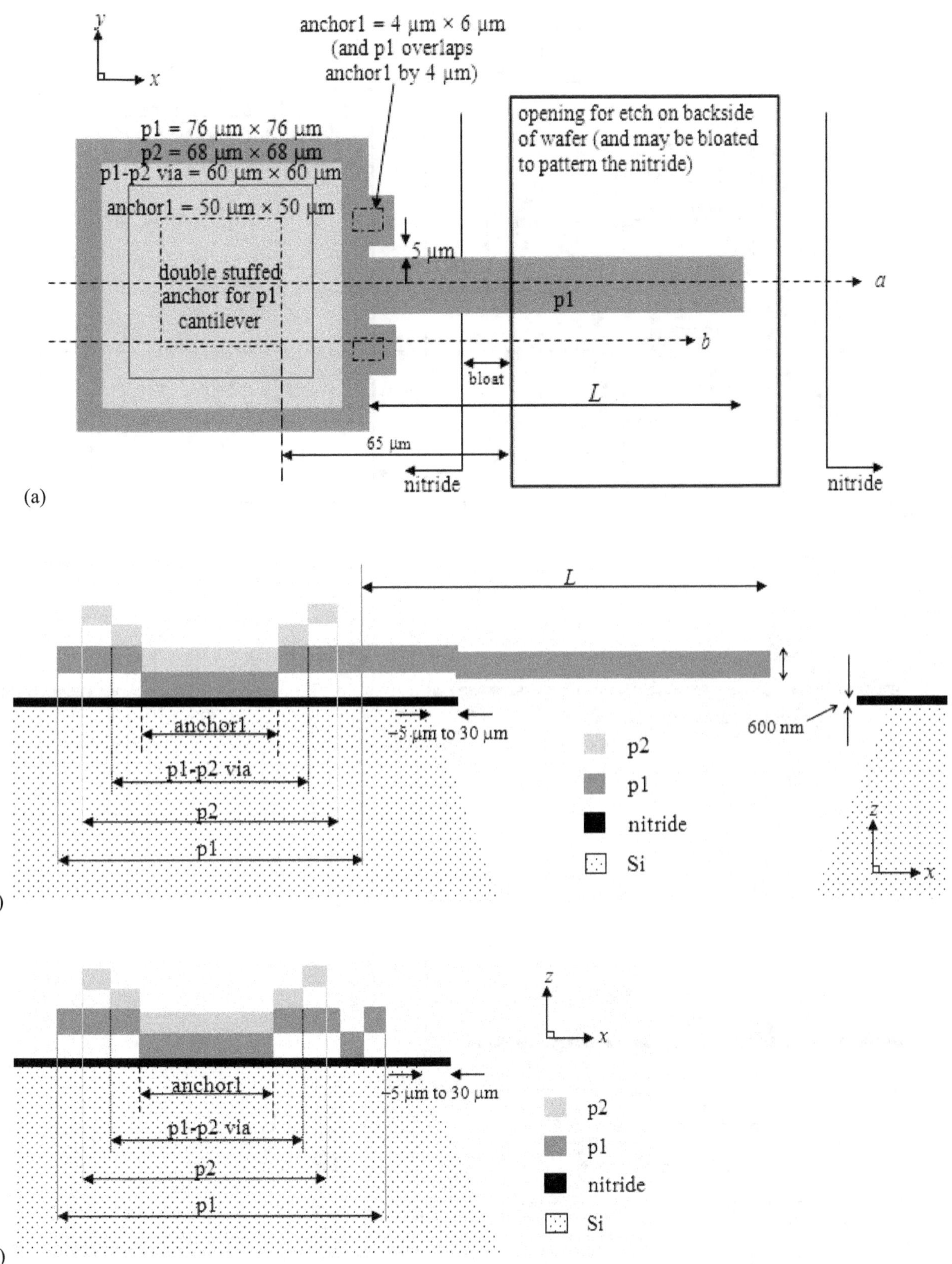

Figure YM3. For a p1 cantilever test structure on a surface-micromachined SRM 2495 chip (with a backside etch) shown in Fig. 2 (a) a design rendition, (b) a cross section along Trace a in (a), and (c) a cross section along Trace b in (a).

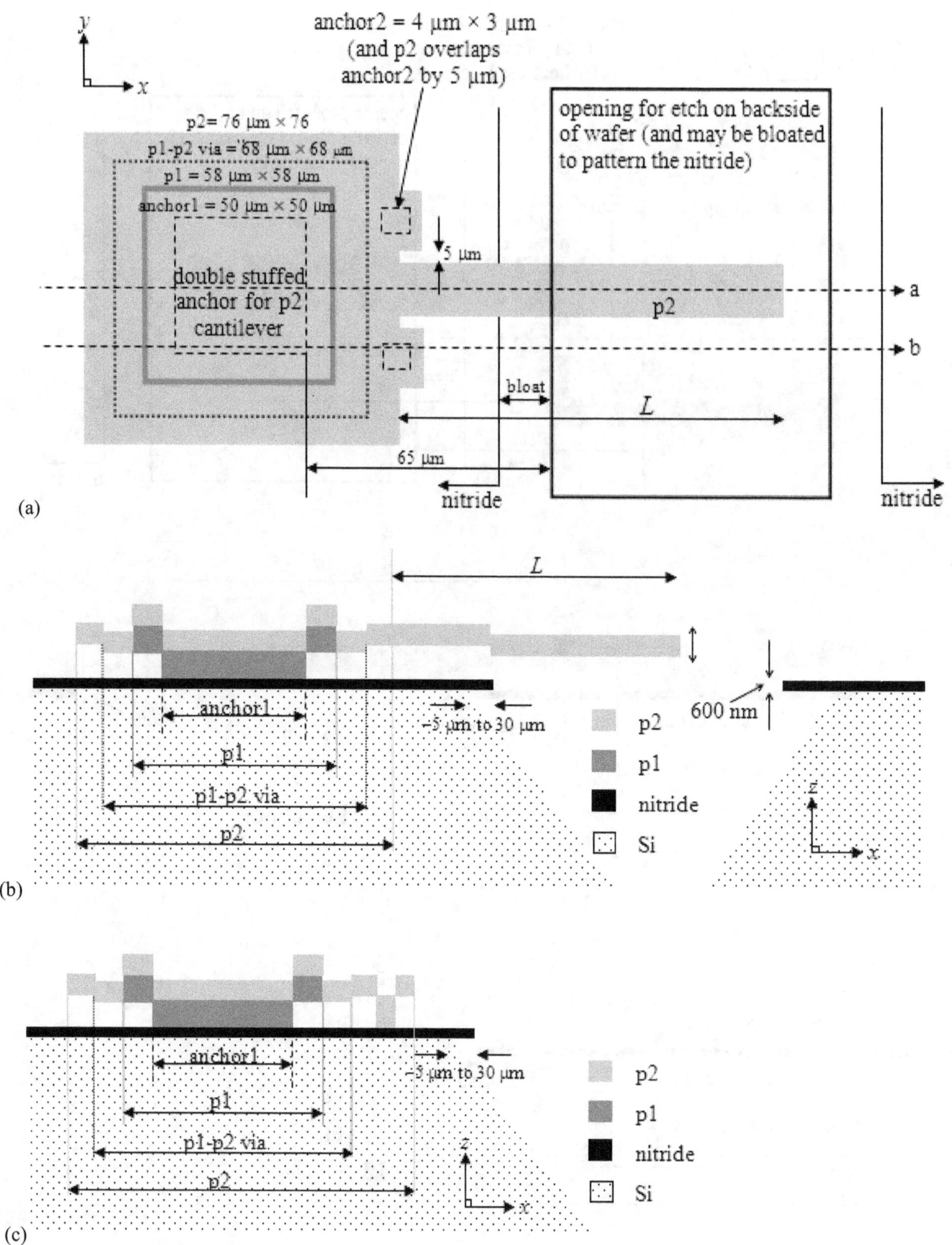

Figure YM4. *For a p2 cantilever test structure on a surface-micromachined SRM 2495 chip (with a backside etch) shown in Fig. 2 (a) a design rendition, (b) a cross section along Trace a in (a), and (c) a cross section along Trace b in (a).*

The specifications for the cantilevers shown in Figs. YM1(a and b) for SRM 2494 and SRM 2495, respectively, are given in Table YM1.

Table YM1. Cantilever Specifications for Young's Modulus Measurements

	For SRM 2494	For SRM 2495	
Structural Layer	SiO_2	poly1	poly2
W_{can} **(μm)**	28	20	20
L_{can} **(μm)**	200, 248, 300, 348, and 400	100, 150, 200, 250, 300, 350, 400, 450, and 500	100, 150, 200, 250, 300, 350, 400, 450, and 500
t **(μm)**	≈ 2.743	≈ 2.0	≈ 1.5
Orientation[a]	$0°$ and $180°$	$0°$ and $90°$	$0°$ and $90°$
Quantity of Beams	three of each length and for each orientation (making 30 beams)	three of each length and for each orientation (making 54 beams)	three of each length and for each orientation (making 54 beams)
E_{init} **(GPa)**	70	160	160
ρ **(g/cm^3)**	2.2	2.33	2.33

[a] A $0°$ orientation implies that the length of the beam is parallel to the x-axis of the test chip, the axes of which are shown in Fig. 1 and again in Fig. YM2 for SRM 2494 (and in Fig. 2 and again in Figs. YM3 and YM4 for SRM 2495), with the connection point of the cantilever having a smaller x-value than the x-values associated with the suspended portion of the cantilever.

For SRM 2494: On SRM 2494, all oxide cantilevers shown in Fig. YM1(a) are designed with both a $0°$ orientation and a $180°$ orientation. As seen in this figure, the length of a cantilever (in micrometers) is given at the top of each column of cantilevers following the column number (i.e., 1 to 5). These design lengths (and the design width) are specified in Table YM1. There are three cantilevers designed at each length for each orientation. Therefore, there are 15 oxide cantilevers with a $0°$ orientation and 15 oxide cantilevers with a $180°$ orientation. The fixed-fixed beams designed in the bottom portion of the Young's modulus grouping of test structures shown in Fig. YM1(a) will not be used for SRM measurements.

As specified in Sec. 1.4.1 for SRM 2494, the exposed silicon, as shown in Fig. YM2(a), is isotropically etched in XeF_2 to release the cantilever, as shown in Fig. YM2(b), by removing the silicon around and beneath the cantilever. The dimensional markers are instrumental in firming up the support region. They also can be used to measure the small amount of SiO_2 that has also been etched in XeF_2,[16] however the tip of the cantilever will also be etched a comparable amount such that the length of the cantilever should remain the same. The etch stop, also shown in this figure, helps to inhibit the etch away from the test structure to shield neighboring structures from the etch. It consists of an n-implant designed to surround the active area. As seen in Fig. YM2(c), there is undercutting of the beam.

For SRM 2494, the dimensions were chosen such that $5 \ \mu m \leq W_{can} \leq 40 \ \mu m$, $W_{can} > t$, and $L_{can} >> t$ where $t=2.743 \ \mu m$, as determined by the electro-physical technique [6] for a previous processing run. (The on-line version of Data Analysis Sheet T.1 [13], as reproduced in Appendix 6, can be used to calculate t.) In addition, the cantilever dimensions were chosen to achieve a) an estimated resonance frequency ($f_{caninit}$) between 10 kHz and 75 kHz using Eq. (YM5) to be presented in Sec. 2.3, b) a Q value above 30 using Eq. (YM8) to be presented in Sec. 2.3, and c) a value less than 2 % for p_{diff} as given by the following equation [10,23]:

$$p_{diff} = \left(1 - \frac{f_{dampedn}}{f_{undampedn}}\right)100\% = \left(1 - \sqrt{1 - \frac{1}{4Q^2}}\right)100\%, \qquad (YM1)$$

where $f_{dampedn}$ and $f_{undampedn}$ are given in Sec. 2.3. See Table YM2 for the calculations of $f_{caninit}$, Q, and p_{diff} for the chosen dimensions.

---

[16] The design dimension from the dimensional marker to the exposed silicon is 16 μm, as shown in Fig. YM2(a).

Table YM2. Calculations of $f_{caninit}$, Q, and p_{diff} for SRM 2494

L_{can} (μm)	$f_{caninit}$ (kHz)	Q	p_{diff} (%)
200	62.5	148.0	0.0006
248[a]	40.6	96.3	0.0013
300	27.8	65.8	0.0029
348[a]	20.6	48.9	0.0052
400	15.6	37.0	0.0091

[a] These values were chosen in order to design on a 0.8 μm grid to simplify the interface with the fabrication service [8].

Also, to ensure that the resonance frequency of the cantilever is not altered by squeeze film or other damping phenomena, [17] the cantilever should be suspended high enough above the underlying layer such that its motion is not altered by the underlying layer. In other words, the gap, d, between the bottom of the suspended cantilever and the top of the underlying layer should conform to the following lower bound [26]

$$d \geq \frac{W_{can}}{3} \; . \tag{YM2}$$

Therefore, for SRM 2494, with W_{can}=28 μm, d should be at least 9.3 μm, which is verified during the pre-package inspection (see Sec. 1.5).

The oxide cantilever consists of four SiO$_2$ layers. The thickness of this beam is calculated using Data Analysis Sheet T.1. See Sec. 8 for specifics. Because the beam is made up of different layers of SiO$_2$ that are prepared in different ways, the layers may have different properties ranging over those given in [30] thus deviating from a single-layered cantilever model. Also there may be remaining debris in the attachment corners of the cantilevers to the beam support. Due to these non-idealities in the geometry and composition of the cantilever and/or beam support [and including the undercutting of the beam shown in Fig. YM2(c)] an effective Young's modulus is reported in the SRM Certificate presented in Sec. 2.6.

For SRM 2495: For SRM 2495, there is approximately four times more chip area available for test structures than there is on SRM 2494. Therefore, less strict criteria were used in determining the cantilever dimensions. Still, the dimensions were chosen such that 5 μm $\leq W_{can} \leq$ 40 μm, $W_{can} > t$, and $L_{can} >> t$ where t=2.0 μm and t=1.5 μm are nominal values provided by the chip fabricator for the poly1 and poly2 thicknesses, respectively [9]. However, the poly1 cantilever dimensions were chosen to achieve a) an estimated resonance frequency ($f_{caninit}$) between 10 kHz and 275 kHz [18] using Eq. (YM5) to be presented in Sec. 2.3, b) a Q value above 10 using Eq. (YM8) to be presented in Sec 2.3, and c) a value less than 2 % for p_{diff} using Eq. (YM1). For the chosen dimensions (with W_{can}=20 μm), Table YM3 provides the calculations of $f_{caninit}$, Q, and p_{diff}. It should be noted, however, that the maximum displacement (i.e., 2.2 μm) of the PZT at 100 V can be obtained from DC to 100 kHz. Therefore, choosing the poly1 cantilevers in Table YM3 with $L_{can} \geq$ 200 μm is recommended. The poly2 cantilevers with 150 μm $\leq L_{can} \leq$ 400 μm are recommended.

In the Young's modulus grouping of test structures on SRM 2495, as shown in Fig. YM1(b), in particular for the cantilevers in this grouping, the p1 cantilever pad design shown in Fig. YM3(a) includes both p1 and p2. If the p2 is not included in the anchor, the cross section would look like that shown in Fig. YM3(b) but without the p2. Without the p2, the attachment point of the p1 cantilever to the anchor would not be considered rigid (or fixed) and would result in a smaller value for the resonance frequency than the resonance freqeuncy for an ideal cantilever with fixed boundary conditions. By including p2 in the anchor design, the p1 and p2 fuse during the fabrication process to make a more rigid and reliable attachment point.

To make an even more rigid attachment point, in the p1 cantilever pad design shown in Fig. YM3(a), the p1 layer is also anchored to the nitride on either side of the cantilever. If this is not done, the cantilever can be viewed as a cantilever with two widths. By anchoring the p1 on either side of the cantilever, it becomes a cantilever with a single width (W_{can}=20 μm), and can be treated as a beam fixed at one end [10].

---

[17] Damping phenomena (such as squeeze film damping) lead to amplitude-dependent resonance frequencies and shifts in the natural frequency of the system, which may limit the accuracy of the technique. The damping may not be present in bulk-micromachining processes because it is dependent upon the depth of the cavity and the vicinity of the sides of the cavity to the beam.

[18] Less than the maximum allowable frequency of the PZT.

Table YM3. Calculations of $f_{caninit}$, Q, and p_{diff} for Cantilevers on SRM 2495

L_{can} (μm)	Poly1			Poly2		
	$f_{caninit}$ (kHz)	Q	p_{diff} (%)	$f_{caninit}$ (kHz)	Q	p_{diff} (%)
100	267.7	349.8	0.0001	200.8	196.8	0.0003
150	119.0	155.5	0.0005	89.2	87.4	0.0016
200	66.9	87.4	0.0016	50.2	49.2	0.0052
250	42.8	56.0	0.0040	32.1	31.5	0.0126
300	29.7	38.9	0.0083	22.3	21.9	0.0262
350	21.9	28.6	0.0153	16.4	16.1	0.0485
400	16.7	21.9	0.0262	12.5	12.3	0.0827
450	13.2	17.3	0.0419	9.9	9.7	0.1325
500	10.7	14.0	0.0639	8.0	7.9	0.2020

Also, as seen in Fig. YM3(b), a flat cantilever is not fabricated. There is an approximate 600 nm vertical transition (or kink) in the cantilever. As shown in Fig. YM3(a), an opening is created on the backside of the wafer for a backside etch. This etch removes the material beneath the cantilevers to ensure the existence of cantilevers that have not adhered to the top of the underlying layer and to ensure there are no squeeze film or other damping phenomena. [19] Earlier in the fabrication process, the nitride layer is patterned using a mask similar to that used to create the openings in the backside of the wafer, however, all the features were bloated (for example, by 20 μm). As a result, the polysilicon cantilevers traverse an approximate 600 nm fabrication step over the nitride as seen in Fig. YM3(b) and Fig. YM5. For single layered p1 and p2 pad designs for p1 and p2 cantilevers, respectively, fabricated on a 2010 processing run (run #93 [9]), this step is approximtely 40 μm from the anchor lip (or 45 μm from the anchor when the opening for the backside etch is designed 65 μm from the anchor). For the double stuffed pad designs shown in Figs. YM3(a) and YM4(a) for the p1 and p2 cantilevers, respectively, on the SRM 2495 chips fabricated on a 2011 processing run (run #95 [9]), this step is approximtely 25 μm from the anchor lip (or 38 μm from the anchor when the opening for the backside etch is designed 65 μm from the anchor). In other words, the amount of bloat is expected to change for different processing runs. Although the backside etch is assumed to eliminate squeeze film and other damping phenomena, the p1 cantilever layer is typically still suspended above the nitride layer for this short distance between anchor1 and the kink in the cantilever, thus potentially introducing a residual damping effect. For a finite element model (FEM) simulation of a 500 μm long cantilever with W_{can}=20 μm and t=2 μm, a modeled 600 nm vertical transition (65 μm from the anchor) decreased the resonance frequency by 5 Hz.

Similar comments can be made concerning the p2 cantilever shown in Fig. YM4.

Non-idealities in the geometry of the cantilever and beam support (such as mentioned above) are responsible for an "effective" Young's modulus value being reported on the SRM Certificate as presented in Sec. 2.6.

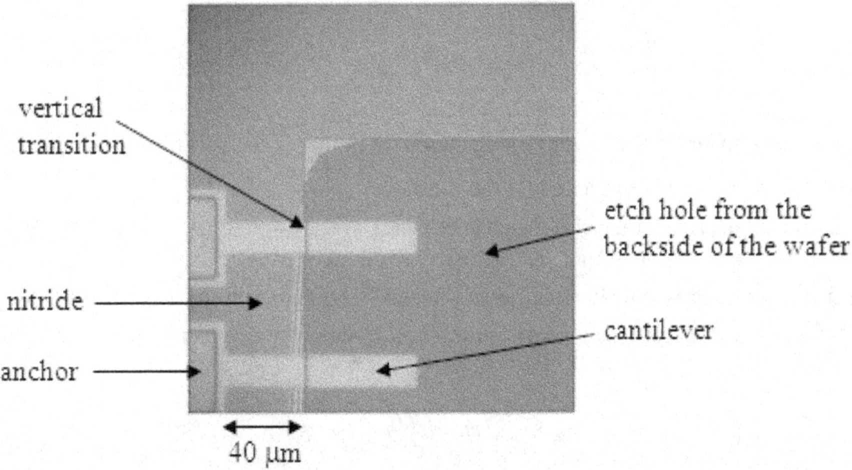

Figure YM5. A photograph of two p1 cantilevers on the 2010 processing run #93 (after the backside etch yet before the release of the beams) which reveals the abrupt vertical transition along the beams associated with a fabrication step over nitride. The pad designs in this figure consist of a single layer of p1.

---

[19] Unless the measurement is performed in a vacuum, damping phenomena are expected in surface micromachining processes without the use of a backside etch.

2.2 Calibration Procedure for Young's Modulus Measurements

For Young's modulus measurements, the time base of the instrument can be calibrated as described here.

Calibration of the time base: To calibrate the time base of the instrument used to obtain the resonance frequency measurements, the following steps are taken:

1. The instrument manufacturer is contacted to ensure that the appropriate signal(s) are measured. In most cases, only the maximum frequency (from which all other signals are derived) needs to be measured, and we will consider this case. Therefore, with the instrument set at the maximum frequency, $f_{instrument}$, as specified by the manufacturer, with a calibrated frequency meter take at least three measurements and record the average of the measurements as f_{meter} and the standard deviation of the measurements as σ_{meter}.

2. Given f_{meter}, record the certified uncertainty of the frequency meter, u_{certf}, from the frequency meter's certificate. (It may be considered negligible such that u_{certf} can be set to 0 Hz.) For use in Sec. 2.4, calculate the uncertainty of a frequency measurement using the following equation:

$$u_{cmeter} = \sqrt{\sigma_{meter}^2 + u_{certf}^2} \ . \tag{YM3}$$

3. The calibration factor, cal_f, is determined using the following equation:

$$cal_f = \frac{f_{meter}}{f_{instrument}} \ . \tag{YM4}$$

The frequency measurements are multiplied by cal_f to obtain calibrated values.

2.3 Young's Modulus Measurement Procedure

Young's modulus measurements are taken from cantilever test structures such as shown in Figs. YM2, YM3, and YM4. To obtain a Young's modulus measurement, the following steps are taken [1]:

1. An estimate for the fundamental resonance frequency of a cantilever, $f_{caninit}$, is obtained using the following equation (a derivation of which is presented in [10]):[20]

$$f_{caninit} = \sqrt{\frac{E_{init}t^2}{38.330\rho L_{can}^4}} \ , \tag{YM5}$$

where E_{init} is the initial estimate for the Young's modulus value of the thin film layer, t is the thickness, ρ is the density, and L_{can} is the suspended length of the cantilever.

2. Measurements are taken at frequencies which encompass $f_{caninit}$ and an excitation-magnitude versus frequency plot is obtained from which the resonance frequency is found.

3. For a given cantilever, three uncalibrated measurements of resonance frequency are obtained (namely, f_{meas1}, f_{meas2}, and f_{meas3}). If these are damped measurements, when calibrated using the following equation:

$$f_{dampedn} = f_{measn} cal_f \ , \tag{YM6}$$

they are called $f_{damped1}$, $f_{damped2}$, and $f_{damped3}$, respectively, where the trailing n in the subscript of f_{measn} and $f_{dampedn}$ is 1, 2, or 3. If f_{meas1}, f_{meas2}, and f_{meas3} are undamped measurements (e.g., if the measurements are performed in a vacuum), they are multiplied by cal_f to become $f_{undamped1}$, $f_{undamped2}$, and $f_{undamped3}$, respectively.

---

[20] By inserting the inputs into the correct locations on the appropriate NIST Web page [13], the calculations can be performed on-line in a matter of seconds.

4. For each damped resonance frequency ($f_{damped1}$, $f_{damped2}$, and $f_{damped3}$), a corresponding undamped resonance frequency ($f_{undamped1}$, $f_{undamped2}$, and $f_{undamped3}$, respectively), is calculated using the equation below:

$$f_{undampedn} = \frac{f_{dampedn}}{\sqrt{1 - 1/\left(4Q^2\right)}},$$

(YM7)

where the trailing n in the subscript of $f_{dampedn}$ and $f_{undampedn}$ is *1*, *2*, or *3* and where Q is the oscillatory quality factor of the cantilever as given by the following equation [23]

$$Q = \left[\frac{W_{can}\sqrt{E_{init}\rho}}{24\mu}\right]\left(\frac{t}{L_{can}}\right)^2,$$

(YM8)

where W_{can} is the suspended cantilever width and μ is the viscosity of the ambient surrounding the cantilever (in air, $\mu=1.84\times10^{-5}$ Ns/m^2 at 20 °C).

5. The average calibrated undamped resonance frequency, $f_{undampedave}$, is calculated from the three calibrated undamped resonance frequencies using the following equation:

$$f_{undampedave} = \frac{f_{undamped1} + f_{undamped2} + f_{undamped3}}{3}.$$

(YM9)

6. Given this value for $f_{undampedave}$, f_{can} is calculated using the following equation:

$$f_{can} = f_{undampedave} + f_{correction},$$

(YM10)

where $f_{correction}$ is a resonance frequency correction term, intending to correct for deviations from the ideal cantilever geometry and/or composition. This correction term is included in Data Analysis Sheet YM.3 but is not included in Data Analysis Sheet YM.1 and Data Analysis Sheet YM.2. The correction terms used for SRM 2494 are given in Table 3 in Sec. 1.13.

7. The Young's modulus value, E, is calculated as follows:

$$E = \frac{38.330\rho f_{can}^2 L_{can}^4}{t^2}.$$

(YM11)

Given a Young's modulus variation with length for SRM 2494, as specified in Sec. 2.5, the Young's modulus is modeled for a cantilever with $L=300$ µm (i.e., $f_{correction}=0$ Hz for $L=300$ µm). The frequency correction terms (given in Table 3 in Sec. 1.13) are used for cantilevers with $L=200$ µm and $L=400$ µm. Eq. (YM11) [with $f_{correction}=0$ Hz in Eq. (YM10)] assumes an ideal geometry for a single-layered cantilever and clamped-free boundary conditions (with no undercutting of the beam). A derivation of this text book equation is presented in [10]. The combined standard uncertainty for E, or u_{cE}, is given below in Sec. 2.4.

2.4 Young's Modulus Uncertainty Analysis

In this section, two combined standard uncertainty equations are presented for use with Young's modulus . The first combined standard uncertainty equation is used for the MEMS 5-in-1. It uses the propagation of uncertainty technique [19-21] (a brief overview of which is given in Appendix 8, which results in relative uncertainties that can be of more value to the user than absolute uncertainties. For example, it can be used to determine what parameters Young's modulus is most s ensitive to and how accurate the parameters must be to assure a pre-determined accuracy. The second combined standard uncertainty equation is similar to that used in the MEMS Young's modulus round robin. It uses a technique which adds absolute uncertainties in quadrature.

2.4.1 Young's Modulus Uncertainty Analysis for the MEMS 5-in-1

This section presents the combined standard uncertainty equation used with the MEMS 5-in-1. To obtain this equation, the propagation of uncertainty technique [19-21] for parameters that multiply (see Appendix 8) is applied to Eq. (YM11). The resulting one sigma uncertainty of the value of E, that is σ_E, is given by:

$$\sigma_E = E\sqrt{2^2\left(\frac{\sigma_{thick}}{t}\right)^2 + \left(\frac{\sigma_\rho}{\rho}\right)^2 + 4^2\left(\frac{\sigma_L}{L_{can}}\right)^2 + 2^2\left(\frac{\sigma_{fcan}}{f_{can}}\right)^2}\,, \tag{YM12}$$

where σ_{thick} is the one sigma uncertainty of the value of t, which is found using the electro-physical technique [6] for SRM 2494 and found using the optomechanical technique [7] for SRM 2495. The on-line version of Data Analysis Sheet T.1 [13], as reproduced in Appendix 6, can be used to calculate t and σ_{thick} for SRM 2494. The on-line version of Data Analysis Sheet T.3 [13], as reproduced in Appendix 7, can be used to calculate t and σ_{thick} for SRM 2495. Also in the above equation, σ_ρ is the estimated one sigma uncertainty of the value of ρ, σ_L is the estimated one sigma uncertainty of the value of L_{can}, and σ_{fcan} [as obtained by applying the propagation of uncertainty technique [19-21] for parameters in an additive relationship (see Appendix 8) to Eq. (YM10)] is given by the following equation:

$$\sigma_{fcan} = \sqrt{\sigma^2_{fundampedave} + \sigma^2_{fcorrection}}\,, \tag{YM13}$$

where

$$\sigma_{fundampedave} = \sqrt{\sigma^2_{fundamped} + \sigma^2_{fresol} + \sigma^2_{freqcal}}\,, \tag{YM14}$$

and

$$\sigma_{fcorrection} = \sqrt{\sigma^2_{support} + \sigma^2_{cantilever}}\,, \tag{YM15}$$

where $\sigma_{fundamped}$ (the standard deviation of the calibrated undamped resonance frequency measurements) is given by the following:

$$\sigma_{fundamped} = STDEV\left(f_{undamped\,1}, f_{undamped\,2}, f_{undamped\,3}\right), \tag{YM16}$$

where σ_{fresol} is calculated assuming a uniform distribution using the following equation:

$$\sigma_{fresol} = \frac{f_{resol}\,cal_f}{2\sqrt{3}}\,, \tag{YM17}$$

where f_{resol} is the uncalibrated frequency resolution for the given set of measurement conditions. Also in Eq. (YM14), $\sigma_{freqcal}$ is the one sigma uncertainty of the value of $f_{undampedave}$ due to the frequency calibration as given by the following equation:

$$\sigma_{freqcal} = f_{undampedave}\,\frac{u_{cmeter}}{f_{meter}}\,, \tag{YM18}$$

where f_{meter} is the average of the three measurements taken with the frequency meter in Sec. 2.2 and u_{cmeter} is the uncertainty of the three frequency measurements taken with the frequency meter for which the uncertainty is assumed to scale linearly.

In Eq. (YM15), $\sigma_{support}$ is the uncertainty in the resonance frequency due to a non-ideal support or attachment conditions (such as any undercutting of the beam and remaining debris in the attachment corners of the cantilever to the beam support as mentioned in Sec. 2.1 for SRM 2494) and $\sigma_{cantilever}$ is the uncertainty in the resonance frequency due to geometry or composition deviations from the ideal cantilever (such as the four oxide layers that comprise the cantilever discussed in Sec. 2.1 for SRM 2494 or the vertical transition discussed in Sec. 2.1 for SRM 2495). Table 3 in Sec. 1.13 gives the values of $\sigma_{support}$ and $\sigma_{cantilever}$ used for SRM 2494. Equation (YM15) assumes that $\sigma_{support}$ and $\sigma_{cantilever}$ are uncorrleated. If it is determined that they are correleated, set $\sigma_{support}=0$ Hz and include the uncertainty assoicated with the support into $\sigma_{cantilever}$.

Looking at Eq. (YM12), it is assumed that the one sigma uncertainty of the value of E, that is σ_E, is equal to the combined standard uncertainty, u_{cE}. In this case,

$$u_{cE3} = \sigma_E \, , \tag{YM19}$$

where a number following the subscript "E" in "u_{cE}" indicates the data analysis sheet that is used to obtain the combined standard uncertainty value. Therefore, u_{cE3} implies that Data Analysis Sheet YM.3 is used. Combining Eq. (YM12), Eq. (YM13), Eq. (YM14), Eq. (YM15), and Eq. (YM19) produces the following equation for the combined standard uncertainty:

$$u_{cE3} = E \sqrt{ \begin{array}{l} 4\left(\dfrac{\sigma_{thick}}{t}\right)^2 + \left(\dfrac{\sigma_\rho}{\rho}\right)^2 + 16\left(\dfrac{\sigma_L}{L_{can}}\right)^2 + 4\left(\dfrac{\sigma_{fundamped}}{f_{can}}\right)^2 \\[2ex] + 4\left(\dfrac{\sigma_{fresol}}{f_{can}}\right)^2 + 4\left(\dfrac{\sigma_{freqcal}}{f_{can}}\right)^2 + 4\left(\dfrac{\sigma_{support}}{f_{can}}\right)^2 + 4\left(\dfrac{\sigma_{cantilever}}{f_{can}}\right)^2 \end{array} } \, , \tag{YM20}$$

for use with the MEMS 5-in-1 where each component in Eq. (YM20) was obtained using a Type B analysis, except where noted. The parameters specified in the denominators of the various ratios in Eq. (YM12) and Eq. (YM20) are the parameters in the Young's modulus equation given in Eq. (YM11). The parameters specified in the numerators are various standard deviations of the parameter in the denominator, which makes each ratio a relative uncertainty. Table YM4 gives example values for each of the uncertainty components as well as the combined standard uncertainty value, u_{cE3}. Using the relative uncertainty values, as given in Table YM4 for this approach, allows one to more easily determine the most sensitive parameters as well as allowing one to determine how accurate the parameters must be to assure a pre-determined accuracy.

The expanded uncertainty for Young's modulus, U_E, is calculated using the following equation:

$$U_E = k u_{cE3} = 2 u_{cE3} \, , \tag{YM21}$$

where the k value of 2 approximates a 95 % level of confidence.

Reporting results [9-21]: Since it can be assumed that the estimated values of the uncertainty components are either approximately uniformly or Gaussianly distributed with approximate combined standard uncertainty u_{cE3}, the Young's modulus value is believed to lie in the interval $E \pm u_{cE3}$ (expansion factor $k=1$) representing a level of confidence of approximately 68 %.

Table YM4. *Example young's Modulus Uncertainty Values Using the MEMS 5-in-1 Approach as Given in Eq. (YM20)*

uncertainty		Type A or B	uncertainty values
1. $\dfrac{\sigma_{thick}}{t}$ (using t=2.743 μm and σ_{thick}=0.058 μm)		B	0.0211
2. $\dfrac{\sigma_{\rho}}{\rho}$ (using ρ=2.2 g/cm^3 and σ_{ρ}=0.05 g/cm^3)		B	0.023
3. $\dfrac{\sigma_{L}}{L_{can}}$ (using L_{can}=300 μm and σ_{L}=0.2 μm)		B	0.00067
4. $\dfrac{\sigma_{fcan}}{f_{can}} = \sqrt{\left(\dfrac{\sigma_{fundamped}}{f_{can}}\right)^2 + \left(\dfrac{\sigma_{fresol}}{f_{can}}\right)^2 + \left(\dfrac{\sigma_{freqcal}}{f_{can}}\right)^2 + \left(\dfrac{\sigma_{support}}{f_{can}}\right)^2 + \left(\dfrac{\sigma_{cantilever}}{f_{can}}\right)^2}$ (using f_{meas1} =26.82625 kHz, f_{meas2} =26.8351 kHz, f_{meas3} =26.8251 kHz, cal_f= 1, and Q= 65.8, such that $f_{undampedave}$= 26.8296 kHz and using $f_{correction}$= 0 kHz [a])		–	0.0002
4a.	$\dfrac{\sigma_{fundamped}}{f_{can}}$ (using W_{can}=28 μm, E_{init}=70 GPa, and μ=1.84×10^{-5} Ns/m^2)	A	0.0002
4b.	$\dfrac{\sigma_{fresol}}{f_{can}}$ (using f_{resol}=1.25 Hz)	B	0.000013
4c.	$\dfrac{\sigma_{freqcal}}{f_{can}}$ (using $\sigma_{freqcal}$=0.0 Hz) [a]	B	0.0
4d.	$\dfrac{\sigma_{support}}{f_{can}}$ (support and attachment assumed to be ideal, $\sigma_{support}$=0 Hz) [a]	B	0.0
4e.	$\dfrac{\sigma_{cantilever}}{f_{can}}$ (using $\sigma_{cantilever}$=0 Hz) [a]	B	0.0
$\dfrac{\sigma_{E}}{E} = \sqrt{4\left(\dfrac{\sigma_{thick}}{t}\right)^2 + \left(\dfrac{\sigma_{\rho}}{\rho}\right)^2 + 16\left(\dfrac{\sigma_{L}}{L_{can}}\right)^2 + 4\left(\dfrac{\sigma_{fcan}}{f_{can}}\right)^2}$		–	0.048
$u_{cE3} = \sigma_E$ (using E = 65.35 GPa)		–	3.1 GPa

[a] This was assumed to be zero to more appropriately compare this data set with the Young's modulus round robin data set in Table YM5.

2.4.2 Previous Young's Modulus Uncertainty Analyses

For the second combined standard uncertainty equation (similar to that used in the MEMS Young's modulus round robin [10,31-32]), seven sources of uncertainty are identified with all other sources considered negligible. The seven sources of uncertainty are the uncertainty of the thickness (u_{thick}), the uncertainty of the density (u_ρ), the uncertainty of the cantilever length (u_L), the uncertainty of the average resonance frequency (u_{freq}), the uncertainty due to the frequency resolution (u_{fresol}), the uncertainty due to damping (u_{damp}), and the uncertainty due to the frequency calibration ($u_{freqcal}$). As such, the combined standard uncertainty equation for u_{cE} with seven sources of uncertainty is as follows:

$$u_{cE2} = \sqrt{u_{thick}^2 + u_\rho^2 + u_L^2 + u_{freq}^2 + u_{fresol}^2 + u_{damp}^2 + u_{freqcal}^2} \,, \qquad \text{(YM22)}$$

where a number following the subscript "E" in "u_{cE}" indicates the data analysis sheet that is used to obtain the combined standard uncertainty value. Therefore, u_{cE2} implies that Data Analysis Sheet YM.2 [13] is used. [The equation for u_{cE1}, which uses Data Analysis Sheet YM.1, does not include the last uncertainty component, $u_{freqcal}$, in Eq. (YM22) and is the actual combined standard uncertainty equation used in the MEMS Young's modulus round robin.] In determining the combined standard uncertainty, a statistical Type A evaluation is used for u_{freq}. The other sources of uncertainty are obtained using a Type B evaluation [19-21] (i.e., one that uses means other than the statistical Type A analysis). Table YM5 gives example values for each of the uncertainty components in Eq. (YM22) assuming that $u_{freqcal}$=0 GPa such that the combined standard uncertainty value u_{cE1} is also given in Table YM5. Note that the resulting values for u_{cE3} (3.1 GPa in Table YM4) and u_{cE1} (3.2 GPa in Table YM5) are comparable.

Table YM5. Example Young's Modulus Uncertainty Values, From a Round Robin Bulk-Micromachined CMOS Chip (Assuming E_{init}=70 GPa), Comparable to Those Obtained in Table YM4

source of uncertainty or descriptor		uncertainty values
1. u$_{thick}$	thickness	2.8 GPa (using t=2.743 μm and σ_{thick}=0.058 μm)
2. u$_\rho$	density	1.5 GPa (using ρ=2.2 g/cm^3 and σ_ρ=0.05 g/cm^3)
3. u$_L$	cantilever length	0.17 GPa (using L_{can}=300 μm and σ_l=0.2 μm)
4. u$_{freq}$	average resonance frequency	0.027 GPa (using f_{meas1} =26.82625 kHz, f_{meas2} =26.8351 kHz, and f_{meas3} =26.8251 kHz)
5. u$_{fresol}$	frequency resolution	0.0018 GPa (using f_{resol}=1.25 Hz)
6. u$_{damp}$	damping	0.0004 GPa (using W_{can}=28 μm and σ_W=0.1 μm and using μ=1.84×10^{-5} Ns/m^2 and σ_μ=0.01×10^{-5} Ns/m^2)
u$_{cE1}$[a]	combined standard uncertainty for Young's modulus	3.2 GPa $= \sqrt{u_{thick}^2 + u_\rho^2 + u_L^2 + u_{freq}^2 + u_{fresol}^2 + u_{damp}^2}$

[a] This u_{cE1} uncertainty (times 3) is plotted in Fig. YM6 with the *repeatability* data point corresponding to the first cantilever with length of 300μm.

The first uncertainty component in Eq. (YM22) and listed in Table YM5 and Table YM6 is u_{thick}. (Table YM6 provides the equations for the uncertainty components.) The uncertainty for u_{thick} is determined from the calculated minimum and maximum Young's modulus values (namely, E_{min} and E_{max}, respectively) as derived using the extremes of values expected for the cantilever thickness. The values for E_{min} and E_{max} are given in the second and third columns, respectively, of Table YM6 where σ_{thick} is the one sigma uncertainty of the value of t, as specified in Sec. 2.4.1. With 99.7 % nominal probability of coverage, assuming a Gaussian distribution (and assuming u_ρ, u_L, u_{freq}, u_{fresol}, u_{damp}, and $u_{freqcal}$ equal zero), the value for E lies between E_{min} and E_{max}. Therefore, u_{thick} is calculated using the formula given in the fifth column of Table YM6.

Table YM6. Previous Young's Modulus Uncertainty Equations [10.31-32]

	E_{min}	E_{max}	G or U[a] / A or B[b]	equation
1. u_{thick}	$\dfrac{38330\rho f_{can}^2 L_{can}^4}{(t+3\sigma_{thick})^2}$	$\dfrac{38330\rho f_{can}^2 L_{can}^4}{(t-3\sigma_{thick})^2}$	G / B	$u_{thick} = \dfrac{E_{max} - E_{min}}{6}$
2. u_ρ	$\dfrac{38.330\left(\rho - 3\sigma_\rho\right)f_{can}^2 L_{can}^4}{t^2}$	$\dfrac{38.330\left(\rho + 3\sigma_\rho\right)f_{can}^2 L_{can}^4}{t^2}$	G / B	$u_\rho = \dfrac{E_{max} - E_{min}}{6}$
3. u_L	$\dfrac{38.330\rho f_{can}^2 \left(L_{can} - 3\sigma_L\right)^4}{t^2}$	$\dfrac{38.330\rho f_{can}^2 \left(L_{can} + 3\sigma_L\right)^4}{t^2}$	G / B	$u_L = \dfrac{E_{max} - E_{min}}{6}$
4. u_{freq}	$\dfrac{38.330\rho \left(f_{can} - 3\sigma_{freq}\right) L_{can}^4}{t^2}$	$\dfrac{38.330\rho \left(f_{can} + 3\sigma_{freq}\right) L_{can}^4}{t^2}$	G / A	$u_{freq} = \dfrac{E_{max} - E_{min}}{6}$
5. u_{fresol}	$\dfrac{38.330\rho \left(f_{can} - \dfrac{f_{resol}cal_f}{2}\right)^2 L_{can}^4}{t^2}$	$\dfrac{38.330\rho \left(f_{can} + \dfrac{f_{resol}cal_f}{2}\right)^2 L_{can}^4}{t^2}$	U / B	$u_{fresol} = \dfrac{E_{max} - E_{min}}{2\sqrt{3}}$
6. u_{damp}	$\dfrac{38.330\rho \left(f_{can} - 3\sigma_{fQ}\right) L_{can}^4}{t^2}$	$\dfrac{38.330\rho \left(f_{can} + 3\sigma_{fQ}\right) L_{can}^4}{t^2}$	G / B	$u_{damp} = \dfrac{E_{max} - E_{min}}{6}$
7. $u_{freqcal}$	$\dfrac{38.330\rho \left(f_{can} - 3\sigma_{freqcal}\right) L_{can}^4}{t^2}$	$\dfrac{38.330\rho \left(f_{can} + 3\sigma_{freqcal}\right) L_{can}^4}{t^2}$	G / B	$u_{freqcal} = \dfrac{E_{max} - E_{min}}{6}$

[a] "G" indicates a Gaussian distribution and "U" indicates a uniform distribution

[b] Type A or Type B analysis

In the same way, using the formulas in Table YM6, the remaining uncertainty components in Eq. (YM22) are calculated, where σ_ρ and σ_L are given in Sec. 2.4.1, where σ_{freq} is the standard deviation of $f_{undamped1}$, $f_{undamped2}$, and $f_{undamped3}$ [also called $\sigma_{fundamped}$ in Eq. (YM16) in Sec. 2.4.1], and where σ_{fQ} is the one sigma uncertainty of the value of f_{can} due to damping as mostly detailed in [10.31-32].[21] [If undamped resonance frequencies (e.g., if the measurements were performed in a vacuum) were recorded as f_{meas1}, f_{meas2}, and f_{meas3}, then u_{damp} is set equal to 0.0 Pa.]

2.5 Young's Modulus Round Robin Results

The round robin *repeatability* and *reproducibility* results are given in this section for Young's modulus measurements. The *repeatability* measurements are performed using the same test method, in the same laboratory (NIST), by the same operator, with the same equipment, in the shortest practical period of time (nominally, within a day). These measurements are done on random test structures. For the *reproducibility* measurements, at least six independent data sets (each using a different piece of equipment or equipment setup) must be obtained following the same test method before the results can be recorded in the precision and bias statement of a SEMI standard test method. These measurements are also done on random test structures.

---

[21] For Data Sheet YM.2 [13], the approach in [10] is slightly modified to include the uncertainty due to the frequency calibration

The *repeatability* data were taken in one laboratory using a dual beam vibrometer (see Sec. 1.1.1 for specifications of the vibrometer used in this procedure). The round robin test chips were processed using a bulk-micromachined CMOS process, similar to that used for SRM 2494 depicted in Fig. 1. A total of 48 Young's modulus values were obtained from twelve different cantilevers four times, with each Young's modulus value determined from the average of three resonance frequency measurements. Of these values, 16 were from four different cantilevers with L=200 μm, 16 from four different cantilevers with L=300 μm, and 16 from four different cantilevers with L=400 μm.

For the *reproducibility* data, eight participants were identified. [22] Each participant was supplied with a round robin test chip and asked to obtain three Young's modulus values; one from an oxide cantilever with a design length of 200 μm, one from an oxide cantilever with a design length of 300 μm, and one from an oxide cantilever with a design length of 400 μm. (The participant could choose to measure any one of five cantilevers of the given length as long as it passed a visual inspection.) Each Young's modulus value was determined from the average of three resonance frequency measurements from the cantilever as specified in Sec. 2.3 using an instrument that meets the manufacturer's alignment and calibration criteria. Following SEMI standard test method MS4 for Young's modulus measurements [32] the measurements were taken then recorded on Data Analysis Sheet YM.1 [13].

The eight participants used a variety of instruments to obtain Young's modulus. These included a single beam vibrometer, a dual beam vibrometer, and a stroboscopic interferometer (consult Sec. 1.1.1 for details associated with these instruments). In addition, thermal excitation measurements [33-35] were obtained on the same chip as PZT excitation measurements and the results are included for comparison purposes.

Tables YM7 and YM8 present the Young's modulus *repeatability* and *reproducibility* results, respectively. In these tables, n indicates the number of calculated Young's modulus values. The average (namely E_{ave}) of the *repeatability* or *reproducibility* measurement results is listed next, followed by the standard deviation (σ_E) of these measurements. Then, the $\pm 2\sigma_E$ limits are given followed by the average of the *repeatability* or *reproducibility* combined standard uncertainty values (u_{cE1ave}).

The Young's modulus *repeatability* and *reproducibility* results are plotted in Fig. YM6. At the top of this figure, E_{ave} and $3u_{cE1ave}$ are specified for the *repeatability* data. The values for $E_{ave} \pm 3u_{cE1ave}$ are also plotted in this figure with both the *repeatability*[23] and *reproducibility* data. As an observation, all of the *reproducibility* results fall comfortably between the *repeatability* bounds of $E_{ave} \pm 3u_{cE1ave}$.

Table YM7. Young's Modulus Repeatability Data
(One Participant, One Laboratory, One Instrument, One Chip, Twelve Different Cantilevers)

	L_{can}=200 μm	L_{can}=300 μm	L_{can}=400 μm	L_{can}=200 μm to L_{can}=400 μm
n	16	16	16	48
E_{ave}	59.8 GPa	65.4 GPa	67.5 GPa	64.2 GPa
σ_E	0.40 GPa	0.17 GPa	0.38 GPa	3.3 GPa
$\pm 2\sigma_E$ **limits**	±0.81 GPa (±1.4 %)	±0.33 GPa (±0.51 %)	±0.76 GPa (±1.1 %)	±6.6 GPa (±10 %)
u_{cE1ave}	2.9 GPa (4.9 %)	3.2 GPa (4.8 %)	3.3 GPa (4.8 %)	3.1 GPa (4.8 %)

---

[22] The term participant refers to a single data set from a unique combination of measurement setup and researcher. In other words, a single researcher with multiple, unique instruments (e.g., a dual beam vibrometer and a single beam vibrometer) or different forms of excitation (e.g., PZT and thermal excitation) could serve as multiple "participants."

[23] Table YM5 specifies the value of each of the uncertainty components comprising u_{cE1} for a Young's modulus measurement where the $3u_{cE1}$ uncertainty bars for this measurement are associated with the *repeatability* data point for the first cantilever in Fig. YM6 with a length of 300μm.

Table YM8. *Young's Modulus Reproducibility Data*
(Eight Participants, Five Laboratories, Seven Instruments, Four Chips)

	L_{can}=200 μm	L_{can}=300 μm	L_{can}=400 μm	L_{can}=200 μm to L_{can}=400 μm
n	8	8	8	24
E_{ave}	58.7 GPa	63.7 GPa	66.0 GPa	62.8 GPa
σ_E	1.3 GPa	1.8 GPa	1.4 GPa	3.4 GPa
$\pm 2\sigma_E$ limits	±2.6 GPa (±4.4 %)	±3.5 GPa (±5.5 %)	±2.9 GPa (±4.4 %)	±6.9 GPa (±11 %)
u_{cElave}	2.8 GPa (4.9 %)	3.1 GPa (4.8 %)	3.2 GPa (4.9 %)	3.0 GPa (4.9 %)

In Fig. YM6, the *repeatability* data are grouped according to the cantilever length with the L=200 μm data plotted first, followed by the L=300 μm data, then the L=400 μm data. The same is done with the *reproducibility* data where for each participant, the L=200 μm data are plotted first, followed by the L=300 μm data, then the L=400 μm data. Both the *repeatability* data and the *reproducibility* data indicate a length dependency. In Fig. YM6, the *repeatability* data show a clustering of the data at each length. In other words, the absolute value of the ±2 σ_E limits at each length (which are plotted in this figure along with E_{ave} for each length) are all less than 1.5 %, which is much less than the 10 % value (as given in Table YM7) when all the lengths are considered. This suggests that when comparing Young's modulus values extracted by different measurement instruments or excitation methods, the cantilevers should have the same length. This length dependency can be due to a number of things including debris in the attachment corners of the cantilevers to the beam support, which would cause larger errors for shorter length cantilevers. This can be a topic for future investigation where a) the physical form and chemical composition of the cantilever is checked to see if it matches the assumptions used in the calculations and b) finite element methods are used to determine if the length dependency is due to the attachment conditions (including debris in the attachment corners of the cantilever and any undercutting of the beam). Therefore, at this point, we can only state that, given the existing cantilevers, we can only report an "effective" value for Young's modulus.

As seen in Fig. YM6, round robin participant #1, participant #2, and participant #3 took measurements on the same chip (chip #1) using a dual beam vibrometer, a single beam vibrometer, and a stroboscopic interferometer, respectively. The results indicate that comparable results were obtained from these instruments.

Round robin participant #4, participant #5, and participant #6 took measurements on the same chip (chip #2) with participant #5 using thermal excitation to obtain the required data and participant #4 and participant #6 using PZT excitation. Fig. YM6 shows no significant difference in the results.

Round robin participant #7 and participant #8 took data from chip #3 and chip #4, respectively.

No information can be presented on the bias of the procedure in the test method for measuring Young's modulus because there is not a certified MEMS material for this purpose. Many values for Young's modulus for various materials have been published with an attempt to consolidate this information in [30]. For a silicon dioxide film, the Young's modulus values reported in [30] range from 46 GPa to 92 GPa. The average *repeatability* value reported in Table YM7 of 64.2 GPa obtained from the oxide cantilever that consists of field oxide, two depositied oxides and a glass layer falls comfortably within this range.

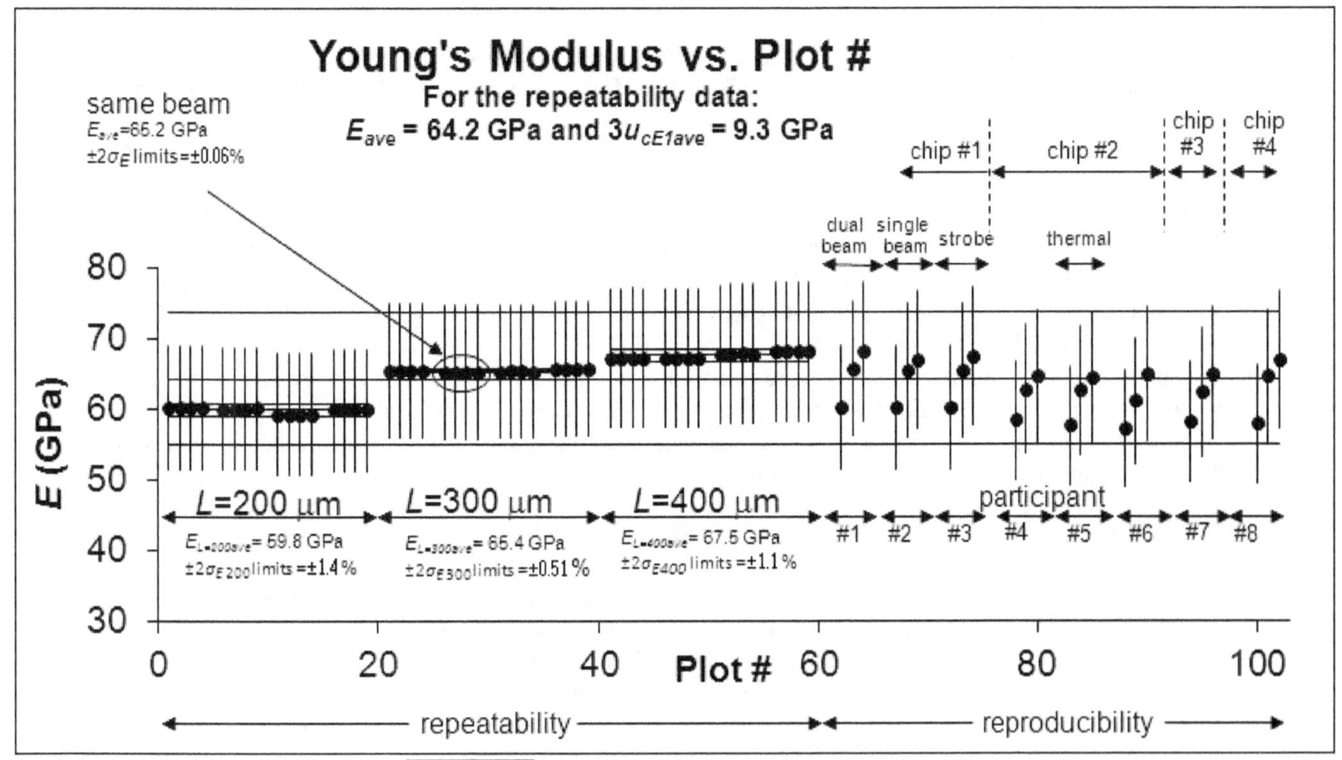

Figure YM6. *Young's modulus round robin results[24]*

2.6 Using the MEMS 5-in-1 to Verify Young's Modulus Measurements

To compare your Young's modulus measurements with NIST measurements, you will need to fill out Data Analysis Sheet YM.3; this data analysis sheet is accessible via the URL specified in the reference [13] a reproduction of which is given in Appendix 1. After obtaining an estimate for the resonance frequency, calibrating the instrument, locating the cantilever test structure, taking the measurements, and performing the calculations, the data on the completed form can be compared with the data on the SRM Certificate and the completed data analysis sheet supplied with the MEMS 5-in-1. Details of the procedure are given below.

Obtain an estimate for the resonance frequency: To determine an estimate for the resonance frequency of the cantilever (that may be a bit on the high side especially for shorter length cantilevers) do the following:
1. Access Data Analysis Sheet YM.3,
2. Supply inputs for:
    a. ρ (input #5),
    b. t (input #9),
    c. E_{init} (input #12), and
    d. L_{can} (input #19),
3. Press the "Calculate Estimates" button that appears before the Preliminary Estimates Table (Table 5) on the data analysis sheet, and
4. The value for $f_{caninit}$ (output #38) is the estimate for the resonance frequency.

Calibrate the instrument: Calibrate the time base of the instrument as given in Sec. 2.2. Obtain the inputs for Table 1 in Data Analysis Sheet YM.3.

Locate the cantilever: In the first grouping of test structures on the MEMS 5-in-1 chips shown in Fig. 1 and Fig. 2 for SRM 2494 and SRM 2495, respectively, Young's modulus measurements are made. Cantilever and fixed-fixed beam test structures are provided for this purpose; however, we will only be concerned with the cantilevers, such as shown in Fig. YM2 for SRM 2494 and as shown in Figs. YM3 and YM4 for SRM 2495. Specifications for the cantilevers in the Young's modulus grouping of test structures in Figs. YM1(a and b) for SRM 2494 and SRM 2495, respectively, are given in Table YM1.

---

[24] *Republished with permission from Semiconductor Equipment and Materials International, Inc. (SEMI) © 2011.*

Data Analysis Sheet YM.3 requires measurements from one cantilever on the MEMS 5-in-1 chip. The specific cantilever to be measured can be deduced from the data entered on the NIST-supplied Data Analysis Sheet YM.3 that accompanies the SRM.

For the Young's modulus grouping of test structures for SRM 2494, as shown in Fig. YM1(a) the target test structure can be found as follows:

1. The input L_{can} (i.e., input #19 on Data Analysis Sheet YM.3) specifies the length of the cantilever. The length of the cantilever (in micrometers) is given at the top of each column of cantilevers in Fig. YM1(a) following the column number (i.e., 1 to 5), therefore L_{can} can be used to locate the column in which the target cantilever resides.
2. The input *whichcan* (i.e., input #20) specifies which cantilever in the column to measure (i.e., the "first," "second," "third," etc.) regardless of the orientation.
3. The input *orient* (i.e., input #18) can be used as a form of verification. The cantilevers are designed at both a 0° and a 180° orientation with the cantilevers having a 0° orientation being the first, second, and third cantilevers in each column and the cantilevers with a 180° orientation being the fourth, fifth, and sixth cantilevers in each column. Therefore, either 0° or 180° will be selected for *orient*.

For the Young's modulus grouping of test structures for SRM 2495, as shown in Fig. YM1(b) the target test structure can be found as follows:

1. The input *mat* (i.e., input #4) specifies the composition of the cantilever, which should be either "poly1" or "poly2" since there are two arrays of poly1 cantilevers and two arrays of poly2 cantilevers.
2. The input *orient* (i.e., input #18) can be used to locate the appropriate array since one of the two arrays of a given composition has a 0° orientation and the other has a 90° orientation.
3. The input L_{can} (i.e., input #19) can be used to locate the appropriate length cantilever within the array. Within each array, the cantilevers are arranged by increasing length with the shortest cantilevers (L_{can}=100 μm) at the top (or leftmost) part of the array and the longest cantilevers (L_{can}=500 μm) at the bottom (or rightmost) part of the array.
4. The input *whichcan* (i.e., input #20) is used to identify which of the three identically designed cantilevers (the "first," "second," or "third") is the one to measure within the array. The length is specified to the left of or below the anchor of the second of the three identical cantilevers.

Take the measurements: For Data Analysis Sheet YM.3, the uncalibrated frequency resolution (as specified by the software) and three uncalibrated resonance frequency measurements are required for the cantilever. Obtain these measurements using the highest magnification objective that is available and feasible (e.g., a 20$\times$ objective) following the steps in SEMI standard test method MS4 [1] for measuring Young's modulus.

Perform the calculations: Enter the data into Data Analysis Sheet YM.3 as follows:

1. Press one of the "Reset this form" buttons. (One of these buttons is located near the top of the data analysis sheet and the other is located near the middle of the data analysis sheet.)
2. Fill out Table 1 and Table 2. (Table 4 is needed for residual stress and stress gradient calculations, as indicated in Sec. 7.3.)
3. Press one of the "Calculate and Verify" buttons to obtain the results for the cantilever. (One of these buttons is located near the top of the data analysis sheet and the other is located near the middle of the data analysis sheet.)
4. Verify the data by checking to see that all the pertinent boxes in the verification section at the bottom of the data analysis sheet say "ok". If one or more of the boxes say "wait," address the issue, if necessary, by modifying the inputs and recalculating.
5. Print out the completed data analysis sheet to compare both the inputs and outputs with those on the NIST-supplied data analysis sheet.

Compare the measurements: The MEMS 5-in-1 is accompanied by a Certificate. This Certificate specifies an effective Young's modulus value, E, and the expanded uncertainty, U_E, (with k=2) intending to approximate a 95 % level of confidence. It is your responsibility to determine an appropriate criterion for acceptance, such as given below:

$$D_E = \left| E_{(customer)} - E \right| \le \sqrt{U_{E(customer)}^2 - U_E^2} \,, \tag{YM23}$$

where D_E is the absolute value of the difference between your Young's modulus value, $E_{(customer)}$, and the Young's modulus value on the SRM Certificate, E, and where $U_{E(customer)}$ is your expanded uncertainty value and U_E is the expanded uncertainty on the SRM Certificate. If your measured value for Young's modulus (as obtained in the newly filled out Data Analysis Sheet

YM.3) satisfies your criterion for acceptance and there are no pertinent "wait" statements at the bottom of your Data Analysis Sheet YM.3, you can consider yourself to be appropriately measuring Young's modulus according to the SEMI MS4 Young's modulus standard test method [1] according to your criterion for acceptance.

An effective Young's modulus is reported for SRM 2494, as shown in Fig. 1, and for SRM 2495, as shown in Fig. 2, due to non-idealities associated with the geometry and/or composition of the cantilever and/or the beam support as specified in Sec. 2.1 and Sec. 2.5. When you use SEMI standard test method MS4 with your own test structures, you must be cognizant of the geometry and composition of your cantilever because this test method assumes an ideal geometry and composition, implying that you would be obtaining an "effective" Young's modulus value if the geometry and/or composition of your cantilever deviates from the ideal.

Any questions concerning the measurements, analysis, or comparison can be directed to mems-support@nist.gov.

3 Grouping 2: Residual Strain

Residual strain is defined in a MEMS process as the amount of deformation (or displacement) per unit length constrained within the structural layer of interest after fabrication yet before the constraint of the sacrificial layer (or substrate) is removed (in whole or in part) [2]. It is a measurement of the strain the parts of a microsystem undergo before they relax after the removal of the stiff oxides that surround them during manufacturing. ASTM standard test method E 2245 [2] on residual strain measurements is an aid in the design and fabrication of MEMS devices [28-29].

This section on residual strain is not meant to replace but to supplement the ASTM standard test method E 2245 [2], which more completely presents the scope, significance, terminology, apparatus, and test structure design as well as the calibration procedure, measurement procedure, calculations, precision and bias data, etc. In this section, the NIST-developed residual strain test structures on SRM 2494 and SRM 2495, as shown in Fig. 1 and Fig. 2, respectively, in the Introduction are given in Sec. 3.1. Sec. 3.2 discusses the calibration procedure for the residual strain measurements, and Sec. 3.3 discusses the residual strain measurement procedure. Following this, the uncertainty analysis is presented in Sec. 3.4, the round robin results are presented in Sec. 3.5, and Sec. 3.6 describes how to use the MEMS 5-in-1 to verify residual strain measurements.

3.1 Residual Strain Test Structures

Residual strain measurements are taken in the second grouping of test structures, as shown in Fig. RS1(a) for SRM 2494 depicted in Fig. 1 and as shown in Fig. RS1(b) for SRM 2495 depicted in Fig. 2.

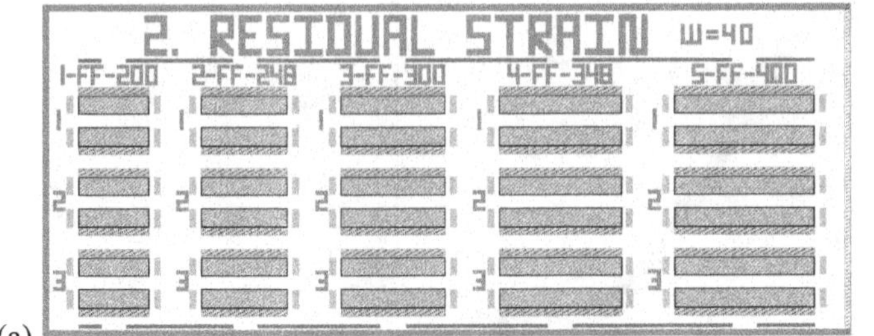

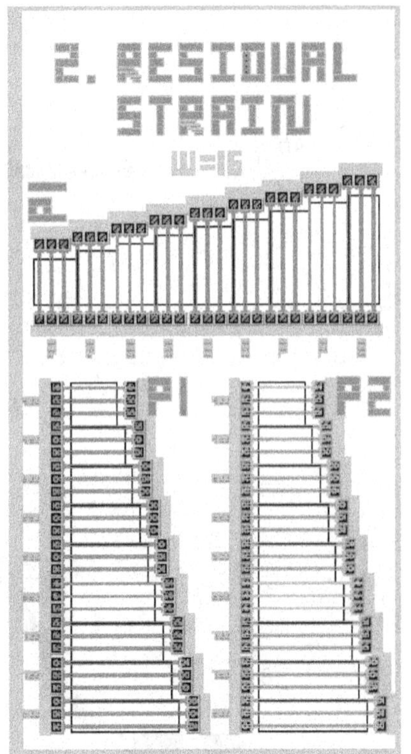

(a)   (b)

Figure RS1. The residual strain grouping of test structures on (a) SRM 2494, fabricated on a multi-user 1.5 µm CMOS process [8] followed by a bulk-micromachining etch, as depicted in Fig. 1 and (b) SRM 2495, fabricated using a polysilicon multi-user surface-micromachining MEMS process [9] with a backside etch, as depicted in Fig. 2.

Residual strain measurements are obtained from fixed-fixed beam test structures. A fixed-fixed beam test structure in the residual strain grouping of test structures, as shown in Figs. RS1(a and b), can be seen in Fig. RS2(a) and Fig. RS3(a), respectively, for the bulk-micromachined SRM 2494 chip and the surface-micromachined SRM 2495 chip with a backside etch. Applicable data traces taken from these test structures are given in Figs. RS2(b and c) and Figs. RS3(b and c), respectively.

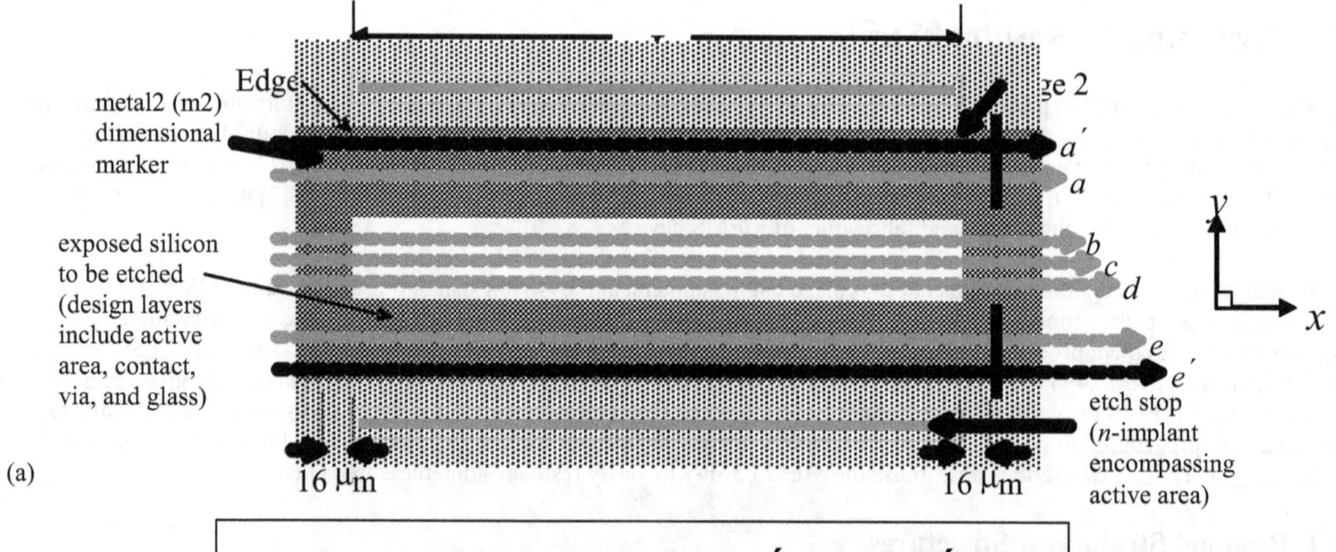

(a)

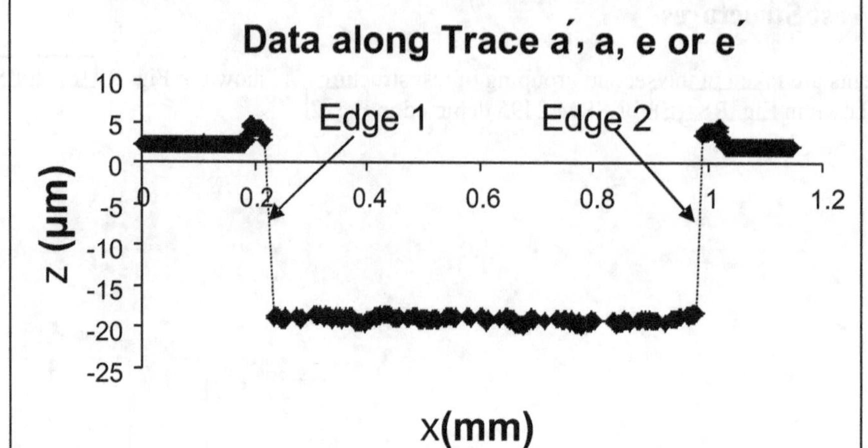

(b)

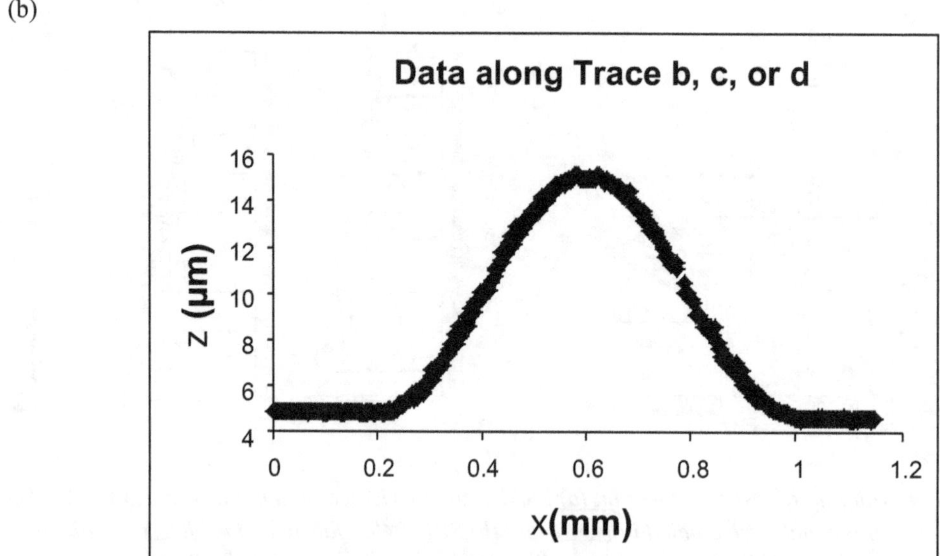

(c)

Figure RS2. For a fixed-fixed beam test structure on SRM 2494, (a) a design rendition, (b) an example of a 2-D data trace used to determine L in (a),[25] and (c) an example of a 2-D data trace taken along the length of the fixed-fixed beam in (a).[26]

---

[25] *Copyright, ASTM International, 100 Barr Harbor Drive, West Conshohocken, PA 19428, USA. Reproduced via permissions with ASTM International.*
[26] Ibid.

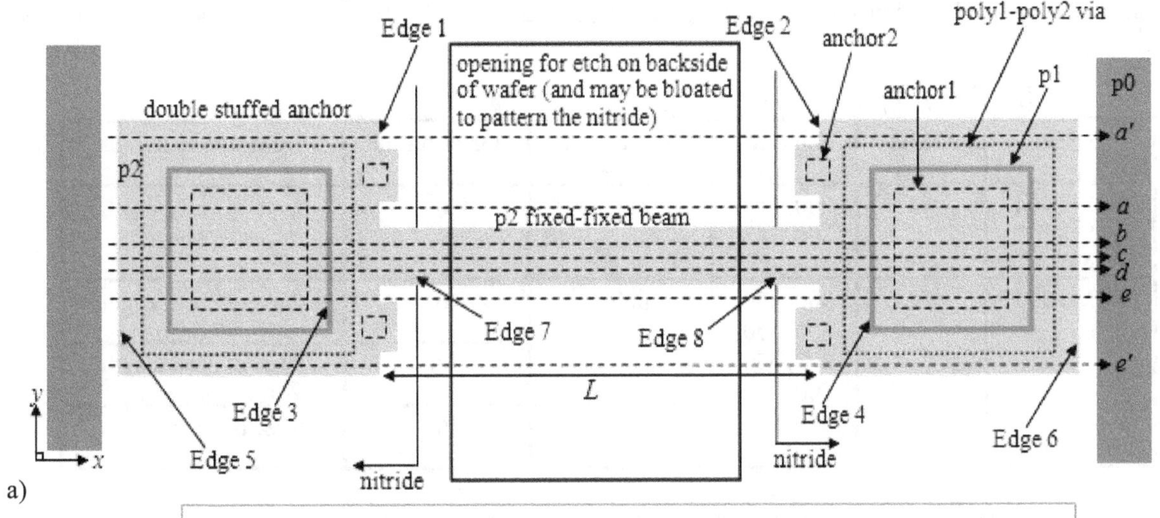

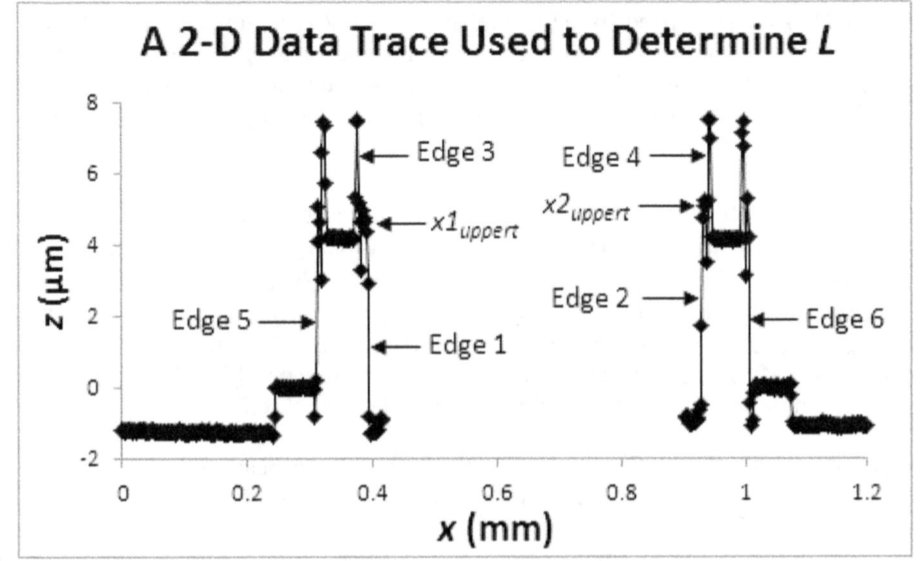

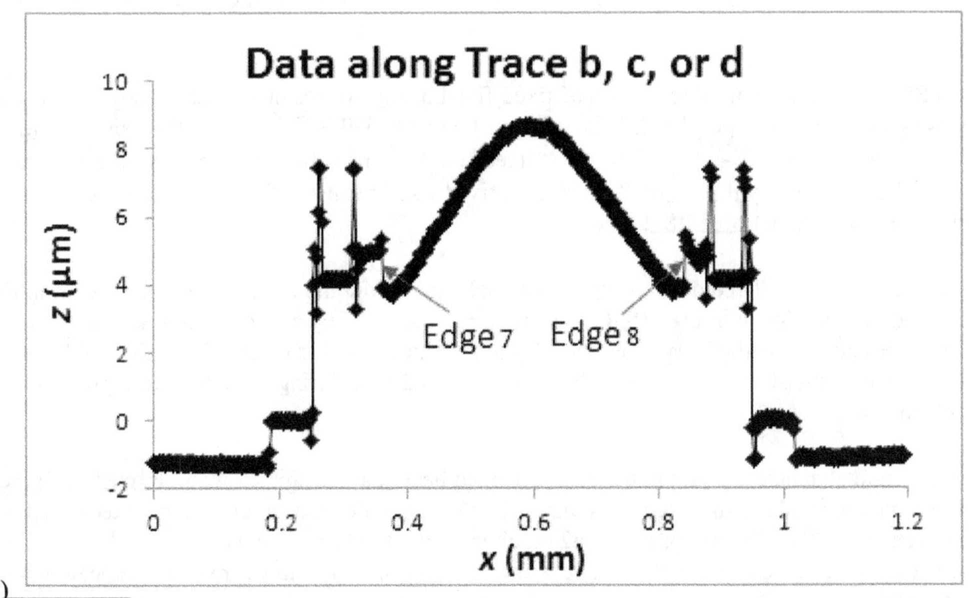

Figure RS3. *For a p2 fixed-fixed beam test structure, (a) a design rendition on SRM 2495, (b) an example of a 2-D data trace used to determine L, and (c) an example of a 2-D data trace taken along the length of a fixed-fixed beam.*

The specifications for the fixed-fixed beams shown in Figs. RS1(a and b) for SRM 2494 and SRM 2495, respectively, are given in Table RS1.

Table RS1. Fixed-Fixed Beam Configurations for Residual Strain Measurements

SRM	Width (μm)	Length (μm)	Structural Layer	Orientation	Quantity of Beams
SRM 2494	40	200, 248, 300, 348, 400	oxide	$0°$	three of each length (or 15 beams)
SRM 2495	16	400, 450, 500, 550, 600, 650, 700, 750, 800	poly1	$0°$	three of each length (or 27 beams)
				$90°$	three of each length (or 27 beams)
			poly2	$0°$	three of each length (or 27 beams)

For SRM 2494: On SRM 2494, all oxide fixed-fixed beams shown in Fig. RS1(a) are designed with a $0°$ orientation. As seen in this figure, the length of a fixed-fixed beam (in micrometers) is given at the top of each column of fixed-fixed beams following the column number (i.e., 1 to 5) and the letters "FF" to indicate a fixed-fixed beam. These design lengths (and the design width) are specified in Table RS1. There are three fixed-fixed beams designed at each length. Therefore, there are 15 oxide fixed-fixed beams with a $0°$ orientation.

As specified in Sec. 1.4.1, the exposed silicon, as shown in Fig. RS2(a), is isotropically etched in XeF_2 to release the fixed-fixed beam by removing the silicon around and beneath the beam. The dimensional markers are instrumental in firming up the support region. They also can be used to measure the small amount of SiO_2 that has also been etched in XeF_2,[27] which would have the effect of modifying the in-plane length of the fixed-fixed beam. The etch stop, also shown in this figure, helps to inhibit the etch away from the test structure to shield neighboring structures from the etch. It consists of an *n*-implant designed to surround active area. Figs. YM2(a, b, and c) show a cantilever test structure on SRM 2494. As can be seen in Fig. YM2(c), there is undercutting of the cantilever. This is the case for fixed-fixed beams as well.

An oxide fixed-fixed beam consists of four SiO_2 layers. The thickness of these beams is calculated using Data Analysis Sheet T.1. See Sec. 8 for specifics. Even though the beam is made up of four layers of SiO_2, the layers may not have the same properties. Also there may be remaining debris in the attachment corners of the fixed-fixed beam to the beam support. Due to these non-idealities in the geometry and composition of the fixed-fixed beam and/or beam support [and including the undercutting of the beam shown in Fig. YM2(c)] an effective residual strain is reported on the SRM Certificate presented in Sec. 3.6.

For SRM 2495: On SRM 2495, there are three arrays of fixed-fixed beams for residual strain measurements, as shown in Fig. RS1(b). Two of these arrays consist of poly1 fixed-fixed beams (as indicated by a "P1" symbol) and one array consists of poly2 fixed-fixed beams (as indicated by a "P2" symbol). The fixed-fixed beams within the top poly1 array have a $90°$ orientation, and the remaining poly1 array along with the poly2 array have fixed-fixed beams with a $0°$ orientation. The design dimensions of the fixed-fixed beams are given in Table RS1.

Fig. RS3(a) shows one of the p2 fixed-fixed beams in the residual strain grouping of test structures shown in Fig. RS1(b). The p2 fixed-fixed beam pad designs shown in Fig. RS3(a) are similar to the pad design shown for the cantilever given in Fig. YM4(a). The pad includes both p1 and p2. Without the p1, the attachment points of the p2 fixed-fixed beam to each anchor are not rigid (or fixed). By including p1 in the anchor design, the p1 and p2 fuse during the fabrication process to make a more rigid and reliable attachment point.

To make an even more rigid attachment point, in the p2 fixed-fixed beam pad design shown in Fig. RS3(a), the p2 layer is also anchored to the nitride on either side of the fixed-fixed beam. If this is not done, each end of the fixed-fixed beam can be viewed as having two beam widths. By anchoring the p2 on either side of the fixed-fixed beam (and at each end of the fixed-fixed beam), it becomes a fixed-fixed beam with a single width ($W=16$ μm), and can be treated as a beam with fixed-fixed boundary conditions [14].

---

[27] The design dimension from the dimensional marker to the exposed silicon is 16 μm, as shown in Fig. RS2(a).

Also, as seen in Fig. YM4(b), a flat cantilever is not fabricated. There is an approximate 600 nm vertical transition (or kink) in the cantilever. For fixed-fixed beams, there are two vertical transistions. As shown in Fig. RS3(a), an opening is created on the backside of the wafer for a backside etch. This etch removes the material beneath the fixed-fixed beams to ensure the existence of fixed-fixed beams that have not adhered to the top of the underlying layer. Earlier in the fabrication process, the nitride layer is patterned using a mask similar to that used to create the openings in the backside of the wafer, however, all the features may be bloated by an amount that is expected to change for different processing runs. As a result, the polysilicon beams traverse two approximate 600 nm fabrication steps over the nitride, as can be seen for one step in the cantilever test structures in Fig. YM4(b) and Fig. YM5. For the double-stuffed pad designs for the p1 and p2 fixed-fixed beams on the SRM 2495 chips fabricated on a 2011 processing run (run #95 [9]) this step is approximtely 25 μm from each anchor lip (or 38 μm from the nearest anchor when the opening for the backside etch is designed 65 μm from the anchor). These non-idealities in the geometry of the fixed-fixed beams are responsible for an "effective" residual strain value being reported on the SRM Certificate as presented in Sec. 3.6.

3.2 Calibration Procedures for Residual Strain Measurements

For SRM residual strain measurements, the interferometric microscope is calibrated in the z-direction as specified in Sec. 5.2 for step height calibrations as used with the MEMS 5-in-1. For discussing earlier version of the uncertainty equations presented in Sec. 3.4.2, Eq. (SH2) in Sec. 5.2 is also used. Therefore, six measurements are taken along the certified area of the physical step height standard before the data session and six measurements are taken along the certified area after the data session. The interferometric microscope is calibrated in the x- and y-directions as given in Sec. 6.2 for in-plane length calibrations. These calibration procedures are the same as those for strain gradient and in-plane length measurements, as indicated in Sec. 4.2 and Sec. 6.2, respectively.

3.3 Residual Strain Measurement Procedure

Residual strain measurements are taken from a fixed-fixed beam test structure such as shown in Fig. RS3(a). To obtain a residual strain measurement, the following steps are taken for SRM 2495 (consult the standard test method [2] for additional details and for modifications to these steps for a bulk-micromachined test structure on SRM 2494):

1. Seven 2-D data traces, such as shown in Fig. RS3(a) are extracted from a 3-D data set.

2. From Traces a', a, e, and e', the uncalibrated values from Edge 1 and Edge 2 (namely, for $x1_{uppert}$ and $x2_{unnert}$) along with the corresponding values for $n1_t$ and $n2_t$, respectively, are obtained (as defined and specified in Sec. 6.3 for in-plane length measurements). The trailing subscript "t" indicates the data trace (a', a, e, or e') being examined. These 16 values are entered into Data Analysis Sheet RS.3 along with the uncalibrated y values associated with Traces a' and e' (namely, ya' and ye').

3. The uncalibrated endpoints ($x1_{ave}$ and $x2_{ave}$) of the measured in-plane length of the fixed-fixed beam are calculated using the equations below:

$$x1_{ave} = \frac{x1_{uppera'} + x1_{uppera} + x1_{uppere} + x1_{uppere'}}{4} \text{ , and} \tag{RS1}$$

$$x2_{ave} = \frac{x2_{uppera'} + x2_{uppera} + x2_{uppere} + x2_{uppere'}}{4} \ . \tag{RS2}$$

4. As specified in Sec. 6.3, the misalignment angle, α, is shown in Fig. L5(a), where L_{meas} and L_{align} in this figure are calculated using Eqs. (L2) through (L4). The misalignment angle is typically determined using the two outermost data traces [a' and e' in this case, as seen in Figs. RS3(a) and L5(b)] and is calculated to be either α_1 or α_2 using either Δ_{x1} or Δ_{x2}, respectively [as seen in Fig. L5(b)]. The equations for Δ_{x1} and Δ_{x2} are:

$$\Delta_{x1} = x1_{uppera'} - x1_{uppere'} \text{ , and} \tag{RS3}$$

$$\Delta_{x2} = x2_{uppera'} - x2_{uppere'} \ . \tag{RS4}$$

The equation for α is as follows:

$$\alpha = \tan^{-1}\left[\frac{\Delta x\; cal_x}{\Delta y\; cal_y}\right] \text{, where} \tag{RS5}$$

$$\Delta y = y_{a'} - y_{e'} \; . \tag{RS6}$$

In addition,

$$\text{if } n1'_a + n1'_e \leq n2'_a + n2'_e \text{, then } \alpha = \alpha_1 \text{ and } \Delta x = \Delta x1, \tag{RS7}$$

$$\text{and if } n1'_a + n1'_e > n2'_a + n2'_e \text{, then } \alpha = \alpha_2 \text{ and } \Delta x = \Delta x2. \tag{RS8}$$

5. The 2-D data along the fixed-fixed beam from Traces b, c, and d, as shown in Fig. RS3(c) for one data trace, are used to obtain three independent measurements of the curved length of the fixed-fixed beam. This is done for one data trace as follows:

   a. Eliminate the data values at both ends of the trace that will not be included in the modeling. This would include all data values outside and including Edges 1 and 2 in Figs. RS3(a and b). If not already eliminated, the data values less then $x1_{ave}$ and the data values greater than $x2_{ave}$ should also be eliminated. In addition, for the 2-D data trace given in Fig. RS3(c), the data values outside and including Edges 7 and 8, if present, should be eliminated. [For the test structure shown in Fig. RS3(a), a backside etch is used. As a result for the process used, the beam traverses two approximate 600 nm fabrication steps over the nitride used in conjunction with the backside etch (as specified in Sec. 3.1). One of these steps can be seen in the cantilever test structures in Fig. YM4(b) and Fig. YM5. Due to these fabrication steps, the data values outside and including Edges 7 and 8 are eliminated.]

   b. Divide the remaining data into two data sets (as shown in Fig. RS4) if there is a peak (or valley) within the length of the curved structure. [The division should occur at the x value corresponding to the maximum (or minimum) z value. This data point should be included in both data sets.]

   c. Choose three representative data points (sufficiently separated) within each data set. The three uncalibrated points within the first data set are called (x_{1F}, z_{1F}), (x_{2F}, z_{2F}), and (x_{3F}, z_{3F}). And, the three uncalibrated points within the second data set are called (x_{1S}, z_{1S}), (x_{2S}, z_{2S}), and (x_{3S}, z_{3S}), such that $x_{3F} = x_{1S}$ and $z_{3F} = z_{1S}$ as specified above such that only five points are obtained from the data trace. In choosing these points x_{3F} is typically the x value corresponding to the maximum (or minimum) z value, (x_{2F}, z_{2F}) and (x_{2S}, z_{2S}) are located near the inflection points, and x_{1F} is slightly larger than $x1_{ave}$ (or an estimate of $x7_{ave}$ if Edge 7 is present) and x_{3S} is slightly smaller than $x2_{ave}$ (or an estimate of $x8_{ave}$ if Edge 8 is present). The five uncalibrated data points are entered into Data Analysis Sheet RS.3.

   d. To account for the misalignment angle, α, as shown in Fig. RS5, and the x-calibration factor, cal_x, the values obtained above for $x1_{ave}, x_{1F}, x_{2F}, x_{3F} = x_{1S}, x_{2S}, x_{3S}$, and $x2_{ave}$ become $f, g, h, i, j, k,$ and l, respectively, along the v-axis (the axis used to measure the length of the fixed-fixed beam) as also shown in Fig. RS5. The uncalibrated z-values of the data points along the beam remain the same, which assumes there is no curvature of the fixed-fixed beam across the width of the fixed-fixed beam. Therefore, the calibrated data points along the beam become $(g, z_{1F}\, cal_z)$, $(h, z_{2F}\, cal_z)$, $(i, z_{3F}\, cal_z)$ or $(i, z_{1S}\, cal_z)$, $(j, z_{2S}\, cal_z)$, and $(k, z_{3S}\, cal_z)$. The equations for $f, g, h, i, j, k,$ and l are given below:

$$f = x1_{ave}\, cal_x \; , \tag{RS9}$$

$$g = \left(x_{1F}\, cal_x - f\right)\cos\alpha + f \; , \tag{RS10}$$

$$h = \left(x_{2F}cal_x - f\right)\cos\alpha + f \,, \tag{RS11}$$

$$i = \left(x_{3F}cal_x - f\right)\cos\alpha + f = \left(x_{1S}cal_x - f\right)\cos\alpha + f \,, \tag{RS12}$$

$$j = \left(x_{2S}cal_x - f\right)\cos\alpha + f \,, \tag{RS13}$$

$$k = \left(x_{3S}cal_x - f\right)\cos\alpha + f \,, \text{ and} \tag{RS14}$$

$$l = \left(x2_{ave}cal_x - f\right)\cos\alpha + f \,. \tag{RS15}$$

e. The in-plane length of the fixed-fixed beam, L, as shown in Fig. RS5, is calculated using the following equation:

$$L = L_{align} + L_{offset} \tag{RS16}$$

$$= l - f + L_{offset}$$

$$= \left(x2_{ave}cal_x - x1_{ave}cal_x\right)\cos\alpha + L_{offset}$$

where L_{offset} is entered into Data Analysis Sheet RS.3 as the in-plane length correction term for the given type of in-plane length measurement on similar structures when using similar calculations and for the given magnification of the given interferometric microscope. See Sec. 6.3 for additional details. One endpoint, $v1_{end}$, of the in-plane length, L, is given below:

$$v1_{end} = f - \frac{1}{2}L_{offset} = x1_{ave}\,cal_x - \frac{1}{2}L_{offset} \,, \tag{RS17}$$

and the other endpoint, $v2_{end}$, of L is as follows:

$$v2_{end} = l + \frac{1}{2}L_{offset} = \left(x2_{ave}\,cal_x - f\right)\cos\alpha + f + \frac{1}{2}L_{offset} \tag{RS18}$$

$$= \left(x2_{ave}\,cal_x - x1_{ave}\,cal_x\right)\cos\alpha + x1_{ave}\,cal_x + \frac{1}{2}L_{offset}\,.$$

These endpoints are used next in the determination of the curved length of the fixed-fixed beam.

f. Using g, h, i, j, and k, as shown in Fig. RS5, two cosine functions (as seen in Fig. RS6) are used to model the out-of-plane shape of the fixed-fixed beam and obtain the curved length, L_c, of the beam [with the v-values of the endpoints given in Eqs. (RS17) and (RS18)].

6. The curved length of the fixed-fixed beam, L_c, for the given data trace, is compared with the in-plane length for a calculation of residual strain, ε_{rt}, using the following equation [14]:

$$\varepsilon_{rt} = \left[\left(L - L_0\right)/L_0\right]\left(1 + \delta_{\varepsilon rcorrection}\right), \tag{RS19}$$

where $L_0 = \left[12\,L_c\left(L_c\,L_e'\,/\,L\right)^2\right]/\left[12(L_c\,L_e'\,/\,L)^2 - \pi^2\,t^2\right]$, so that $\tag{RS20}$

$$\varepsilon_{rt} = \left\{ \{L - [12L_c(L_cL_e'/L)^2]/[12(L_cL_e'/L)^2 - \pi^2 t^2]\} / \{[12L_c(L_cL_e'/L)^2]/[12(L_cL_e'/L)^2 - \pi^2 t^2]\} \right\} (1 + \delta_{\varepsilon rcorrection}) \,, \quad (RS21)$$

where L_0 is the length with zero force applied, L_e' is the effective length of the fixed-fixed beam when the forces P_c and P_+ are applied, t is the thickness of the beam as obtained using Data Analysis Sheet T.1 for SRM 2494 and Data Analysis Sheet T.3 for SRM 2495, and where $\delta_{\varepsilon rcorrection}$ is a relative residual strain correction term intending to correct for deviations from the ideal fixed-fixed beam geometry and/or composition as discussed in more detail in the next step. For a more complete description of these lengths, the forces mentioned, and the derivation of this equation (without the correction term), consult reference [14]. The subscript "t" in ε_{rt} refers to the data trace used to obtain this residual strain value.

7. The relative residual strain correction term, $\delta_{\varepsilon rcorrection}$, is intended to correct for deviations from the ideal fixed-fixed beam geometry and/or composition. This includes deviations from the ideal in both the beam support and the beam itself (such as vertical transitions along the beam as discussed in Sec. 3.1 for SRM 2495). For SRM 2494 and 2495, it is currently assumed that $\delta_{\varepsilon rcorrection} = 0$ and an effective value for residual strain is entered on the SRM Certificates.

8. The resulting residual strain value, ε_r, is the average of the residual strain values obtained from Traces b, c, and d, as given below:

$$\varepsilon_r = \frac{\varepsilon_{rb} + \varepsilon_{rc} + \varepsilon_{rd}}{3} \,. \quad\quad (RS22)$$

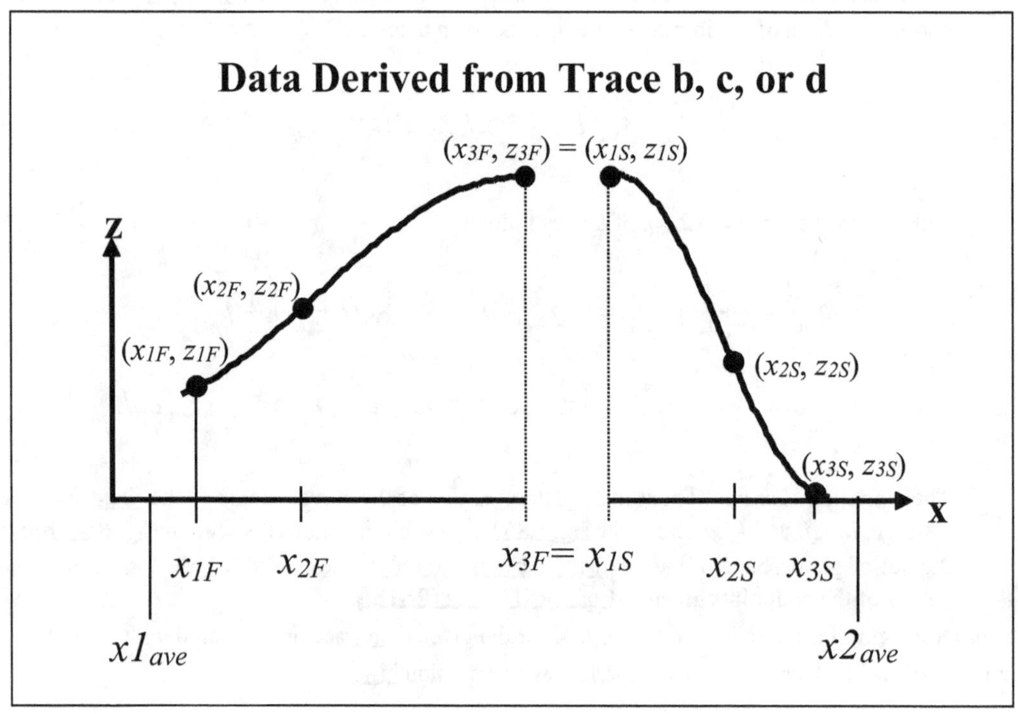

Figure RS4. *Two data sets derived from an abbreviated data trace along a fixed-fixed beam. The data in the figure above have been exaggerated.*

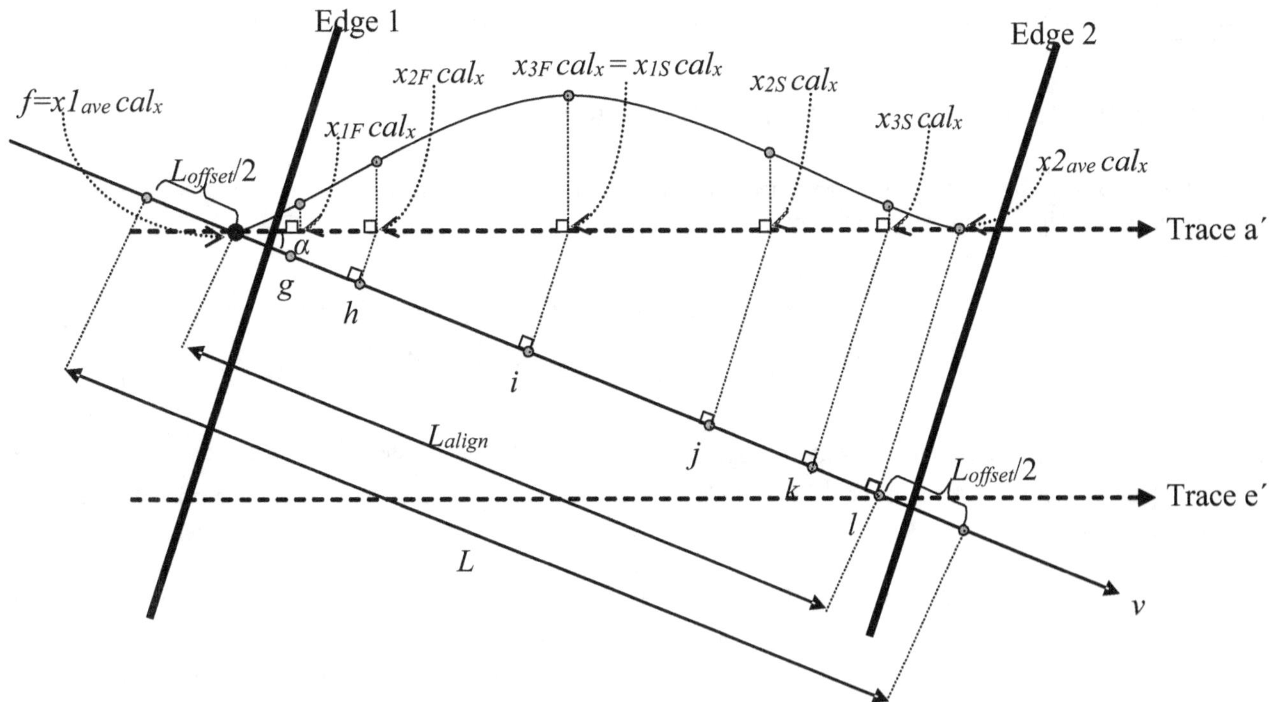

Figure RS5. *Sketch used to derive the appropriate v-values (f, g, h, i, j, k, and l) along the length of the beam*

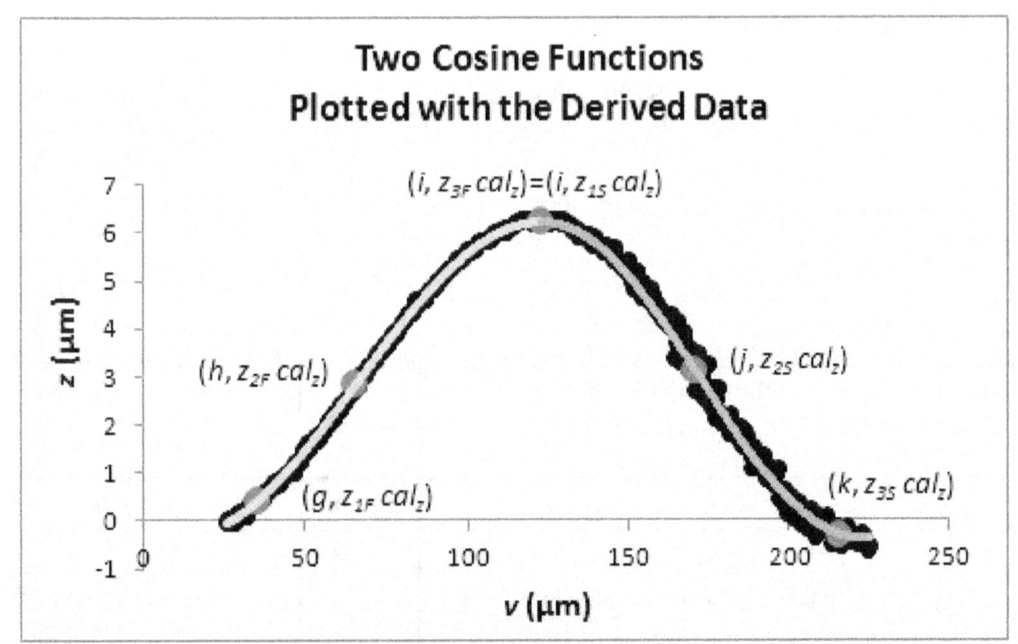

Figure RS6. *A comparison plot of the model with the derived data for an upward bending fixed-fixed beam*

3.4 Residual Strain Uncertainty Analysis

In this section, uncertainty equations are presented for use with residual strain. The first uncertainty equation (presented in Sec. 3.4.1) is used for the MEMS 5-in-1. The equations used in the round robin experiment and other previous work are presented in Sec. 3.4.2.

3.4.1 Residual Strain Uncertainty Analysis for the MEMS 5-in-1

For the MEMS 5-in-1, the combined standard uncertainty, $u_{c\varepsilon r3}$, for residual strain measurements with 13 uncertainty components is given by the following equation:

$$u_{c\varepsilon r3} = \sqrt{\begin{array}{l} u_W^2 + u_L^2 + u_{zres}^2 + u_{xcal}^2 + u_{xres}^2 + u_{Rave}^2 + u_{noise}^2 + u_{cert}^2 \\ + u_{repeat(shs)}^2 + u_{drift}^2 + u_{linear}^2 + u_{correction}^2 + u_{repeat(samp)}^2 \end{array}} , \qquad \text{(RS23)}$$

with additional sources of uncertainty considered negliglible. A number following the subscript "ε_r" in "$u_{c\varepsilon r}$" indicates the data analysis sheet that is used to obtain the combined standard uncertainty value. Therefore, $u_{c\varepsilon r3}$ implies that Data Analysis Sheet RS.3 [2,13] is used.

In Eq. (RS23), u_W is the uncertainty due to variations across the width of the beam, u_L is due to the measurement uncertainty of L (without including the calibration uncertainty), u_{zres} is the uncertainty due to the resolution of the interferometer in the z-direction, u_{xcal} is the uncertainty due to the calibration in the x-direction, and u_{xres} is the uncertainty due to the resolution of the interferometric microscope in the x-direction. Next, u_{Rave} is the uncertainty due to the sample's surface roughness, u_{noise} is the uncertainty due to interferometric noise, u_{cert} is due to the uncertainty of the value of the physical step height standard, $u_{repeat(shs)}$ is the uncertainty due to the *repeatability* of a measurement taken on the physical step height standard, u_{drift} is the uncertainty due to the amount of drift during the data session, and u_{linear} is the uncertainty due to the deviation from linearity of the data scan. Then, in Eq. (RS23), $u_{correction}$ is the uncertainty in the residual strain correction term due to non-ideal support (or attachment conditions) and/or geometry and/or composition deviations from the ideal fixed-fixed beam geometry and composition. Then, $u_{repeat(samp)}$ is the uncertainty of residual strain *repeatability* measurements taken on fixed-fixed beams processed similarly to the one being measured.

Calculations for each of the uncertainty components in Eq. (RS23) are presented below in sequence, with Table RS2 giving a brief tabular summary of how each uncertainty component is obtained. This table can be referenced as each component is discussed.

Table RS2. *Determination of the Residual Strain Uncertainty Components in Eq. (RS23) for the MEMS 5-in-1* [2]

Uncertainty Component	Method to Obtain ε_{r-high} and ε_{r-low}, if applicable	G or U[a] /A or B[b]	Equation
1. u_W	–	G / A	$u_W = STDEV(\varepsilon_{rb}, \varepsilon_{rc}, \varepsilon_{rd})$
2. u_L	using $L_{minuL} = L - 3u_{cLnoxcal}$ for L and $L_{maxuL} = L + 3u_{cLnoxcal}$ for L $u_{cLnoxcal} = \sqrt{u_{cL0}^2 - u_{Lxcal}^2}$ endpoints for L_{minuL}: $vl_{end} + \dfrac{3}{2}u_{cLnoxcal}$	G / B	$u_{Lt} = \dfrac{\lvert \varepsilon_{r-high} - \varepsilon_{r-low}\rvert}{6}$ $u_L = \dfrac{u_{Lb} + u_{Lc} + u_{Ld}}{3}$

	$$v2_{end} - \frac{3}{2}u_{cLnoxcal}$$ endpoints for L_{maxuL}: $$v1_{end} - \frac{3}{2}u_{cLnoxcal}$$ $$v2_{end} + \frac{3}{2}u_{cLnoxcal}$$		
3. u_{zres}	using $d=(1/2)z_{res}$ in Table RS3	U / B	$$u_{zrest} = \frac{\left|\varepsilon_{r-high} - \varepsilon_{r-low}\right|}{2\sqrt{3}}$$ $$u_{zres} = \frac{u_{zresb} + u_{zresc} + u_{zresd}}{3}$$
4. u_{xcal}	using cal_{xmin} for cal_x where $$cal_{xmin} = cal_x - 3\sigma_{xcal}cal_x / ruler_x$$ and cal_{xmax} for cal_x where $$cal_{xmax} = cal_x + 3\sigma_{xcal}cal_x / ruler_x$$	G / B	$$u_{xcalt} = \frac{\left|\varepsilon_{r-high} - \varepsilon_{r-low}\right|}{6}$$ $$u_{xcal} = \frac{u_{xcalb} + u_{xcalc} + u_{xcald}}{3}$$
5. u_{xres}	using $d=(1/2)x_{res}(cal_x)\cos(\alpha)$ in Table RS4	U / B	$$u_{xrest} = \frac{\left|\varepsilon_{r-high} - \varepsilon_{r-low}\right|}{2\sqrt{3}}$$ $$u_{xres} = \frac{u_{xresb} + u_{xresc} + u_{xresd}}{3}$$
6. u_{Rave}	using $d=3\sigma_{Rave}$ in Table RS3 where $\sigma_{Rave} = \frac{1}{6}R_{ave}$	G / B	$$u_{Ravet} = \frac{\left|\varepsilon_{r-high} - \varepsilon_{r-low}\right|}{6}$$ $$u_{Rave} = \frac{u_{Raveb} + u_{Ravec} + u_{Raved}}{3}$$
7. u_{noise}	using $d=3\sigma_{noise}$ in Table RS3 where $\sigma_{noise} = \frac{1}{6}(R_{tave} - R_{ave})$	G / B	$$u_{noiset} = \frac{\left|\varepsilon_{r-high} - \varepsilon_{r-low}\right|}{6}$$ $$u_{noise} = \frac{u_{noiseb} + u_{noi\,sec} + u_{noised}}{3}$$
8. u_{cert}	using $d=3(z_{xx}-z_{1F})\sigma_{cert}/cert$ in Table RS5 where z_{xx} is the column heading[c]	G / B	$$u_{certt} = \frac{\left|\varepsilon_{r-high} - \varepsilon_{r-low}\right|}{6}$$ $$u_{cert} = \frac{u_{certb} + u_{certc} + u_{certd}}{3}$$
9. $u_{repeat(shs)}$	using $d=3(z_{xx}-z_{1F})\sigma_{6same}/\overline{z}_{6same}$ in Table RS5 where z_{xx} is the column heading[c]	G / B	$$u_{repeat(shs)t} = \frac{\left|\varepsilon_{r-high} - \varepsilon_{r-low}\right|}{6}$$ $$u_{repeat(shs)} = (u_{repeat(shs)b} + u_{repeat(shs)c} + u_{repeat(shs)d})/3$$
10. u_{drift}	using $d=(z_{xx}-z_{1F})z_{drift}cal_z/(2\,cert)$ in Table RS5 where z_{xx} is the column heading[c]	U / B	$$u_{driftt} = \frac{\left|\varepsilon_{r-high} - \varepsilon_{r-low}\right|}{2\sqrt{3}}$$

			$$u_{drift} = \frac{u_{driftb} + u_{driftc} + u_{driftd}}{3}$$
11. u_{linear}	using $d=z_{linear}$ in Table RS5	U / B	$$u_{lineart} = \frac{\left| \varepsilon_{r-high} - \varepsilon_{r-low} \right|}{2\sqrt{3}}$$ $$u_{linear} = \frac{u_{linearb} + u_{linearc} + u_{lineard}}{3}$$
12. $u_{correction}$	–	G / B	$$u_{correctiont} = \left| \delta_{\varepsilon rcorrection} \varepsilon_{rt} \right| / 3$$ $$u_{correction} = (u_{correctionb} + u_{correctionc} + u_{correctiond})/3$$
13. $u_{repeat(samp)}$	–	G / A	$$u_{repeat(samp)t} = \sigma_{repeat(samp)} \left| \varepsilon_{rt} \right|$$ $$u_{repeat(samp)} = (u_{repeat(samp)b} + u_{repeat(samp)c} + u_{repeat(samp)d})/3$$

[a] "G" indicates a Gaussian distribution and "U" indicates a uniform distribution.

[b] Type A or Type B analysis

[c] For ease of presentation, z_{xx} and z_{1F} in this table are considered calibrated. (Actually, these values are uncalibrated as presented earlier in this SP 260. Therefore the uncalibrated values should be multiplied by cal_z before use in this table.)

The uncertainty value for u_W is the standard deviation of the residual strain values as obtained from three data traces (b, c, and d) across the width of the beam, as given in Table RS2.

The uncertainty component for u_L is found after calculating the residual strain in two different ways for Traces b, c, and d. First, the residual strain is found assuming that L_{minuL} is the in-plane length of the fixed-fixed beam and second, assuming that L_{maxuL} is the in-plane length of the fixed-fixed beam. The equations for L_{minuL} and L_{maxuL} are given in Table RS2, where $u_{cLnoxcal}$ is the combined standard uncertainty for in-plane length, u_{cL0}, as obtained in Sec. 6.4.1 but without the in-plane length x-calibration component u_{Lxcal} (called u_{xcal} in Sec. 6.4). [The square of the x-calibration component is subtracted from the square of u_{cL0} under the square root sign in Table RS2 since an uncertainty contribution due to the x-calibration is incorporated in the residual strain component (also called u_{xcal}) as discussed in a following paragraph.] In obtaining the residual strain value with L_{minuL} then L_{maxuL} as the in-plane length, the curved length of the fixed-fixed beam is also calculated. The two extreme residual strain values (ε_{r-low} and ε_{r-high}) are identified for each data trace. Then, u_{Lt} is calculated, assuming a Gaussian distribution, using the equation given in Table RS2. The average of the three values obtained for u_{Lt} is equated with u_L.

The uncertainty equation for u_{zres} is found from the residual strain calculations using the different sets of inputs given in Table RS3. Here, for each data trace (b, c, and d), the inputed z-values along the top of the beam are varied plus or minus half z_{res}, where z_{res} is the calibrated resolution of the interferometer in the z-direction. The two extreme residual strain values (ε_{r-low} and ε_{r-high}) are identified for each data trace. Then, u_{zrest} is calculated, assuming a uniform distribution, using the equation given in Table RS2. The average of the three values obtained for u_{zrest} is equated with u_{zres}.

The uncertainty equation for u_{xcal} is found by finding the residual strain value for each data trace (b, c, and d) with cal_{xmin} as the x-calibration factor and then finding the residual strain value for each data trace with cal_{xmax} as the x-calibration factor, where cal_{xmin} and cal_{xmax} are determined using the equations given in Table RS2. This component, u_{xcal}, includes the uncertainty of the calibration in the x-direction for the in-plane length as discussed in a preceding paragraph for the component u_L. The two extreme residual strain values, ε_{r-low} and ε_{r-high}, are identified for each data trace. Then, u_{xcalt} is calculated, assuming a Gaussian distribution, using the equation given in Table RS2. The average of the three values obtained for u_{xcalt} is equated with u_{xcal}.

The uncertainty equation for u_{xres} is found from the residual strain calculations using the seven different sets of inputs given in Table RS4. Here, for each data trace (b, c, and d) the inputted x-values along the top of the beam are varied $\pm(1/2)x_{res}(cal_x)\cos(\alpha)$ where x_{res} is the uncalibrated resolution of the interferometric microscope in the x-direction. The two extreme residual strain values, $\varepsilon_{r\text{-}low}$ and $\varepsilon_{r\text{-}high}$, are identified for each data trace. Then, u_{xrest} is calculated, assuming a uniform distribution, using the equation given in Table RS2. The average of the three values obtained for u_{xrest} is equated with u_{xres}.

Table RS3. Seven Sets of Inputs[a] for Residual Strain Calculations to Determine u_{zrest}, u_{Ravet}, u_{noiset}, and u_{samp}[b]

	z_{1F}	z_{2F}	$z_{3F}=z_{1S}$	z_{2S}	z_{3S}
1	z_{1F}	z_{2F}	z_{3F}	z_{2S}	z_{3S}
2	$z_{1F}+d$	z_{2F}	$z_{3F}{}^-d$	z_{2S}	$z_{3S}+d$
3	$z_{1F}{}^-d$	z_{2F}	$z_{3F}{}^+d$	z_{2S}	$z_{3S}{}^-d$
4	$z_{1F}+d$	$z_{2F}+d$	$z_{3F}{}^-d$	$z_{2S}+d$	$z_{3S}+d$
5	$z_{1F}+d$	$z_{2F}{}^-d$	$z_{3F}{}^-d$	$z_{2S}{}^-d$	$z_{3S}+d$
6	$z_{1F}{}^-d$	$z_{2F}+d$	$z_{3F}{}^+d$	$z_{2S}+d$	$z_{3S}{}^-d$
7	$z_{1F}{}^-d$	$z_{2F}{}^-d$	$z_{3F}{}^+d$	$z_{2S}{}^-d$	$z_{3S}{}^-d$

[a] For ease of presentation, the values for z_{1F}, z_{2F}, z_{3F}, z_{1S}, z_{2S}, and z_{3S} in this table are assumed to be calibrated.

[b] In this table, $d=(1/2)z_{res}$ to determine u_{zrest}, $d=3\sigma_{Rave}$ to determine u_{Ravet}, $d=3\sigma_{noise}$ to deterimine u_{noiset}, and in Sec. 3.4.2 $d=3\sigma_{samp}$ to determine u_{samp}.

Table RS4. Seven Sets of Inputs[a] for Residual Strain Calculations to Determine u_{xrest}

	g	h	i	j	k
1	g	h	i	j	k
2	$g+d$	h	i	j	k^-d
3	g^-d	h	i	j	k^+d
4	$g+d$	$h+d$	i	j^-d	k^-d
5	$g+d$	h^-d	i	$j+d$	k^-d
6	g^-d	$h+d$	i	j^-d	k^+d
7	g^-d	h^-d	i	$j+d$	k^+d

[a] In this table, $d = (1/2)x_{res}(cal_x)\cos(\alpha)$.

The uncertainty equation for u_{Rave} is found from a determination of σ_{Rave}, where σ_{Rave} is calculated to be one-sixth the value of R_{ave}. R_{ave} is defined as the calibrated surface roughness of a flat and leveled surface of the sample material calculated to be the average of three or more measurements, each measurement of which is taken from a different 2-D data trace. For each data trace (b, c, and d), the data points obtained along the top of the beam are then varied as specified in Table RS3 (with $d=3\sigma_{Rave}$) and the residual strain determined for the different sets of inputs. The two extreme residual strain values, $\varepsilon_{r\text{-}low}$ and $\varepsilon_{r\text{-}high}$, are then identified for each data trace. The interval from $\varepsilon_{r\text{-}low}$ to $\varepsilon_{r\text{-}high}$ is assumed to encompass 99 % of the measurements. Then, u_{Ravet} is calculated, assuming a Gaussian distribution, using the equation given in Table RS2. The average of the three values obtained for u_{Ravet} is equated with u_{Rave}.

The uncertainty equation for u_{noise} is found from a determination of σ_{noise}, where σ_{noise} is the standard deviation of the noise measurement, calculated to be one-sixth the value of R_{tave} minus R_{ave}, where R_{tave} is the calibrated peak-to-valley roughness of a flat and leveled surface of the sample material calculated to be the average of three or more measurements, each measurement of which is taken from a different 2-D data trace. For each data tace (b, c, and d), the data points obtained along the top of the beam are then varied as specified in Table RS3 (with $d=3\sigma_{noise}$) and the residual strain determined for the different sets of inputs. The two extreme residual strain values, $\varepsilon_{r\text{-}low}$ and $\varepsilon_{r\text{-}high}$, are then identified for each data trace. The interval from $\varepsilon_{r\text{-}low}$ to $\varepsilon_{r\text{-}high}$ is assumed to encompass 99 % of the measurements. Then, u_{noiset} is calculated, assuming a Gaussian distribution, using the equation given in Table RS2. The average of the three values obtained for u_{noiset} is equated with u_{noise}.

The uncertainty equation for u_{cert} is found from the residual strain calculations using the inputs in Table RS5 [with $d=3(z_{xx}-z_{1F})\sigma_{cert}/cert$ for each data trace (b, c, and d) where σ_{cert} is the certified one sigma uncertainty of the certified physical step height standard]. Here, σ_{cert} is assumed to scale linearly with height. Given the three different residual strain values for the inputs specified in Table RS5, ε_{r-low} and ε_{r-high} are identified for each data trace. The interval from ε_{r-low} to ε_{r-high} is assumed to encompass 99 % of the measurements. Then, u_{certt} is calculated, assuming a Gaussian distribution, using the equation given in Table RS2. The average of the three values obtained for u_{certt} is equated with u_{cert}.

Table RS5. Three Sets of Inputs[a] for Residual Strain Calculations to Determine
u_{certt}, $u_{repeat(shs)t}$, u_{driftt}, $u_{lineart}$, and u_{zcal}.[b]

	z_{1F}	z_{2F}	$z_{3F}=z_{1S}$	z_{2S}	z_{3S}
1	z_{1F}	z_{2F}	z_{3F}	z_{2S}	z_{3S}
2	z_{1F}	$z_{2F}+d$	$z_{3F}+d$	$z_{2S}+d$	$z_{3S}+d$
3	z_{1F}	$z_{2F}-d$	$z_{3F}-d$	$z_{2S}-d$	$z_{3S}-d$

[a] For ease of presentation, the values for z_{1F}, z_{2F}, z_{3F}, z_{1S}, z_{2S}, and z_{3S} are assumed to be calibrated in this table and in the applicable equations for d.

[b] In this table, $d=3(z_{xx}-z_{1F})\sigma_{cert}/cert$ to determine u_{certt} where z_{xx} is the column heading,

$d=3(z_{xx}-z_{1F})\sigma_{6same}/\bar{z}_{6same}$ to determine $u_{repeat(shs)t}$, $d=(z_{xx}-z_{1F})z_{drift}cal_z/(2\ cert)$ to determine u_{driftt},

$d=z_{linear}$ to determine $u_{lineart}$, and in Sec. 3.4.2 $d=3(z_{xx}-z_{1F})\sigma_{zcal}/cert$ to determine u_{zcal}.

The uncertainty equation for $u_{repeat(shs)}$ is found from the residual strain calculations using the inputs in Table RS5 [with

$d=3(z_{xx}-z_{1F})\sigma_{6same}/\bar{z}_{6same}$ for each data trace (b, c, and d) where σ_{6same} is the maximum of two uncalibrated values (σ_{same1} and

σ_{same2}) and $\bar{z}_{6same}$ is the uncalibrated average of the six calibration measurements from which σ_{6same} is found. (See Sec. 5.2 for specifics.)] Here, σ_{6same} is assumed to scale linearly with height. Given the three different residual strain values for the inputs specified in Table RS5, ε_{r-low} and ε_{r-high} are identified for each data trace. The interval from ε_{r-low} to ε_{r-high} is assumed to encompass 99 % of the measurements. Then, $u_{repeat(shs)t}$ is calculated, assuming a Gaussian distribution, using the equation given in Table RS2. The average of the three values obtained for $u_{repeat(shs)t}$ is equated with $u_{repeat(shs)}$.

The uncertainty equation for u_{drift} is found from the residual strain calculations using z_{drift}, which is calculated as follows: the uncalibrated average of the six calibration measurements taken before the data session at the same location on the physical step height standard ($\bar{z}_{same1}$) is determined, and the uncalibrated average of the six calibration measurements taken after the data session at this same location ($\bar{z}_{same2}$) is determined. Then, z_{drift} is calculated as the positive difference of these two values. Here, z_{drift} is assumed to scale linearly with height. For each data trace (b, c, and d), the input values to the residual strain calculations are then varied as specified in Table RS5 [with $d=(z_{xx}-z_{1F})z_{drift}cal_z/(2\ cert)$ where $cert$ is the certified value of the physical step height standard]. Given the three different residual strain values for the inputs specified in Table RS5, ε_{r-low} and ε_{r-high} are identified for each data trace. Then, u_{driftt} is calculated, assuming a uniform distribution, using the equation given in Table RS2. The average of the three values obtained for u_{driftt} is equated with u_{drift}.

The uncertainty equation for u_{linear} is found from the residual strain calculations using z_{linear}, which is the difference in height between two points times z_{lin}, where z_{lin} is the percent quoted (typically less than 3 %) by the interferometer manufacturer for the maximum deviation from linearity of the data scan over the total scan range or as determined in Sec. 1.1.2.2. For each data trace (b, c, and d), the input values to the residual strain calculations are varied as specified in Table RS5 with $d=z_{linear}$, where z_{linear} is given by the following equation:

$$z_{linear}=(z_{xx}-z_{1F})z_{lin},\qquad\text{(RS24)}$$

with z_{xx} being the column heading in the table and where z_{xx} and z_{1F} are considered calibrated values for this discussion. Given the three different residual strain values for the inputs specified in Table RS5, $\varepsilon_{r\text{-}low}$ and $\varepsilon_{r\text{-}high}$ are identified for each data trace. Then, $u_{lineart}$ is calculated, assuming a uniform distribution, using the equation given in Table RS2. The average of the three values obtained for $u_{lineart}$ is equated with u_{linear}.

In Eq. (RS23), $u_{correction}$ is calculated using the equations given in Table RS2.

In Eq. (RS23), $u_{repeat(samp)}$ is the uncertainty of residual strain *repeatability* measurements taken on fixed-fixed beams processed similarly to the one being measured. For each data trace, $u_{repeat(samp)t}$ is given by the following equation:

$$u_{repeat(samp)t} = \sigma_{repeat(samp)} \left| \varepsilon_{rt} \right| , \qquad (RS25)$$

and the average of the three values obtained for $u_{repeat(samp)t}$ is equated with $u_{repeat(samp)}$. In the above equation, the residual strain relative *repeatability* standard deviation, $\sigma_{repeat(samp)}$, is found from at least twelve 3-D data sets of a given fixed-fixed beam from which twelve values of ε_r are calculated as given in Sec. 3.3. The standard deviation of the twelve or more measurements of ε_r divided by the avearge of these measurements is equated with $\sigma_{repeat(samp)}$. Table 3 in Sec. 1.13 specifies that the residual strain relative *repeatability* standard deviation, $\sigma_{repeat(samp)}$, for fixed-fixed beams fabricated on a bulk micromachined process similar to that used to fabricate SRM 2494, is 2.49 %.

In determining the combined standard uncertainty, a Type B evaluation [19-21] (i.e., one that uses means other than the statistical Type A analysis) is used for each source of uncertainty, except where noted in Table RS2.

The expanded uncertainty for residual strain, $U\varepsilon_r$, is calculated using the following equation:

$$U_{\varepsilon r} = k u_{c\varepsilon r3} = 2 u_{c\varepsilon r3} , \qquad (RS26)$$

where the k value of 2 approximates a 95 % level of confidence.

Reporting results [19-21]. Since it can be assumed that the estimated values of the uncertainty components are either approximately uniformly or Gaussianly distributed with approximate combined standard uncertainty $u_{c\varepsilon r3}$, the residual strain is believed to lie in the interval $\varepsilon_r \pm u_{c\varepsilon r3}$ (expansion factor k=1) representing a level of confidence of approximately 68 %.

3.4.2 Previous Residual Strain Uncertainty Analyses

In this section, two uncertainty equations are presented; one that was used in the round robin experiment and one that was used before Eq. (RS23). For these equations the residual strain is assumed to be the residual strain value obtained from Trace c without a correction term, with the in-plane length, L, and the combined standard uncertainty for L (namely, u_{cL}) calculated using Eqs. (L17) and (L18), respectively, in Sec. 6.4.2. Also, it is assumed that $\alpha = 0$.

The uncertainty equation used in the round robin experiment uses eight sources of uncertainty with all other sources of uncertainty considered negligible. This residual strain combined standard uncertainty equation (as calculated in Data Analysis Sheet RS.1 [13] and in ASTM standard test method E 2245-05 [36]) with eight sources of uncertainty is as follows:

$$u_{c\varepsilon r1} = \sqrt{u_W^2 + u_L^2 + u_{zres}^2 + u_{samp}^2 + u_{zcal}^2 + u_{xcal}^2 + u_{xres}^2 + u_{xresL}^2} . \qquad (RS27)$$

The number following the subscript "ε_r" in "$u_{c\varepsilon r}$" indicates the data analysis sheet that is used to obtain the combined standard uncertainty value. Therefore, $u_{c\varepsilon r1}$ implies that Data Analysis Sheet RS.1 is used. In Eq. (RS27), u_W, u_L, u_{zres}, u_{xcal}, and u_{xres} are defined in Sec. 3.4.1; however with slightly different calculations. Also, u_{samp} is the uncertainty due to the sample's peak-to-valley surface roughness as measured with the interferometer, u_{zcal} is the uncertainty of the calibration in the z-direction, and

u_{xresL} is the uncertainty due to the resolution of the interferometric microscope in the x-direction as pertains to the in-plane length measurement.

Calculations for each of the uncertainty components in Eq. (RS27) are presented below in sequence, with Table RS6 giving a brief tabular summary of how each uncertainty component is obtained. This table can be referenced as each component is discussed.

Table RS6. Determination of Some Residual Strain Uncertainty Components in Eq. (RS27) and Eq. (RS28) [11,36]

Uncertainty Component	Method to Obtain $\varepsilon_{r\text{-high}}$ and $\varepsilon_{r\text{-low}}$	G or U[a] / A or B[b]	Equation
1. u_W	using Trace b, Trace c, and Trace d	U / B	$u_W = \dfrac{\left\| \varepsilon_{r-high} - \varepsilon_{r-low} \right\|}{2\sqrt{3}}$
2. u_L	using $L_{minuL} = (x2_{lower} - x1_{lower})cal_x$ for L and $L_{maxuL} = (x2_{upper} - x1_{upper})cal_x$ for L	G / B	$u_L = \dfrac{\left\| \varepsilon_{r-high} - \varepsilon_{r-low} \right\|}{6}$
3. u_{zres}	using $d = (1/2)z_{res}$ in Table RS3	U / B	$u_{zres} = \dfrac{\left\| \varepsilon_{r-high} - \varepsilon_{r-low} \right\|}{2\sqrt{3}}$
4. u_{samp}	using $d = 3\sigma_{samp}$ in Table RS3	G / B	$u_{samp} = \dfrac{\left\| \varepsilon_{r-high} - \varepsilon_{r-low} \right\|}{6}$
5. u_{zcal}	using $d = 3(z_{xx} - z_{1F})\sigma_{zcal}/cert$ in Table RS5 where z_{xx} is the column heading[c]	G / B	$u_{zcal} = \dfrac{\left\| \varepsilon_{r-high} - \varepsilon_{r-low} \right\|}{6}$
6. u_{xcal}	using cal_{xmin} for cal_x where $cal_{xmin} = cal_x - 3\sigma_{xcal}cal_x/ruler_x$ and cal_{xmax} for cal_x where $cal_{xmax} = cal_x + 3\sigma_{xcal}cal_x/ruler_x$	G / B	$u_{xcal} = \dfrac{\left\| \varepsilon_{r-high} - \varepsilon_{r-low} \right\|}{6}$
7. u_{xres}	using $d = (1/2)x_{res}(cal_x)$ in Table RS4	U / B	$u_{xres} = \dfrac{\left\| \varepsilon_{r-high} - \varepsilon_{r-low} \right\|}{2\sqrt{3}}$
8. u_{xresL}	using $d = (1/2)x_{res}(cal_x)$ in Table RS7	U / B	$u_{xresL} = \dfrac{\left\| \varepsilon_{r-high} - \varepsilon_{r-low} \right\|}{2\sqrt{3}}$
9. $u_{repeat(shs)}$	using $d = (z_{xx} - z_{1F})z_{repeat(shs)}/(2\bar{z}_6)$ in Table RS5 where z_{xx} is the column heading[c]	U / B	$u_{repeat(shs)} = \dfrac{\left\| \varepsilon_{r-high} - \varepsilon_{r-low} \right\|}{2\sqrt{3}}$

[a] "G" indicates a Gaussian distribution and "U" indicates a uniform distribution.

[b] Type A or Type B analysis

[c] For ease of presentation, z_{xx} and z_{1F} in this table are considered calibrated. (Actually, these values are uncalibrated as presented earlier in this SP 260. Therefore the uncalibrated values should be multiplied by cal_z before use in this table.)

The uncertainty equation for u_W is found using the residual strain results from three data traces (b, c, and d) across the width of the beam. The two extreme residual strain values (ε_{r-low} and ε_{r-high}) are obtained. Assuming a uniform probability distribution, u_W is calculated using the formula given in Table RS6.

The uncertainty equation for u_L is found after calculating the residual strain in two different ways using the data from Trace c. First, the residual strain is found assuming that L_{minuL} is the in-plane length of the fixed-fixed beam and second, assuming that

L_{maxuL} is the in-plane length of the fixed-fixed beam. The equations for L_{minuL} and L_{maxuL} are given in Table RS6. Consult Sec. 6.3 and Sec. 6.4.2 for specifics associated with the x-values used in these equations. Assuming that L_{minuL} then L_{maxuL} is the in-plane length precipitates recalculations of the curved length of the fixed-fixed beam. Then, u_L is calculated, assuming a Gaussian distribution, using the equation given in Table RS6.

The uncertainty equation for u_{zres} is found from the residual strain calculations using the different sets of inputs given in Table RS3. Here, for Trace c, the inputed z-values along the top of the beam are varied plus or minus half z_{res}, where z_{res} is the calibrated resolution of the interferometer in the z-direction. The two extreme residual strain values (ε_{r-low} and ε_{r-high}) are identified. Then, u_{zres} is calculated, assuming a uniform distribution, using the equation given in Table RS6.

The uncertainty equation for u_{samp}[28] is found from a determination of R_{tave}, the calibrated peak-to-valley roughness of a flat and leveled surface of the sample material calculated to be the average of three or more measurements, each measurement of which is taken from a different 2-D data trace. Then, the standard deviation, σ_{samp}, of this measurement is calculated to be one-sixth the value of R_{tave}. For Trace c, the data points obtained along the top of the fixed-fixed beam are then varied as specified in Table RS3 (with $d=3\sigma_{samp}$) and the residual strain determined for the different sets of inputs. Given the resulting residual strain values, ε_{r-low} and ε_{r-high} are identified. The interval from ε_{r-low} to ε_{r-high} is assumed to encompass 99 % of the measurements. Then, u_{samp} is calculated, assuming a Gaussian distribution, using the equation given in Table RS6.

The method of calibration of the interferometer in the z-direction affects the determination of the uncertainty component u_{zcal}.[29] In view of the method of calibration in the z-direction as referred to in Sec. 3.2, the uncertainty equation for u_{zcal} is found from the residual strain calculations using the three different sets of inputs given in Table RS5 [with $d=3(z_{xx}-z_{1F})\sigma_{zcal}/cert$]. Here, σ_{zcal} is the calibrated standard deviation of the twelve step height measurements taken along the certified portion of the physical step height standard before and after the data session and is assumed to scale linearly with height. Due to the fact that the difference in height being measured for residual strain measurements is small, u_{zcal} is almost negligible. Also, due to the method of calibration and other factors, u_{zcal} is very much considered an estimate. For Trace c, given the three different residual strain values for the inputs specified in Table RS5, ε_{r-low} and ε_{r-high} are identified. The interval from ε_{r-low} to ε_{r-high} is assumed to encompass 99 % of the measurements. Then, u_{zcal}, is calculated, assuming a Gaussian distribution, using the equation given in Table RS6.

The uncertainty equation for u_{xcal} is found by finding the residual strain value for Trace c with cal_{xmin} as the x-calibration factor and then finding the residual strain value for Trace c with cal_{xmax} as the x-calibration factor, where cal_{xmin} and cal_{xmax} are determined using the equations given in Table RS6. This component, u_{xcal}, includes the uncertainty of the calibration in the x-direction for the in-plane length. The two extreme residual strain values, ε_{r-low} and ε_{r-high}, are identified. Then, u_{xcal} is calculated, assuming a Gaussian distribution, using the equation given in Table RS6.

The uncertainty equation for u_{xres} is found from the residual strain calculations using the seven different sets of inputs given in Table RS4 after setting $\alpha=0$ such that g, h, i, j, and k in Eqs. (RS10) to (RS14) become the calibrated values of x_{1F}, x_{2F}, x_{3F} or x_{1S}, x_{2S}, and x_{3S}, respectively. Here, for Trace c, the inputted x-values along the top of the beam are varied $\pm(1/2)x_{res}(cal_x)$ where x_{res} is the uncalibrated resolution of the interferometric microscope in the x-direction. The two extreme residual strain values, ε_{r-low} and ε_{r-high}, are identified. Then, u_{xres} is calculated, assuming a uniform distribution, using the equation given in Table RS6.

The uncertainty equation for u_{xresL} is found from the residual strain calculations for Trace c from the two sets of inputs given in Table RS7 for $x1_{upper}$, $x1_{lower}$, $x2_{lower}$, and $x2_{upper}$. In this table, $x1_{max}$ is the original value for $x1_{upper}$, $x1_{min}$ is the original value for $x1_{lower}$, $x2_{min}$ is the original value for $x2_{lower}$, and $x2_{max}$ is the original value for $x2_{upper}$. As can be seen in this table, the minimum and maximum in-plane length endpoints are varied $\pm(1/2)x_{res}(cal_x)$. Then, ε_{r-low} and ε_{r-high} are determined and u_{xresL} is calculated, assuming a uniform distribution, using the equation given in Table RS6.

---

[28] In Eq. (RS27), u_{samp} is found from R_{tave} and one-sixth this value (or σ_{samp}). For Eq. (RS23) and Eq. (RS28) (which is presented later), R_{tave} is divided into R_{ave} (for the determination of u_{Rave}) and R_{tave} minus R_{ave} (for the determination of u_{noise}).

[29] Later in this section, using a different calibration method, u_{zcal} is divided into four separate components.

In determining the combined standard uncertainty, a Type B evaluation [19-21] (i.e., one that uses means other than the statistical Type A analysis) is used for each source of uncertainty. Table RS8 gives example values for each of these uncertainty components as well as the combined standard uncertainty value, $u_{c\varepsilon r1}$.

Table RS7. Two Sets of Inputs[a] for Residual Strain Calculations to Determine u_{xresL}[b]

	$\mathbf{x1}_{upper}$	$\mathbf{x1}_{lower}$	$\mathbf{x2}_{lower}$	$\mathbf{x2}_{upper}$
1	$x1_{max}{-}d$	$x1_{min}{-}d$	$x2_{min}{+}d$	$x2_{max}{+}d$
2	$x1_{max}{+}d$	$x1_{min}{+}d$	$x2_{min}{-}d$	$x2_{max}{-}d$

[a] For ease of presentation, the values for $x1_{upper}$, $x1_{lower}$, $x2_{lower}$, $x2_{upper}$, $x1_{max}$, $x1_{min}$, $x2_{min}$, and $x2_{max}$, in this table are assumed to be calibrated.

[b] In this table, $d = (1/2)x_{res}(cal_x)$.

Table RS8. Example Residual Strain Uncertainty Values[a] from a Round Robin Surface-Micromachined Chip

	source of uncertainty or descriptor	uncertainty values
1. $\mathbf{u}_W$	variations across the width of the beam	0.033×10^{-6}
2. $\mathbf{u}_L$	measurement uncertainty of L	0.038×10^{-6}
3. $\mathbf{u}_{zres}$	interferometric resolution in z-direction	0.034×10^{-6}
4. $\mathbf{u}_{samp}$	interferometric peak-to-valley surface roughness	0.393×10^{-6}
5. $\mathbf{u}_{zcal}$	calibration in z-direction	0.052×10^{-6}
6. $\mathbf{u}_{xcal}$	calibration in x-direction	0.123×10^{-6}
7. $\mathbf{u}_{xres}$	interferometric resolution in x-direction	0.273×10^{-6}
8. $\mathbf{u}_{xresL}$	interferometric resolution in x-direction as pertains to the in-plane length measurement	0.013×10^{-6}
$\mathbf{u}_{c\varepsilon r1}$[b]	combined standard uncertainty for residual strain	0.501×10^{-6}

[a] As determined in ASTM standard test method E 2245-05 [36] using Eq. (RS27) for a fixed-fixed beam with a design length of 650 μm and with a 0° orientation.

[b] This value for $u_{c\varepsilon r1}$ was used in the round robin (see Sec. 3.5) and is incorporated into the calculation of $u_{c\varepsilon r1ave}$ as presented in the sixth row of Table RS9, as determined in ASTM standard test method E2245-05 [36] using Eq. (RS27).

An expanded version of the uncertainty calculation presented in Eq. (RS27) is given below, which includes twelve sources of uncertainty:

$$u_{c\varepsilon r2} = \sqrt{u_W^2 + u_L^2 + u_{zres}^2 + u_{xcal}^2 + u_{xres}^2 + u_{xresL}^2 + u_{Rave}^2 + u_{noise}^2 + u_{cert}^2 + u_{repeat(shs)}^2 + u_{drift}^2 + u_{linear}^2} \quad . \qquad \text{(RS28)}$$

This calculation is done using Data Analysis Sheet RS.2 [31]. The first six components (namely, u_W, u_L, u_{zres}, u_{xcal}, u_{xres}, and u_{xresL}) are calculated as they are calculated for use in Eq. (RS27) using only Trace c. By using the above equation [instead of Eq. (RS27)], the z-calibration procedures are the same for the different applicable measurements in this SP 260 for the MEMS 5-in-1 SRMs [by using a procedure such as given in Sec. 5.2 for step height measurements (as used with the MEMS 5-in-1) as opposed to the method referred to in Sec. 3.2 for earlier versions]. Therefore, the component, u_{zcal} in Eq. (RS27) gets replaced with the components u_{cert}, u_{drift}, and u_{linear} as described in Sec. 3.4.1 however using only the data from Trace c and with the component $u_{repeat(shs)}$ as described in the next paragraph. Also, to provide more physical understanding to the resulting uncertainties, the uncertainty component u_{samp} in Eq. (RS27) is replaced with the components u_{Rave} and u_{noise} in Eq. (RS28). These two components are described in Sec. 3.4.1 however using only the data from Trace c.

The uncertainty equation for $u_{repeat(shs)}$ in Eq. (RS28) is found from the residual strain calculations using $z_{repeat(shs)}$, which is calculated to be the maximum of two values; one of which is the positive uncalibrated difference between the minimum and maximum values of the six calibration measurements taken along the certified portion of the physical step height standard before the data session and the other is the positive uncalibrated difference between the minimum and maximum values of the six measurements taken along the physical step height standard after the data session. Here, $z_{repeat(shs)}$ is assumed to scale linearly with height. Using the data from Trace c, the input values to the residual strain calculations are then varied as specified in Table RS5 [with $d=(z_{xx}-z_{1F})z_{repeat(shs)}/(2\bar{z}_6)$ where $\bar{z}_6$ is the uncalibrated average of the six calibration measurements from which $z_{repeat(shs)}$ was found]. For the three different input combinations, ε_{r-low} and ε_{r-high} are identified. Then, $u_{repeat(shs)}$ is calculated, assuming a uniform distribution, using the equation given in Table RS6.

3.5 Residual Strain Round Robin Results

The MEMS Length and Strain Round Robin *repeatability* and *reproducibility* results are given in this section for residual strain measurements. The *repeatability* data were taken in one laboratory using an optical interferometer (see Sec. 1.1.2). Unlike the MEMS 5-in-1 chip shown in Fig. 2 for SRM 2495, a similarly processed surface-micromachined test chip (from run #46 [9] and without the backside etch) was fabricated on which residual strain measurements were taken from poly1 fixed-fixed beam test structures having a 0° orientation and from poly1 fixed-fixed beams having a 90° orientation. An array of the fixed-fixed beam test structures on the round robin test chip with a 0° orientation is shown in Fig. RS7. Each fixed-fixed beam array has design lengths from 400 μm to 800 μm, inclusive, in 50 μm increments. (All the fixed-fixed beams are 10 μm wide.) However, only the design lengths between 600 μm and 750 μm, inclusive, were used in obtaining the *repeatability* data. Therefore, with three beams designed at each length, 24 measurements were taken (12 measurements at each orientation). See Fig. RS8(a) for a design rendition of a p1 fixed-fixed beam test structure on the round robin test chip and Figs. RS8(b and c) for applicable 2-D data traces taken from this fixed-fixed beam test structure.

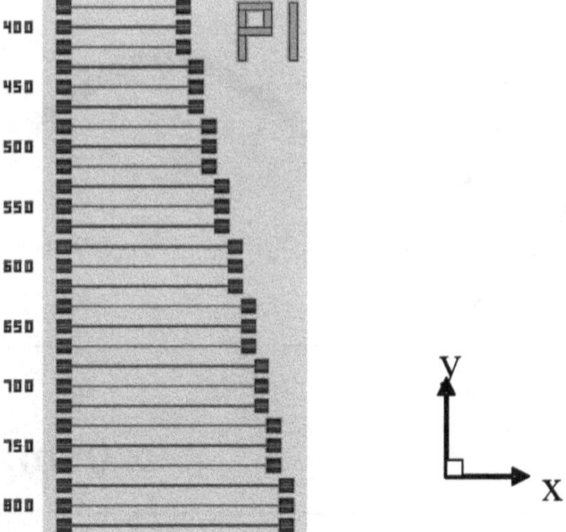

Figure RS7. *An array of fixed-fixed beams on the round robin test chip.*

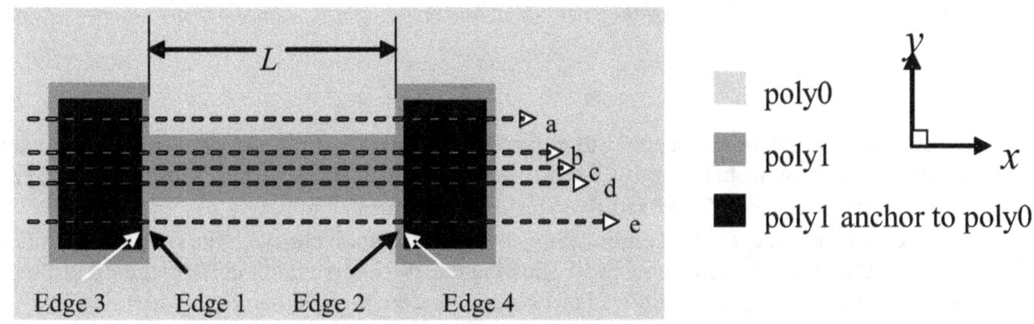

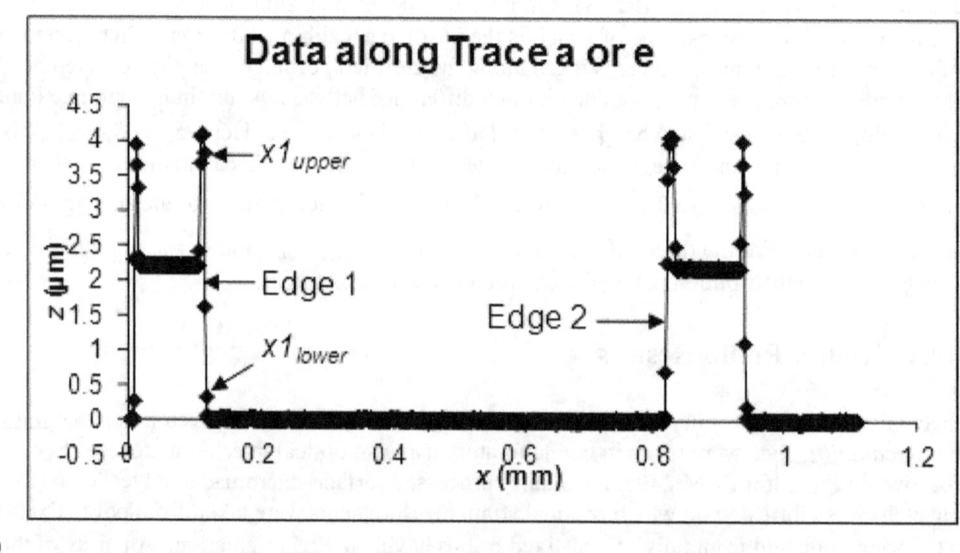

(b)

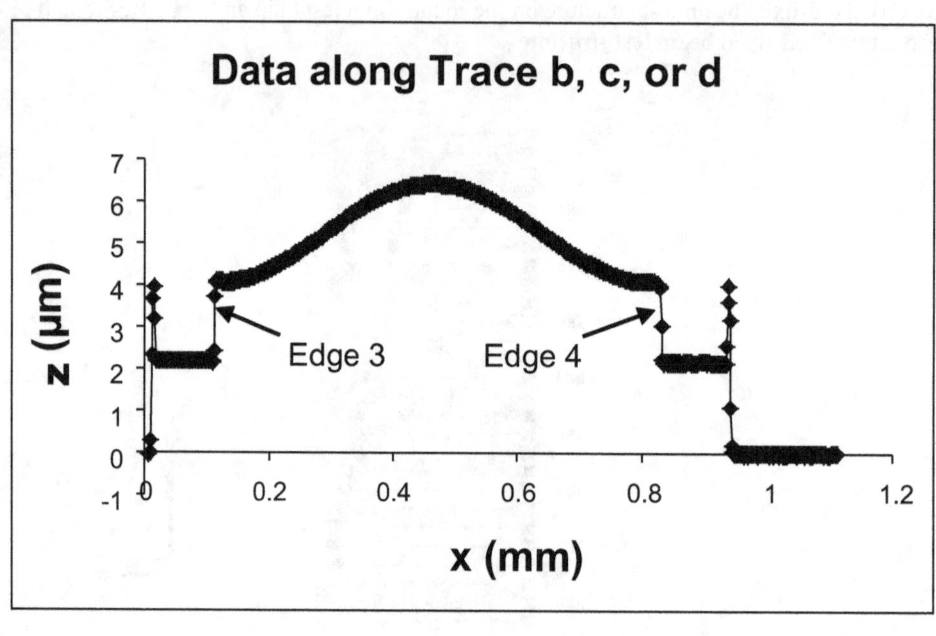

(c)

Figure RS8. *For a fixed-fixed beam test structure on the round robin test chip, (a) a design rendition,*
(b) an example of a 2-D data trace used to determine L in (a),
and (c) an example of a 2-D data trace taken along the length of the fixed-fixed beam in (a).

For the *reproducibility* data, a round robin test chip was passed from laboratory to laboratory. Each participant was asked to obtain a residual strain measurement from two poly1 fixed-fixed beams in an array, such as shown in Fig. RS7, that had either a $0°$ orientation or a $90°$ orientation. One of the fixed-fixed beams was requested to have a design length of 650 μm. The design length for the other fixed-fixed beam could range from 550 μm to 700 μm, inclusive. The results from one of the two fixed-fixed beams is included in the results presented below. Following the 2002 version of ASTM standard test method E 2245 [37] for residual strain measurements, the raw, uncalibrated measurents were recorded on Data Analysis Sheet G (similar to the existing Data Analysis Sheet RS.1 [13]) for measurements of residual strain.

Table RS9 presents the residual strain *repeatability* and *reproducibility* results. In this table, n is the number of measurements followed by the average (namely, ε_{rave}) of the *repeatability* or *reproducibility* measurement results. For the repeatabilitiy measurements only, $\sigma_{repeat(samp)}$ is given next, which is the relative standard deviation of the *repeatability* residual strain measurements. Then, the $\pm 2\sigma_{\varepsilon r}$ limits are given, where $\sigma_{\varepsilon r}$ is the standard deviation of the residual strain measurements, followed by the average of the *repeatability* or *reproducibility* combined standard uncertainty values ($u_{c\varepsilon rave}$) for different calculations.

Table RS9. *Residual Strain Measurement Results*

	Repeatability results L_{des}=600 μm to 750 μm	Reproducibility results L_{des}=550 μm to 700 μm
1. n	24	6[a]
2. ε_{rave}	-41.65×10^{-6}	-44.0×10^{-6}
3. $\sigma_{repeat(samp)}$	5.7 %	–
4. $\pm 2\sigma_{\varepsilon r}$ limits	$\pm 4.7\times10^{-6}$ (± 11 %)	$\pm 8.8\times10^{-6}$ (± 20 %)
5. $u_{c\varepsilon r1ave}$[b]	0.77×10^{-6} (1.8 %)	1.1×10^{-6} (2.4 %)
6. $u_{c\varepsilon r1ave}$[c]	0.53×10^{-6} (1.3 %)	–
7. $u_{c\varepsilon r2ave}$[d]	0.57×10^{-6} (1.4 %)	–
8. $u_{c\varepsilon r3ave}$[e]	2.4×10^{-6} (5.9 %)	–

[a] Two of these measurements were taken from the same instrument by different operators.

[b] Where $u_{c\varepsilon r1}$ is determined in ASTM standard test method E 2245$-$02 [37]. For this calculation, the u_{samp} and u_{zcal} components in the $u_{c\varepsilon r1}$ calculation in Eq. (RS27) are combined into one component. As such, for this component, the limits, assuming a uniform (that is, rectangular) probability distribution, are represented by a ± 20 nm variation in the z-value of the data points. Also in ASTM standard test method E 2245$-$02, u_{zres}= u_{xcal}=u_{xres}=u_{xresL}=0.

[c] Where $u_{c\varepsilon r1}$ is determined in determined in ASTM standard test method E 2245$-$05 [36] using Eq. (RS27).

[d] Where $u_{c\varepsilon r2}$ is determined in determined using Eq. (RS28).

[e] Where $u_{c\varepsilon r3}$ is determined in determined using Eq. (RS23).

Comments concerning the round robin data include the following:

a) *Plots*: In this round robin, random length fixed-fixed beams were measured. As such, there are at least two variables (orientation and length) as discussed below:

i) *Orientation*: Figure RS9 is a plot of $-\varepsilon_r$ versus orientation, which reveals no obvious orientation dependence. The values for ε_{rave} are approximately the same for the two different orientations. It is interesting that the $\pm 2\sigma_{\varepsilon r}$ limits for the data taken from the test structures with a 0° orientation are approximately half the $\pm 2\sigma_{\varepsilon r}$ limits for the data taken from the test structures with a 90° orientation.

ii) *Length*: Figure RS10 is a plot of $-\varepsilon_r$ versus length, which reveals no obvious length dependence.

b) *Precision*: The *repeatability* and *reproducibility* precision data appear in Table RS9. In particular, for the $\pm 2\sigma_{\varepsilon r}$ limits, the *repeatability* data (i.e., ± 11 %) are tighter than the *reproducibility* data (i.e., ± 20 %). This is due to the *repeatability* measurements being taken in the same laboratory using the same instrument by the same operator.

c) *Bias*: No information can be presented on the bias of the procedure in ASTM standard test method E 2245 for measuring residual strain because there is not a certified MEMS material for this purpose.

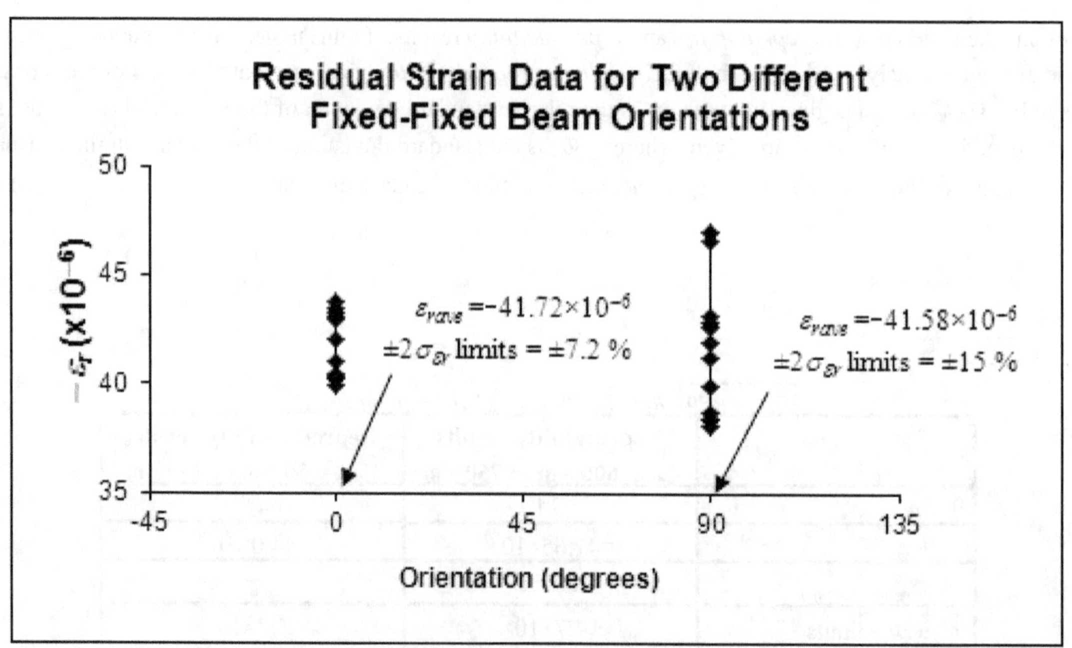

Figure RS9. A plot of $-\varepsilon_r$ versus orientation.[30]

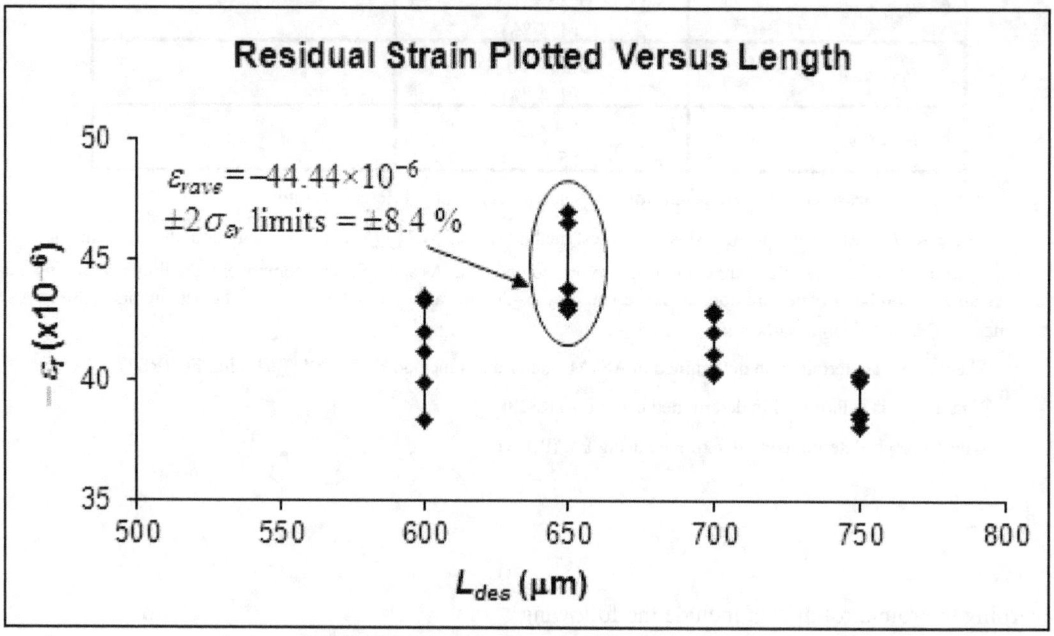

Figure RS10. A plot of $-\varepsilon_r$ versus length.[31]

3.6 Using the MEMS 5-in-1 to Verify Residual Strain Measurements

To compare your in-house residual strain measurements with NIST measurements, you will need to fill out Data Analysis Sheet RS.3. (This data analysis sheet is accessible via the URL specified in the reference [13], a reproduction of which is given in Appendix 2.)   After calibrating the instrument, locating the test structure, taking the measurements, and performing the

[30] *Copyright, ASTM International, 100 Barr Harbor Drive, West Conshohocken, PA 19428, USA. Reproduced via permissions with ASTM International.*
[31] Ibid.

calculations, the data on your completed form can be compared with the data on the SRM Certificate and the completed data analysis sheet supplied with the MEMS 5-in-1. Details of the procedure are given below.

Calibrate the instrument: Calibrate the instrument as specified in Sec. 3.2 for SRM measurements. Obtain the inputs for Table 1 in Data Analysis Sheet RS.3.

Locate the fixed-fixed beam: In the second grouping of test structures, shown in Figs. RS1(a and b), on the MEMS 5-in-1 chips shown in Fig. 1 and Fig. 2 for SRM 2494 and SRM 2495, respectively, residual strain measurements are made. Fixed-fixed beam test structures are provided for this purpose, as shown in Fig. RS2(a) for SRM 2494 and as shown in Fig. RS3(a) for the poly2 cantilevers on SRM 2495. Data Analysis Sheet RS.3 requires measurements from one fixed-fixed beam test structure. The specific test structure to be measured can be deduced from the data entered on the NIST-supplied Data Analysis Sheet RS.3 that accompanies the SRM.

For the residual strain grouping of test structures for SRM 2494, as shown in Fig. RS1(a), the target test structure can be found as follows:
1. The input *design length* (i.e., input #5 on Data Analysis Sheet RS.3, a reproduction of which is given in Appendix 2) specifies the design length of the fixed-fixed beam. The design length of the fixed-fixed beam (in micrometers) is given at the top of each column of test structures in Fig. RS1(a) following the column number (i.e., 1 to 5) and the letters "FF" to indicate a fixed-fixed beam; therefore design length can be used to locate the column in which the target test structure resides. Design lengths for the fixed-fixed beam test structures are given in Table RS1.
2. The input *which beam* (i.e., input #7) specifies which fixed-fixed beam in the column to measure (i.e., the "first," "second," "third," etc.). Since there are three instances of each test structure, the radio button corresponding to "first," "second," or "third" is used to identify the target test structure.

For the residual strain grouping of test structures for SRM 2495, as shown in Fig. RS1(b), the target test structure can be found as follows:
1. The input *material* (i.e., input #3) is used to identify if a fixed-fixed beam in a poly1 array is to be measured or if a fixed-fixed beam in a poly2 array is to be measured. The two poly1 arrays in Fig. RS1(b) have a P1 designation and the one poly2 array has a P2 designation.
2. The input *orientation* (i.e., input #9) specifies the orientation of the fixed-fixed beam array. The fixed-fixed beams in the lower left poly1 array have a $0°$ orientation, and the fixed-fixed beams in the upper poly1 array have a $90°$ orientation. The poly2 array has a $0°$ orientation.
3. The input *design length* (i.e., input #5) specifies the design length of the fixed-fixed beam. The design length of the fixed-fixed beam (in micrometers) is given next to the second of three fixed-fixed beams of the same length, as can barely be seen in Fig. RS1(b). Therefore, *design length* can be used to locate a set of three possible target test structures. Design lengths for the fixed-fixed beam test structures are given in Table RS1.
4. The input *which beam* (i.e., input #7) specifies which fixed-fixed beam in the set of three possible target test structures of the same length in the array to measure (i.e., the "first," "second," "third," etc.). Since there are three instances of each test structure, the radio button corresponding to "first," "second," or "third" is used to identify the target test structure.

Take the measurements: Following the steps in ASTM standard test method E 2245 [2] for residual strain measurements, the fixed-fixed beam is oriented under the interferometric optics as shown in Fig. RS2(a) or Fig. RS3(a) [32], and one 3-D data set is obtained using the highest magnification objective that is available and feasible. (For the given design lengths, the magnification should be at least 10 $\times$.) The data are leveled and zeroed. Traces a$'$, a, e, and e$'$ are obtained. From these data traces, measurements of $x1_{uppert}$, $x2_{uppert}$, $n1_t$, and $n2_t$ from Edge 1 and Edge 2, as shown in Fig. RS2(b) or Fig. RS3(b), are recorded in Data Analysis Sheet RS.3.   Traces a$'$ and e$'$ are used to calculate the misalignment angle, α. The uncalibrated y-values for these traces (namely, ya' and ye') are also recorded in Data Analysis Sheet RS.3.

For SRM 2494, with Data Analysis Sheet RS.3, uncalibrated data points along the fixed-fixed beam for (x_{1F}, z_{1F}), (x_{2F}, z_{2F}), (x_{3F}, z_{3F}), (x_{2S}, z_{2S}), and (x_{3S}, z_{3S}) are requested from Traces b, c, and d, as shown in Fig. RS2(c) and Fig. RS4.

For SRM 2495, there are data restrictions due to non-idealities in the geometry of the fixed-fixed beam, as discussed in Sec. 3.1 and Sec. 3.3. In particular, uncalibrated data points along the fixed-fixed beam for (x_{1F}, z_{1F}), (x_{2F}, z_{2F}), (x_{3F}, z_{3F}), (x_{2S}, z_{2S}),

---

[32] This orientation assumes that the pixel-to-pixel spacing in the x-direction of the interferometric microscope is smaller than or equal to the pixel-to-pixel spacing in the y-direction.

and (x_{3S}, z_{3S}) are requested from Traces b, c, and d, as shown in Fig RS3(c), and these data should be taken between Edges 7 and 8, as also shown in Fig. RS3(c).

Perform the calculations: Enter the data into Data Analysis Sheet RS.3 as follows:
1. Press one of the "Reset this form" buttons. (One of these buttons is located near the top of the data analysis sheet and the other is located near the middle of the data analysis sheet.)
2. Supply inputs to Table 1 through Table 5.
3. Press one of the "Calculate and Verify" buttons to obtain the results from the fixed-fixed beam test structure. (One of these buttons is located near the top of the data analysis sheet and the other is located near the middle of the data analysis sheet.)
4. Verify the data by checking to see that all the pertinent boxes in the verification section at the bottom of the data analysis sheet say "ok". If one or more of the boxes say "wait," address the issue, if necessary, by modifying the inputs and recalculating.
5. Print out the completed data analysis sheet to compare both the inputs and outputs with those on the NIST-supplied data analysis sheet.

Compare the measurements: The MEMS 5-in-1 is accompanied by a Certificate. This Certificate specifies an effective residual strain value, ε_r, for SRM 2494 and SRM 2495 and the expanded uncertainty, U_{ε_r}, (with k=2) intending to approximate a 95 % level of confidence. It is your responsibility to determine an appropriate criterion for acceptance, such as given below:

$$D_{\varepsilon_r} = \left| \varepsilon_{r(customer)} - \varepsilon_r \right| \le \sqrt{U^2_{\varepsilon_r(customer)} - U^2_{\varepsilon_r}} , \qquad (RS29)$$

where D_{ε_r} is the absolute value of the difference between your residual strain value, $\varepsilon_{r(customer)}$, and the residual strain value on the SRM Certificate, ε_r, and where $U_{\varepsilon_r(customer)}$ is your expanded uncertainty value and U_{ε_r} is the expanded uncertainty on the SRM Certificate. If your measured value for residual strain (as obtained in the newly filled out Data Analysis Sheet RS.3) satisfies your criterion for acceptance and there are no pertinent "wait" statements at the bottom of your Data Analysis Sheet RS.3, you can consider yourself to be appropriately measuring residual strain according to the ASTM E 2245 residual strain standard test method [2] according to your criterion for acceptance.

An effective residual strain value is reported for SRMs 2494 and 2495, as shown in Figs 1 and 2, respectively, due to non-idealities associated with the geometry and/or composition of the fixed-fixed beam and/or the beam support as discussed in Sec. 3.1 and Sec. 3.3. When you use ASTM standard test method E 2245 with your own fixed-fixed beam, you must be cognizant of the geometry and composition of your fixed-fixed beam because this test method assumes an ideal geometry and composition, implying that you would be obtaining an "effective" residual strain value if the geometry and/or composition of your fixed-fixed beam deviates from the ideal.

Any questions concerning the measurements, analysis, or comparison can be directed to mems-support@nist.gov.

4   Grouping 3: Strain Gradient

Strain gradient is defined as a through-thickness variation of the residual strain in the structural layer of interest before it is released [3]. ASTM standard test method E 2246 [3] on strain gradient measurements is an aid in the design and fabrication of MEMS devices [28-29]. It can be used to determine the maximum distance that a MEMS component can be suspended say, in air, before it begins to bend or curl.

This section on strain gradient is not meant to replace but to supplement the ASTM standard test method E 2246 [3], which more completely presents the scope, significance, terminology, apparatus, and test structure design as well as the calibration procedure, measurement procedure, calculations, precision and bias data, etc. The NIST-developed strain gradient test structures on SRM 2494 and SRM 2495, as shown in Fig. 1 and Fig. 2, respectively, in the Introduction are given in Sec. 4.1. Sec. 4.2 discusses the calibration procedure for the strain gradient measurements, and Sec. 4.3 discusses the strain gradient measurement procedure. Following this, the uncertainty analysis is presented in Sec. 4.4, the round robin results are presented in Sec. 4.5, and Sec. 4.6 describes how to use the MEMS 5-in-1 to verify strain gradient measurements.

4.1 Strain Gradient Test Structures

Strain gradient measurements are taken in the third grouping of test structures, as shown in Fig. SG1(a) for SRM 2494 depicted in Fig. 1 and as shown in Fig. SG1(b) for SRM 2495 depicted in Fig. 2.

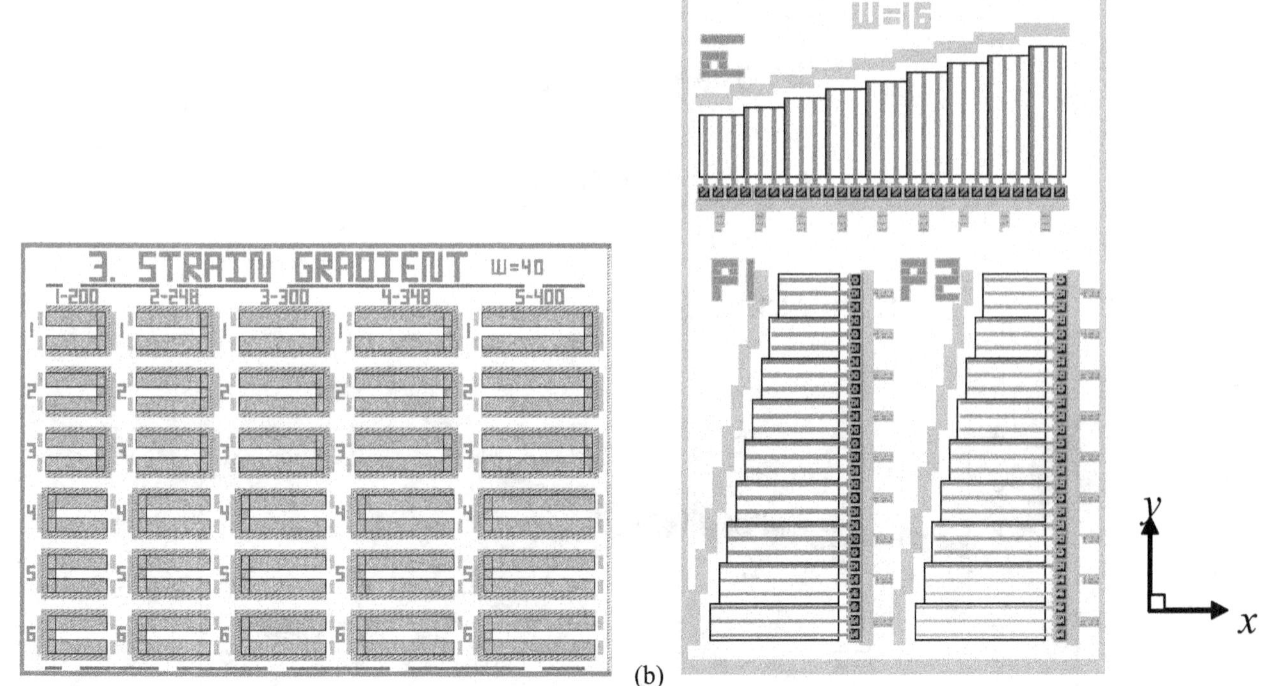

(a) (b)

Figure SG1. The strain gradient grouping of test structures on (a) SRM 2494, fabricated on a multi-user 1.5 µm CMOS process [8] followed by a bulk-micromachining etch, as depicted in Fig. 1 and (b) SRM 2495, fabricated using a polysilicon multi-user surface-micromachining MEMS process [9] with a backside etch, as depicted in Fig. 2.

Strain gradient measurements are obtained from cantilever test structures. A cantilever test structure in the strain gradient grouping of test structures, as shown in Figs. SG1(a and b), can be seen in Fig. SG2(a) and Fig. SG3(a), respectively, for the bulk-micromachined SRM 2494 chip and the surface-micromachined SRM 2495 chip with a backside etch. Applicable data traces taken from these test structures are given in Figs. SG2(b and c) and Figs. SG3(b and c), respectively.

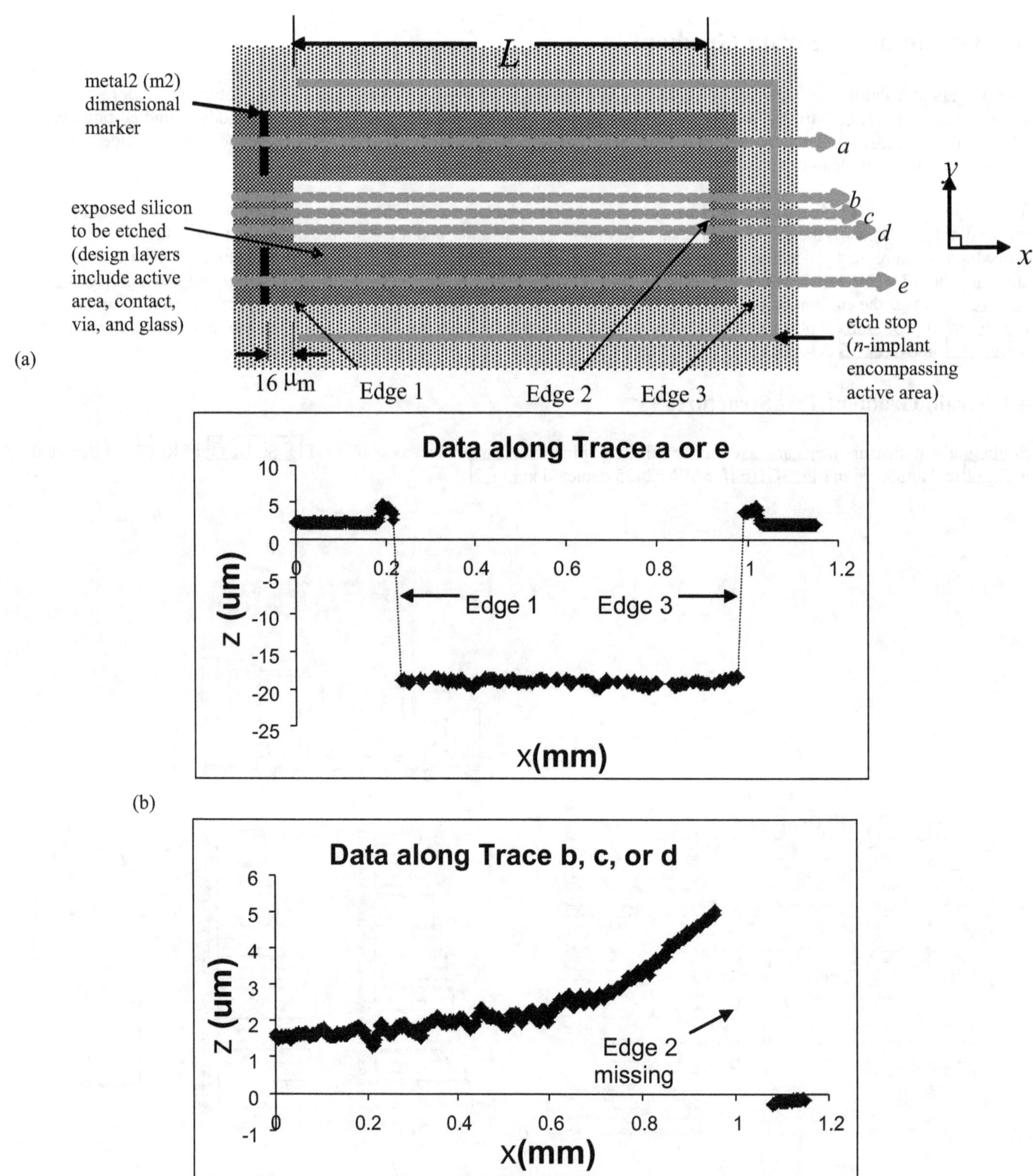

Figure SG2. *For a cantilever test structure on SRM 2494, (a) a design rendition, (b) an example of a 2-D data trace used to locate the attachment point of the cantilever in (a),[33] and (c) an example of a 2-D data trace taken along the length of the cantilever in (a)[34].*

[33] *Copyright, ASTM International, 100 Barr Harbor Drive, West Conshohocken, PA 19428, USA. Reproduced via permissions with ASTM International.*
[34] Ibid.

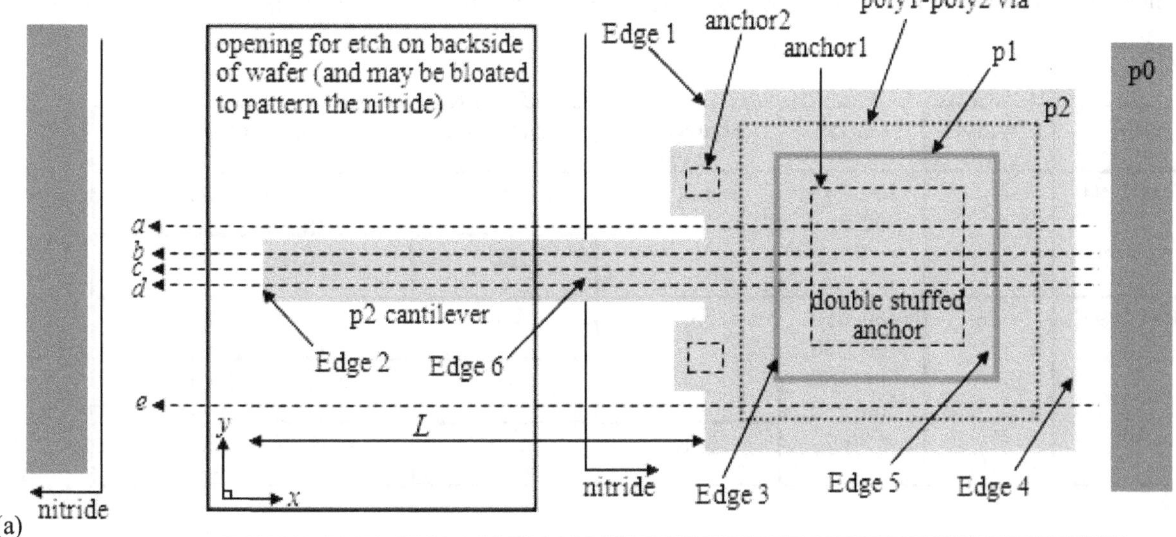

(a)

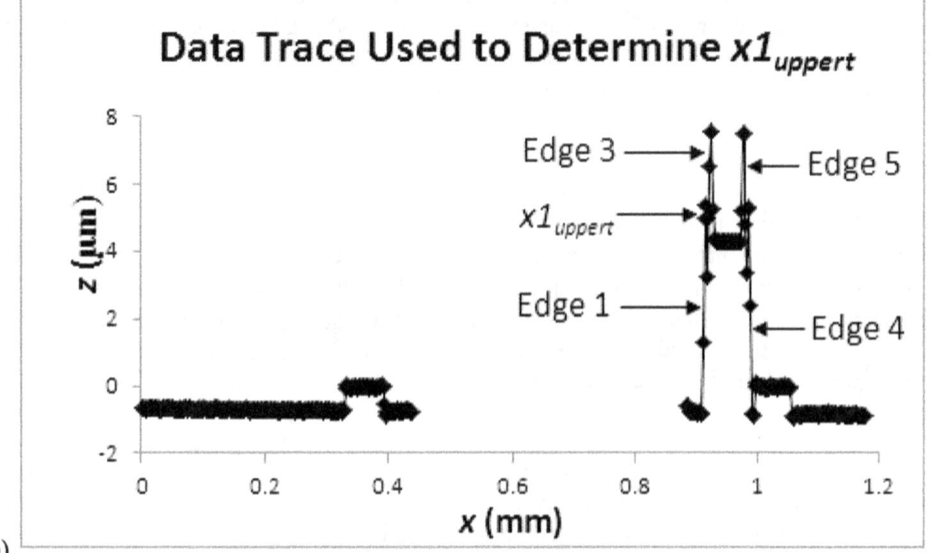

(b)

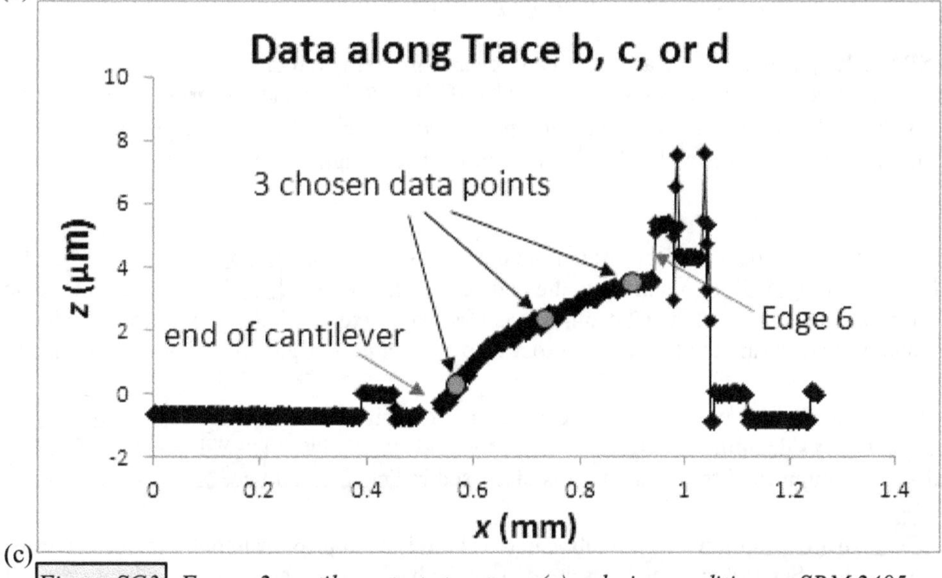

(c)

Figure SG3. *For a p2 cantilever test structure, (a) a design rendition on SRM 2495, (b) an example of a 2-D data trace used to determine $x1_{uppert}$, and (c) an example of a 2-D data trace taken along the length of a cantilever.*

The specifications for the cantilevers shown in Figs. SG1(a and b) for SRM 2494 and SRM 2495, respectively, are given in Table SG1.

Table SG1. Cantilever Configurations for Strain Gradient Measurements

SRM	Width (μm)	Length (μm)	Structural Layer	Orientation	Quantity of Beams
SRM 2494	40	200, 248, 300, 348, 400	oxide	$0°$	three of each length (or 15 beams)
				$180°$	three of each length (or 15 beams)
SRM 2495	16	400, 450, 500, 550, 600, 650, 700, 750, 800	poly1	$90°$	three of each length (or 27 beams)
				$180°$	three of each length (or 27 beams)
			poly2	$180°$	three of each length (or 27 beams)

For SRM 2494: On SRM 2494, all oxide cantilevers shown in Fig. SG1(a) are designed with both a $0°$ orientation and a $180°$ orientation. As seen in this figure, the length of a cantilever (in micrometers) is given at the top of each column of cantilevers following the column number (i.e., 1 to 5). These design lengths (and the design width) are specified in Table SG1. There are three cantilevers designed at each length for each orientation. Therefore, there are 15 oxide cantilevers with a $0°$ orientation and 15 oxide cantilevers with a $180°$ orientation.

As specified in Sec. 1.4.1, the exposed silicon, as shown in Fig. SG2(a), is isotropically etched in XeF_2 to release the cantilever by removing the silicon around and beneath the cantilever. The dimensional markers are instrumental in firming up the support region. They also can be used to measure the small amount of SiO_2 that has also been etched in XeF_2,[35] however the tip of the cantilever will also be etched a comparable amount such that the length of the cantilever should remain the same. The etch stop, also shown in this figure, helps to inhibit the etch away from the test structure to shield neighboring structures from the etch. It consists of an *n*-implant designed to surround the active area. Figs. YM2(a, b, and c) shows a cantilever test structure on SRM 2494. As can be seen in Fig. YM2(c), there is undercutting of the cantilever.

An oxide cantilever consists of four SiO_2 layers. The thickness of these cantilevers is calculated using Data Analysis Sheet T.1. See Sec. 8 for specifics. Even though the beam is made up of four layers of SiO_2, the layers may not have the same properties. Due to this non-ideality in the composition of the cantilever, an effective strain gradient is reported on the SRM Certificate presented in Sec. 4.6. Also, an effective value is reported (as specified in Sec. 4.3) due to excessive curvature of the cantilever making it difficult to obtain data beyond 250 μm along the length of the cantilever.

For SRM 2495: On SRM 2495, there are three arrays of cantilevers for strain gradient measurements, as shown in Fig. SG1(b). Two of these arrays consist of poly1 cantilevers (as indicated by a "P1" symbol) and one array consists of poly2 cantilevers (as indicated by a "P2" symbol). The cantilevers within the top poly1 array have a $90°$ orientation, and the remaining poly1 array along with the poly2 array have cantilevers with a $180°$ orientation. The design dimensions of the cantilevers are given in Table SG1.

Fig. SG3(a) shows one of the p2 cantilevers in the strain gradient grouping of test structures shown in Fig. SG1(b). The p2 cantilever pad design shown in Fig. SG3(a) is similar to the pad design shown in Fig. YM4(a). The pad includes both p1 and p2. Without the p1, the attachment point of the p2 cantilever is less like a fixed boundary condition. By including p1 in the anchor design, the p1 and p2 fuse during the fabrication process to make a more rigid and reliable attachment point.

To make an even more rigid attachment point, in the p2 cantilever pad design shown in Fig. SG3(a), the p2 layer is also anchored to the nitride on either side of the cantilever. This is done to be consistent with the pad designs in the Young's modulus and residual strain groupings of test structures (as discussed in Sec. 2.1 and Sec. 3.1 respectively).

Also, as seen in Fig. YM4(b), a flat cantilever is not fabricated. There is an approximate 600 nm vertical transition (or kink) in the cantilever. As shown in Fig. SG3(a), an opening is created on the backside of the wafer for a backside etch. This etch removes the material beneath the cantilevers to ensure the existence of cantilevers that have not adhered to the top of the

---

[35] The design dimension from the dimensional marker to the exposed silicon is 16 μm, as shown in Fig. SG2(a).

underlying layer. Earlier in the fabrication process, the nitride layer is patterned using a mask similar to that used to create the openings in the backside of the wafer, however, all the features may be bloated by an amount that is expected to change for different processing runs. As a result, the polysilicon cantilevers traverse an approximate 600 nm fabrication step over the nitride, as can be seen in Fig. YM4(b) and Fig. YM5. For the double stuffed pad designs for the p1 and p2 cantilevers on the SRM 2495 chips fabricated on a 2011 processing run (run #95 [9]), this step is approximtely 25 μm from the anchor lip (or 38 μm from the anchor when the opening for the backside etch is designed 65 μm from the anchor). Viable data can be taken along the cantilever without encompassing the kink such that it is not necessary to call the resulting strain gradient value "effective" due to this kink.

4.2 Calibration Procedures for Strain Gradient Measurements

For SRM strain gradient measurements, the interferometric microscope is calibrated in the z-direction as specified in Sec. 5.2 for step height calibrations as used with the MEMS 5-in-1. For discussing earlier versions of the uncertainty equation presented in Sec. 4.4.2, Eq. (SH2) in Sec. 5.2 is also used. Therefore, six measurements are taken along the certified area of the physical step height standard before the data session and six measurements are taken along the certified area after the data session. The interferometric microscope is calibrated in the x- and y-directions as given in Sec. 6.2 for in-plane length calibrations. These calibration procedures are the same as those for residual strain and in-plane length measurements, as indicated in Sec. 3.2 and Sec. 6.2, respectively.

4.3 Strain Gradient Measurement Procedure

Strain gradient measurements are taken from a cantilever test structure, such as shown in Fig. SG3(a). To obtain a strain gradient measurement, the following steps are taken for SRM 2495 (consult the standard test method [3] for additional details and for modifications to these steps for a bulk-micromachined test structure on SRM 2494):

1. Five 2-D data traces, such as shown in Fig. SG3(a) are extracted from a 3-D data set.

2. From Traces a and e, the uncalibrated values typically from Edge 1 (namely, $x1_{uppera}$ and $x1_{unnere}$) along with the corresponding values for $n1_a$ and $n1_e$, respectively, are obtained (as defined and specified in Sec. 6.3 for in-plane length measurements). These four values are entered into Data Analysis Sheet SG.3 along with the uncalibrated y values associated with these traces (namely, y_a and y_e). The main purpose of these entries is to calculate the misalignment angle, α, as shown in Fig. SG4 between Edge 1 and a line drawn perpendicular to Traces a and e. The following equation is used:

$$\alpha = \tan^{-1}\left[\frac{\Delta x \; cal_x}{\Delta y \; cal_y}\right] \text{, where} \tag{SG1}$$

$$\Delta x = x1_{uppera} - x1_{uppere} \text{, and} \tag{SG2}$$

$$\Delta y = y_a - y_e \; . \tag{SG3}$$

    An alternate edge [such as Edge 4 in Fig. SG3(a)] may be used instead of Edge 1 if $(n4_a + n4_e) < (n1_a + n1_e)$.

3. Another purpose of the entries associated with Traces a and e is to ensure that the uncalibrated x-values for the strain gradient data points (obtained in the next step) are all greatrer than $x1_{ave}$ as calculated below (assuming a $0°$ orientation):[36]

$$x1_{ave} = \frac{x1_{uppera} + x1_{uppere}}{2} \; . \tag{SG4}$$

---

[36] Note that the cantilever in Fig. SG3(a) has a $180°$ orientation. For a $180°$ cantilever orientation, all x-values should be multiplied by -1 before entering them into Data Analysis Sheet SG.3 to satisfy this criterion.

(Note that the cantilever in Fig. SG3(a) has a 180° orientation.) Therefore, it is preferable to provide Edge 1 inputs, if possible. If an alternate edge is used, special care must be taken to ensure that all the uncalibrated x-values (obtained in the next step) are greater than an estimate for $x1_{ave}$ (assuming a 0° orientation). However, for the test structure shown in Fig. SG3(a), a backside etch is used. As a result, for the process used, the cantilever traverses an approximate 600 nm fabrication step over the nitride used in conjuntion with the backside etch (as specified in Sec. 4.1) as seen in Fig. YM4(b) and Fig. YM5. Therefore, uncalibrated x-values obtained in the next step should all be greater than an estimate for $x6_{ave}$ for this test structure (assuming a 0° orientation). It should be mentioned that if there is remaining debris in an attachment corner of the cantilever to the beam support (such as may be the case for some cantilevers on SRM 2494), then the uncalibrated x-values obtained in the next step should all be greater than an estimated uncalibrated x-value corresponding to where the cantilever is first free of this debris (assuming a 0° orientation).

4. The 2-D data along the cantilever from Traces b, c, and d, as shown in Fig. SG3(c) for one data trace, are used to obtain three independent measurements of strain gradient. This is done for one data trace as follows:

   a. Eliminate the data values at both ends of the trace that will not be included in the modeling. This would include all data values outside and including Edges 1 and 2 in Fig. SG3(a) and all data values outside and including the data transition [such as, Edge 6 as shown in Fig. SG3(c)].

   b. Choose three representative data points (sufficiently separated) among the remaining data points, as shown in Fig. SG3(c). The three uncalibrated points are called (x_1, z_1), (x_2, z_2), and (x_3, z_3) where the uncalibrated x values (namely, x_1, x_2, and x_3) are all greater than $x1_{ave}$ (or $x6_{ave}$, if Edge 6 is present), assuming a 0° orientation. (For cantilevers with a 180° orientation, as given in Fig. SG3(a), negate the x values of all the data points such that $x1_{ave} < x_1 < x_2 < x_3 < x2_{ave}$.) The three data points are entered into Data Analysis Sheet SG.3.

   c. To account for the misalignment angle, α, as shown in Fig. SG5, and the x-calibration factor, cal_x, the values obtained above for $x1_{ave}$, x_1, x_2, and x_3 become f, g, h, and i, respectively, along the v-axis (the axis assumed to be aligned with respect to the in-plane length of the cantilever) as also shown in Fig. SG5. The uncalibrated z-values of the data points along the cantilever remain the same, which assumes there is no curvature of the cantilever across the width of the cantilever. Therefore, the calibrated data points along the cantilever become $(g, z_1 cal_z)$, $(h, z_2 cal_z)$, and $(i, z_3 cal_z)$. The equations for f, g, h, and i, are given below:

$$f = xl_{ave} cal_x , \tag{SG5}$$

$$g = \left(x_1 cal_x - f \right) \cos\alpha + f , \tag{SG6}$$

$$h = \left(x_2 cal_x - f \right) \cos\alpha + f , \text{ and} \tag{SG7}$$

$$i = \left(x_3 cal_x - f \right) \cos\alpha + f . \tag{SG8}$$

A circular arc (as seen in Fig. SG6 for a cantilever with a 0° orientation) is used to model the out-of-plane shape of the cantilever.

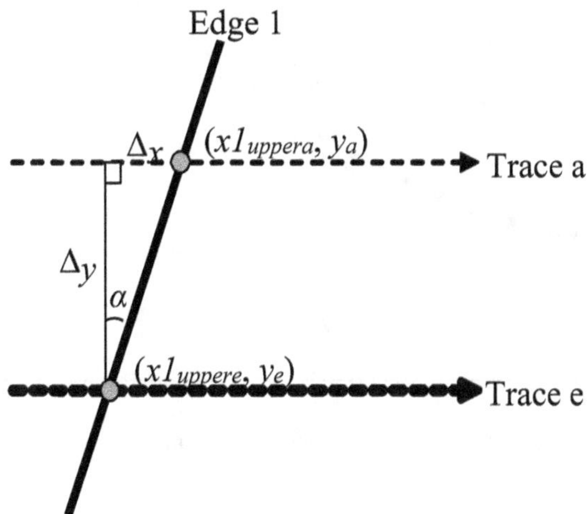

Figure SG4. *Sketch showing the misalignment angle, α, between Edge 1 and a line drawn perpendicular to Traces a and e. In this sketch, it is assumed that the x- and y-values are calibrated.*

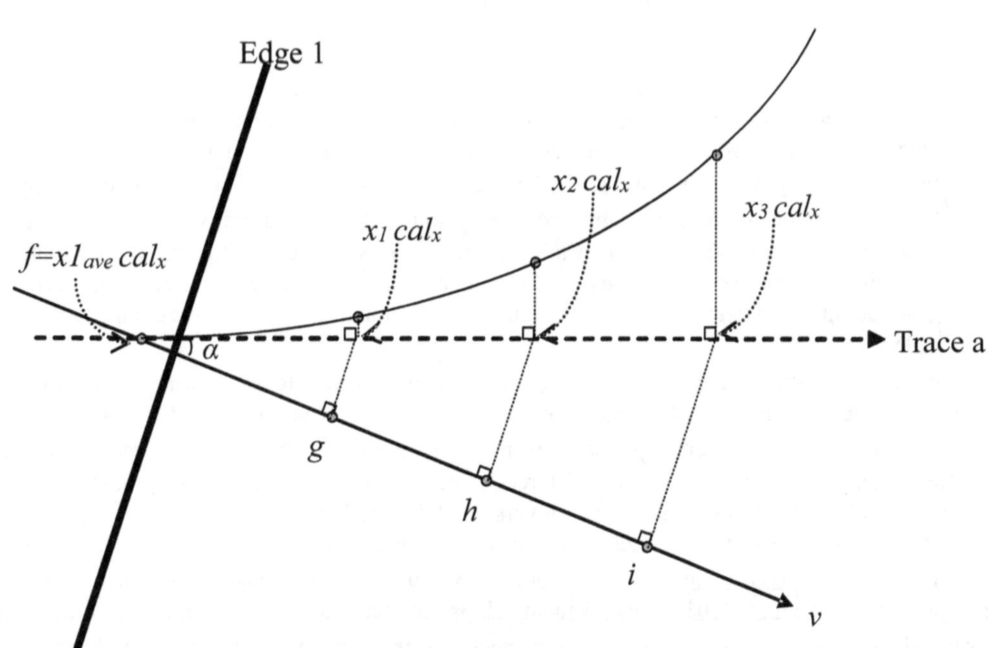

Figure SG5. *Sketch used to derive the appropriate v-values (f, g, h, and i) along the length of the cantilever*

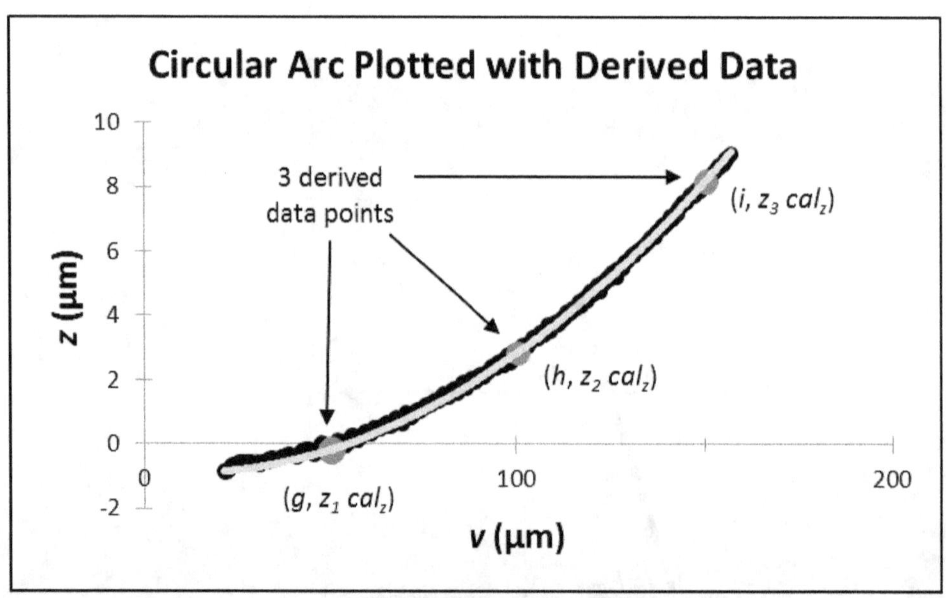

Figure SG6. A circular arc function plotted with derived data for a cantilever with a $0°$ orientation.

    d.   The strain gradient, s_{gt} or s_{g0t} (where the substrcipt "t" refers to the data trace being considered), is calculated using one of the following equations [14]

$$s_{gt} \approx \frac{1}{R_{int}} + s_{gcorrection}, \quad \text{or} \tag{SG9}$$

$$s_{g0t} = \frac{1}{[R_{int} - s(t/2)]} + s_{gcorrectio\ n}, \tag{SG10}$$

where s_{g0t} is the strain gradient when ε_r equals zero, R_{int} is the radius of the circle describing the shape of the topmost surface of the cantilever as measured with the interferometer, t is the thickness of the cantilever, and where $s = 1$ for downward bending cantilevers (or if data were taken from the bottom of an upward bending cantilever) and $s = {}^-1$ for upward bending cantilevers (unless data were taken from the bottom of an upward bending cantilever). Also in the above equation, $s_{gcorrection}$ is a strain gradient correction term intending to correct for any variations associated with length (and also any deviations from the ideal cantilever geometry and/or composition) as discussed in more detail in the next step. For a more complete analysis and a derivation of this equation (without the correction term), consult reference [14].

5.   The strain gradient correction term, $s_{gcorrection}$, is intended to correct for any variations associated with length, and is also assumed to correct for deviations from the ideal cantilever geometry and/or composition. To obtain $s_{gcorrection}$ associated with a given length cantilever in a given process, strain gradient measurements are obtained from different length cantilevers with three different cantilevers measured at each length. Therefore, for SRM 2494, the design lengths of the measured cantilevers are 200 µm, 248 µm, 300 µm, 348 µm, and 400 µm. And, for SRM 2495, the design lengths of the measured cantilevers are 400 µm, 450 µm, 500 µm, 550 µm, 600 µm, 650 µm, 700 µm, 750 µm, and 800 µm. Given a plot of s_g versus the design length, L_{des}, such as obtained in the round robin experiment (see Fig. SG10 in Sec. 4.5) using chips fabricated on a surface-micromachining process, for the longer cantilevers $s_{gcorrection}=0$ or a value that corrects for deviations from the ideal cantilever geometry and/or composition. (For SRM 2495, it is assumed that $s_{gcorrection}=0$ for the longer cantilevers.) For each of the shorter length cantilevers (≤ 500 µm for the data in Fig. SG10) an approximate value of $s_{gcorrection}$ is extracted such that the strain gradient is no longer a function of length. For SRM 2494, due to the excessive curvature of the cantilevers making it difficult to obtain data beyond the first 250 µm along its length, it is assumed that

$s_{gcorrection}$=0 for the shorter length cantilevers and an effective strain gradient is entered on the SRM Certificate. Only the shorter length cantilevers will be used for the SRM 2494 measurements.

6. The resulting strain gradient value, s_g, is the average of the strain gradient values obtained from Traces b, c, and d as given below:

$$s_g = \frac{s_{gb} + s_{gc} + s_{gd}}{3}.$$
(SG11)

4.4 Strain Gradient Uncertainty Analysis

In this section, uncertainty equations are presented for use with strain gradient. The first uncertainty equation (presented in Sec. 4.4.1) is used for the MEMS 5-in-1. The equations used in the round robin experiment and other previous work are presented in Sec. 4.4.2.

4.4.1 Strain Gradient Uncertainty Analysis for the MEMS 5-in-1

For the MEMS 5-in-1, the combined standard uncertainty, u_{csg3}, for strain gradient measurements with twelve uncertainty components is given by the following equation:

$$u_{csg3} = \sqrt{\begin{aligned} & u_W^2 + u_{zres}^2 + u_{xcal}^2 + u_{xres}^2 + u_{Rave}^2 + u_{noise}^2 + u_{cert}^2 + u_{repeat(shs)}^2 \\ & + u_{drift}^2 + u_{linear}^2 + u_{correction}^2 + u_{repeat(samp)}^2 \end{aligned}},$$
(SG12)

with additional sources of uncertainty considered negligible. A number following the subscript "sg" in "u_{csg}" indicates the data analysis sheet that is used to obtain the combined standard uncertainty value. Therefore, u_{csg3} implies that Data Analysis Sheet SG.3 [3,13] is used.

In Eq. (SG12), u_W is the uncertainty due to variations across the width of the beam, u_{zres} is the uncertainty due to the resolution of the interferometer in the z-direction, u_{xcal} is the uncertainty due to the calibration in the x-direction, and u_{xres} is the uncertainty due to the resolution of the interferometric microscope in the x-direction. Next, u_{Rave} is the uncertainty due to the sample's surface roughness, u_{noise} is the uncertainty due to interferometric noise, u_{cert} is due to the uncertainty of the value of the physical step height standard, $u_{repeat(shs)}$ is the uncertainty due to the *repeatability* of a measurement taken on the physical step height standard, u_{drift} is the uncertainty due to the amount of drift during the data session, and u_{linear} is the uncertainty due to the deviation from linearity of the data scan. Then, in Eq. (SG12), $u_{correction}$ is the uncertainty in the strain gradient correction term due to geometry and/or composition deviations from the ideal cantilever and $u_{repeat(samp)}$ is the uncertainty of strain gradient *repeatability* measurements taken on cantilevers processed similarly to the one being measured.

With a few variations, the combined standard uncertainty equations and the calculations for each uncertainty component are similar to those presented for residual strain measurements. Therefore, refer to Sec. 3.4.1 for the general approach. More specifically, refer to Tables SG2, SG3, SG4, and SG5 where mention is made of Tables RS2, RS3, RS4, and RS5, respectively. In addition, replace the words "residual strain" with the words "strain gradient," replace "fixed _fixed beams" with "cantilevers," replace $\varepsilon_{r\text{-}low}$ with $s_{g\text{-}low}$, $\varepsilon_{r\text{-}high}$ with $s_{g\text{-}high}$, z_{xx} with z_x, and replace Eq. (RS23), Eq. (RS24), and Eq. (RS25) with Eq. (SG12), Eq. (SG13), and Eq. (SG14), respectively, where:

$$z_{linear} = (z_x - z_I)z_{lin} \quad \text{(where } z_x \text{ and } z_I \text{ are considered calibrated values for this discussion), and}$$
(SG13)

$$u_{repeat(samp)t} = \sigma_{repeat(samp)}s_{gt}.$$
(SG14)

In determining the combined standard uncertainty, a Type B evaluation [19-21] (i.e., one that uses means other than the statistical Type A analysis) is used for each source of uncertainty, except where noted in Table SG2.

Uncertainty Component	Method to Obtain s_{g-high} and s_{g-low}, if applicable	G or U[a] / A or B[b]	equation
1. u_W	–	G / A	$u_W = STDEV\,(s_{gb}, s_{gc}, s_{gd})$
2. u_{zres}	using $d=(1/2)z_{res}$ in Table SG3	U / B	$u_{zrest} = \dfrac{\left\| s_{g-high} - s_{g-low} \right\|}{2\sqrt{3}}$ $u_{zres} = \dfrac{u_{zresb} + u_{zresc} + u_{zresd}}{3}$
3. u_{xcal}	using cal_{xmin} for cal_x where $cal_{xmin} = cal_x - 3\sigma_{xcal}cal_x / ruler_x$ and cal_{xmax} for cal_x where $cal_{xmax} = cal_x + 3\sigma_{xcal}cal_x / ruler_x$	G / B	$u_{xcalt} = \dfrac{\left\| s_{g-high} - s_{g-low} \right\|}{6}$ $u_{xcal} = \dfrac{u_{xcalb} + u_{xcalc} + u_{xcald}}{3}$
4. u_{xres}	using $d=(1/2)x_{res}(cal_x)\cos(\alpha)$ in Table SG4	U / B	$u_{xrest} = \dfrac{\left\| s_{g-high} - s_{g-low} \right\|}{2\sqrt{3}}$ $u_{xres} = \dfrac{u_{xresb} + u_{xresc} + u_{xresd}}{3}$
5. u_{Rave}	using $d=3\sigma_{Rave}$ in Table SG3 where $\sigma_{Rave} = \dfrac{1}{6} R_{ave}$	G / B	$u_{Ravet} = \dfrac{\left\| s_{g-high} - s_{g-low} \right\|}{6}$ $u_{Rave} = \dfrac{u_{Raveb} + u_{Ravec} + u_{Raved}}{3}$
6. u_{noise}	using $d=3\sigma_{noise}$ in Table SG3 where $\sigma_{noise} = \dfrac{1}{6}\left(R_{tave} - R_{ave}\right)$	G / B	$u_{noiset} = \dfrac{\left\| s_{g-high} - s_{g-low} \right\|}{6}$ $u_{noise} = \dfrac{u_{noiseb} + u_{noisec} + u_{noised}}{3}$
7. u_{cert}	using $d=3(z_x-z_1)\sigma_{cert}/cert$ in Table SG5 where z_x is the column heading[c]	G / B	$u_{certt} = \dfrac{\left\| s_{g-high} - s_{g-low} \right\|}{6}$ $u_{cert} = \dfrac{u_{certb} + u_{certc} + u_{certd}}{3}$
8. $u_{repeat(shs)}$	using $d=3(z_x-z_1)\sigma_{6same}/\bar{z}_{6same}$ in Table SG5 where z_x is the column heading[c]	G / B	$u_{repea(shs)t} = \dfrac{\left\| s_{g-high} - s_{g-low} \right\|}{6}$ $u_{repea(shs)} = (u_{repea(shs)b} + u_{repea(shs)c} + u_{repea(shs)d})/3$
9. u_{drift}	using $d=(z_x-z_1)z_{drift}cal_z/(2\,cert)$ in Table SG5 where z_x is the column heading[c]	U / B	$u_{driftt} = \dfrac{\left\| s_{g-high} - s_{g-low} \right\|}{2\sqrt{3}}$

			$u_{drift} = \dfrac{u_{driftb} + u_{driftc} + u_{driftd}}{3}$
10. u_{linear}	using $d=z_{linear}$ in Table SG5	U / B	$u_{lineart} = \dfrac{\left\| s_{g-high} - s_{g-low} \right\|}{2\sqrt{3}}$ $u_{linear} = \dfrac{u_{linearb} + u_{linearc} + u_{lineard}}{3}$
11. $u_{correction}$	–	G / B	$u_{correction} = u_{correctiont}$ $= \left\| s_{gcorrectia} \right\| / 3$
12. $u_{repeat(samp)}$	–	G / A	$u_{repeat(samp)t} = \sigma_{repeat(samp)} s_{gt}$ $u_{repeat(samp)} = (u_{repeat(samp)b} + u_{repeat(samp)c} + u_{repeat(samp)d})/3$

[a] "G" indicates a Gaussian distribution and "U" indicates a uniform distribution.

[b] Type A or Type B analysis

[c] For ease of presentation, z_x and z_l in this table are considered calibrated. (Actually, these values are uncalibrated as presented earlier in this SP 260. Therefore, the uncalibrated values should be multiplied by cal_z before use in this table.)

Table SG3. Three Sets of Inputs[a] for Strain Gradient Calculations to Determine u_{zrest}, u_{Ravet}, u_{noiset}, and u_{samp}[b]

	z_1	z_2	z_3
1	z_1	z_2	z_3
2	z_1+d	z_2-d	z_3+d
3	z_1-d	z_2+d	z_3-d

[a] For ease of presentation, the values for z_1, z_2, and z_3 in this table are assumed to be calibrated.

[b] In this table, $d=(1/2)z_{res}$ to determine u_{zrest}, $d=3\sigma_{Rave}$ to determine u_{Ravet}, $d=3\sigma_{noise}$ to determine u_{noiset}, and in Sec. 4.4.2 $d=3\sigma_{samp}$ to determine u_{samp}.

Table SG4. Seven Sets of Inputs[a] for Strain Gradient Calculations to Determine u_{xrest}

	g	**h**	**i**
1	g	h	i
2	$g+d$	h	$i-d$
3	$g-d$	h	$i+d$
4	$g+d$	$h+d$	$i-d$
5	$g+d$	$h-d$	$i-d$
6	$g-d$	$h+d$	$i+d$
7	$g-d$	$h-d$	$i+d$

[a] In this table, $d=(1/2)x_{res}(cal_x)\cos(\alpha)$.

$$\text{Table SG5. Three Sets of Inputs}^{a}\text{ for Strain Gradient Calculations}$$
$$\text{to Determine } u_{certt},\ u_{repeat(shs)t},\ u_{driftt},\ u_{lineart},\text{ and } u_{zcal}{}^{b}$$

	z_1	z_2	z_3
1	z_1	z_2	z_3
2	z_1	z_2+d	z_3+d
3	z_1	z_2-d	z_3-d

[a] For ease of presentation, the values for z_1, z_2, and z_3 are assumed to be calibrated in this table and in the applicable equations for d.

[b] In this table, $d=3(z_x-z_1)\sigma_{cert}/cert$ to determine u_{certt} where z_x is the column heading,

$d=3(z_x-z_1)\sigma_{6same}/\bar{z}_{6same}$ to determine $u_{repeat(shs)t}$, $d=(z_x-z_1)z_{drift}\,cal_z/(2\,cert)$ to determine u_{driftt}, $d=z_{linear}$ to determine $u_{lineart}$, and in Sec. 4.4.2 $d=3(z_x-z_1)\sigma_{zcal}/cert$ to determine u_{zcal}.

The expanded uncertainty for strain gradient, U_{sg}, is calculated using the following equation:

$$U_{sg}=ku_{csg3}=2u_{csg3}\ ,\qquad\qquad(\text{SG15})$$

where the k value of 2 approximates a 95 % level of confidence.

Reporting results [9-21]. Since it can be assumed that the estimated values of the uncertainty components are either approximately uniformly or Gaussianly distributed with approximate combined standard uncertainty u_{csg3}, the strain gradient is believed to lie in the interval $s_g\pm u_{csg3}$ (expansion factor $k=1$) representing a level of confidence of approximately 68 %.

4.4.2 Previous Strain Gradient Uncertainty Analyses

In this section, two uncertainty equations are presented; one that was used in the round robin experiment and one that was used before Eq. (SG12). For these equations, the strain gradient is assumed to be the strain gradient value obtained from Trace c without a correction term and it is assumed that $\alpha=0$.

The uncertainty equation used in the round robin experiment uses six sources of uncertainty with all other sources of uncertainty considered negligible. This strain gradient uncertainty equation (as calculated in Data Analaysis Sheet SG.1 [3] and in ASTM standard test method E 2246-05 [38]) with six sources of uncertainty is as follows:

$$u_{csg1}=\sqrt{u_W^2+u_{zres}^2+u_{samp}^2+u_{zcal}^2+u_{xcal}^2+u_{xres}^2}\ .\qquad\qquad(\text{SG16})$$

A number following the subscript "sg" in "u_{csg}" indicates the data analysis sheet that is used to obtain the combined standard uncertainty value. Therefore, u_{csg1} implies that Data Analysis Sheet SG.1 is used. In Eq. (SG16), u_W, u_{zres}, u_{xcal}, and u_{xres} are defined in Sec. 4.4.1. Also in the above equation, u_{samp} is the uncertainty due to the sample's peak-to-valley surface roughness as measured with the interferometer, and u_{zcal} is the uncertainty of the calibration in the z-direction. These six uncertainty components are calculated as specified in Sec. 3.4.2, replacing the words "residual strain" with the words "strain gradient," "fixed-fixed beams" with "cantilevers," ε_{r-low} with s_{g-low}, and ε_{r-high} with s_{g-high}. In addition, refer to Tables SG2, SG3, SG4, SG5, and SG6 where mention is made of Tables RS2, RS3, RS4, RS5, and RS6 respectively.

Table SG6. Determination of Some Strain Gradient Uncertainty Components in Eq. (SG16) and Eq. (SG17) [11,38]

Uncertainty Component	Method to Obtain s_{g-high} and s_{g-low}	G or U[a] / A or B[b]	equation
1. u_W	using Trace b, Trace c, and Trace d	U / B	$u_W = \dfrac{\left\lvert s_{g-high} - s_{g-low} \right\rvert}{2\sqrt{3}}$
2. u_{zres}	using $d=(1/2)z_{res}$ in Table SG3	U / B	$u_{zres} = \dfrac{\left\lvert s_{g-high} - s_{g-low} \right\rvert}{2\sqrt{3}}$
3. u_{samp}	using $d=3\sigma_{samp}$ in Table SG3	G / B	$u_{samp} = \dfrac{\left\lvert s_{g-high} - s_{g-low} \right\rvert}{6}$
4. u_{zcal}	using $d=3(z_x-z_1)\sigma_{zcal}/cert$ in Table SG5 where z_x is the column heading[c]	G / B	$u_{zcal} = \dfrac{\left\lvert s_{g-high} - s_{g-low} \right\rvert}{6}$
5. u_{xcal}	using cal_{xmin} for cal_x where $cal_{xmin} = cal_x - 3\sigma_{xcal}cal_x/ruler_x$ and cal_{xmax} for cal_x where $cal_{xmax} = cal_x + 3\sigma_{xcal}cal_x/ruler_x$	G / B	$u_{xcal} = \dfrac{\left\lvert s_{g-high} - s_{g-low} \right\rvert}{6}$
6. u_{xres}	using $d=(1/2)x_{res}(cal_x)$ in Table SG4	U / B	$u_{xres} = \dfrac{\left\lvert s_{g-high} - s_{g-low} \right\rvert}{2\sqrt{3}}$
7. $u_{repeat(shs)}$	using $d=(z_x-z_1)z_{repeat(shs)}/(2\bar{z}_6)$ in Table SG5 where z_x is the column heading[c]	U / B	$u_{repeat(shs)} = \dfrac{\left\lvert s_{g-high} - s_{g-low} \right\rvert}{2\sqrt{3}}$

[a] "G" indicates a Gaussian distribution and "U" indicates a uniform distribution.

[b] Type A or Type B analysis

[c] For ease of presentation, z_x and z_1 in this table are considered calibrated. (Actually, these values are uncalibrated as presented earlier in this SP 260, therefore, the uncalibrated values should be multiplied by cal_z before use in this table.)

In determining the combined standard uncertainty, a Type B evaluation [19-21] (i.e., one that uses means other than the statistical Type A analysis) is used for each source of uncertainty. Table SG7 gives example values for each of these uncertainty components as well as the combined standard uncertainty value, u_{csg1}.

Table SG7. Example Strain Gradient Uncertainty Values[a]
From a Round Robin Surface-Micromachined Chip

	source of uncertainty or descriptor	uncertainty value (m^{-1})
1. $\mathbf{u}_W$	variations across the width of the beam	0.186
2. $\mathbf{u}_{zres}$	interferometric resolution in z-direction	0.037
3. $\mathbf{u}_{samp}$	interferometric peak-to-valley surface roughness	0.423
4. $\mathbf{u}_{zcal}$	calibration in z-direction	0.012
5. $\mathbf{u}_{xcal}$	calibration in x-direction	0.013
6. $\mathbf{u}_{xres}$	interferometric resolution in x-direction	0.056
$\mathbf{u}_{csg1}$[b]	combined standard uncertainty for strain gradient	0.467

[a] As determined in ASTM standard test method E 2246-05 [38] using Eq. (SG16) for a cantilever with a design length of 550 μm and with a $0°$ orientation.

[b] This value for u_{csg1} was used in the round robin (see Sec. 4.5) and is incorporated into the calculations of $u_{csg1ave}$ as presented in the sixth row of Table SG8, which uses Eq. (SG16).

An expanded version of the uncertainty calculation presented in Eq. (SG16) is given below, which includes ten sources of uncertainty:

$$u_{csg2} = \sqrt{u_W^2 + u_{zres}^2 + u_{xcal}^2 + u_{xres}^2 + u_{Rave}^2 + u_{noise}^2 + u_{cert}^2 + u_{repeat(shs)}^2 + u_{drift}^2 + u_{linear}^2} \quad . \tag{SG17}$$

This calculation is done using Data Analysis Sheet SG.2 [13]. The first four components (u_W, u_{zres}, u_{xcal}, and u_{xres}) are calculated as they are calculated for use in Eq. (SG16) using only Trace c. By using the above equation [instead of Eq. (SG16)], the z-calibration procedures are the same for the different applicable measurements in this SP 260 for the MEMS 5-in-1 SRMs [by using a procedure such as given in Sec. 5.2 for step height measurements (as used with the MEMS 5-in-1) as opposed to the method referred to in Sec. 4.2 for earlier versions]. Therefore, the component, u_{zcal} in Eq. (SG16) gets replaced with the components u_{cert}, u_{drift}, and u_{linear} as described in Sec. 3.4.1 (after making the appropriate substitutions and using only the data from Trace c) and with the component $u_{repeat(shs)}$ as described in Sec. 3.4.2 (after making the appropriate substitutions). Also, to provide more physical understanding to the resulting uncertainties, the uncertainty component u_{samp} in Eq. (SG16) is replaced with the components u_{Rave} and u_{noise} in Eq. (SG17). These two components are described in Sec. 3.4.1 (after making the appropriate substitutions and using only the data from Trace c).

4.5 Strain Gradient Round Robin Results

The MEMS Length and Strain Round Robin *repeatability* and *reproducibility* results are given in this section for strain gradient measurements. The *repeatability* data were taken in one laboratory using an optical interferometer (see Sec. 1.1.2). Unlike the MEMS 5-in-1 chip shown in Fig. 2 for SRM 2495, a similarly processed surface-micromachined test chip (from run #46 [9] and without the backside etch) was fabricated on which strain gradient measurements were taken from poly1 cantilever test structures but having a $0°$ orientation and from poly1 cantilevers having a $90°$ orientation. An array of the cantilevers on the round robin test chip with a $0°$ orientation is shown in Fig. SG7. Each cantilever array has design lengths from 400 μm to 800 μm, inclusive, in 50 μm increments. (All the cantilevers are 10 μm wide.) However, only the design lengths between 400 μm to 750 μm, inclusive, were used in obtaining the *repeatability* data. Therefore, with three beams designed at each length, 48 measurements were taken (24 measurements at each orientation). See Fig. SG8(a) for a design rendition of a p1 cantilever on the round robin test chip and Figs. SG8(b and c) for applicable 2-D data traces taken from this cantilever.

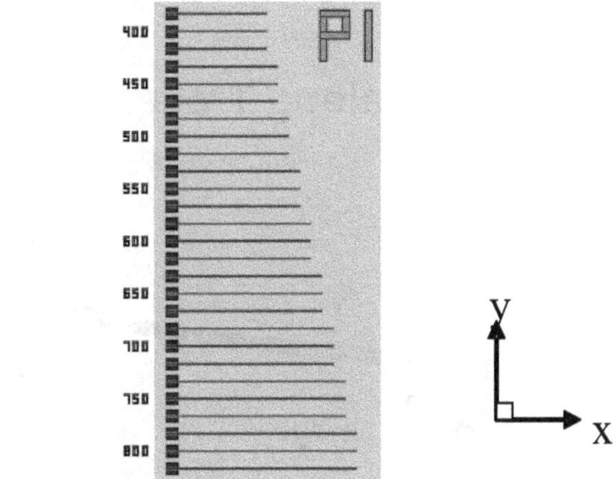

Figure SG7. An array of cantilevers on the round robin test chip.

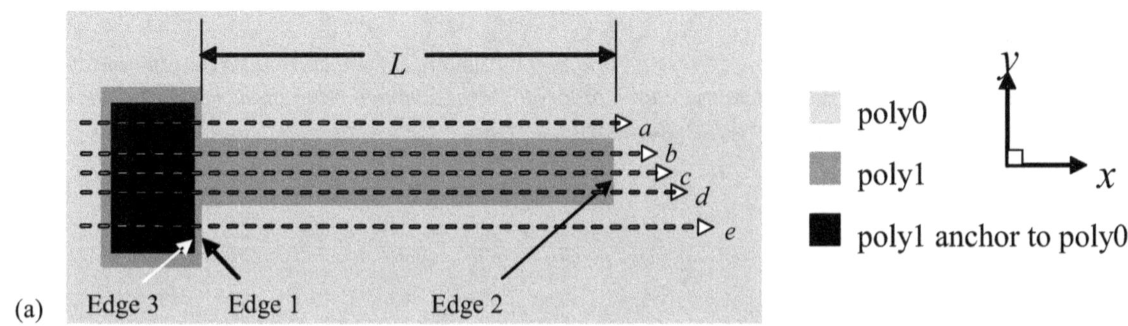

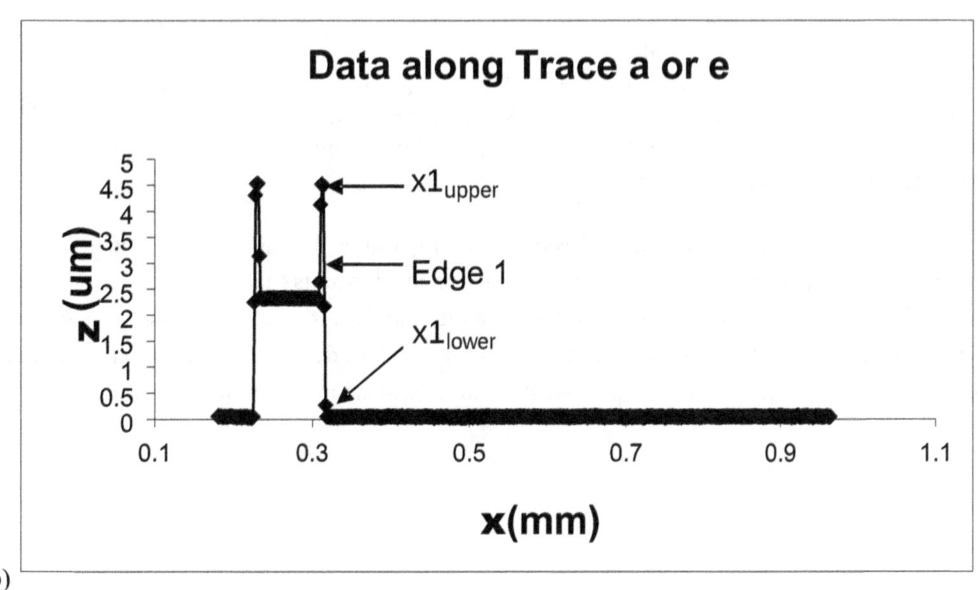

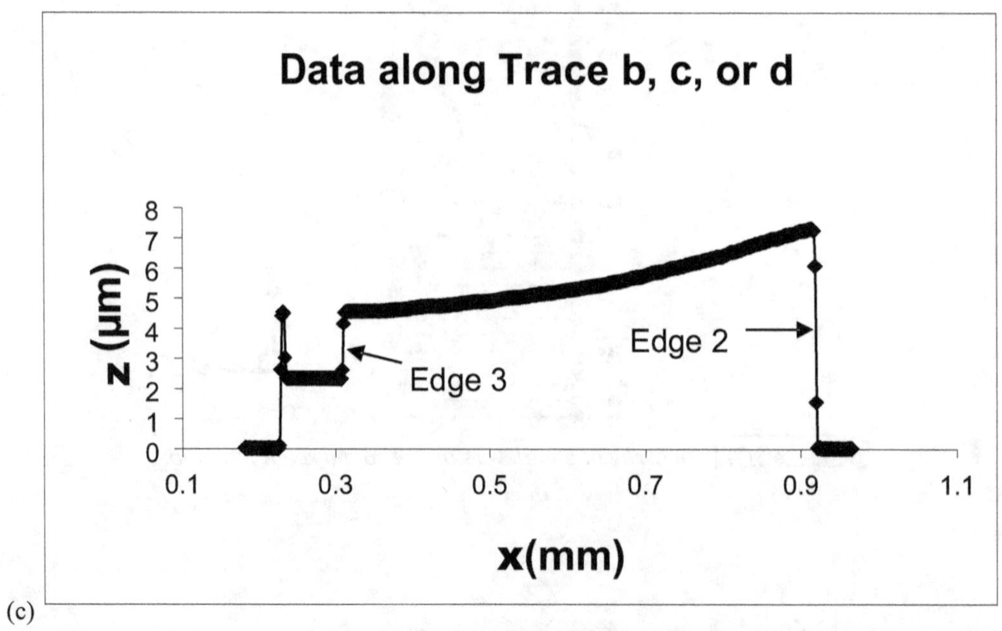

(c)

*Figure SG8. For a cantilever test structure on the round robin test chip, (a) a design rendition,
(b) an example of a 2-D data trace used to locate the attachment point of the cantilever in (a),
and (c) an example of a 2-D data trace taken along the length of the cantilever in (a).*

For the *reproducibility* data,[37] a round robin test chip was passed from laboratory to laboratory. Each participant was asked to obtain a strain gradient measurement from two poly1 cantilevers in an array, such as shown in Fig. SG7, that had either a 0° orientation or a 90° orientation. One of the cantilevers was requested to have a design length of 650 µm. The design length for the other cantilever could range from 500 µm to 650 µm, inclusive. The results from one of the two cantilevers is included in the results presented below. Following the 2002 version of ASTM standard test method E 2246 [39] for strain gradient measurements, the raw, uncalibrated measurements were recorded on Data Analysis Sheet H (similar to the existing Data Analysis Sheet SG.1 [13]) for measurements of strain gradient.

Table SG8 presents the strain gradient *repeatability* and *reproducibility* results. In this table, n is the number of measurements followed by the average of the *repeatability* or *reproducibility* measurement results (namely, s_{gave}). For the *repeatability* measurements only, $\sigma_{repeat(samp)}$ is given, which is the relative standard deviation of the *repeatability* strain gradient measurements. Then, the $\pm 2\sigma_{sg}$ limits are given, where σ_{sg} is the standard deviation of the strain gradient measurements, followed by the average of the *repeatability* or *reproducibility* combined standard uncertainty values (i.e., u_{csgave}) for different calculations.

Table SG8. Strain Gradient Measurement Results

	Repeatability results L_{des}= 500 μm to 650 μm	Repeatability results L_{des}= 400 μm to 750 μm	Reproducibility results L_{des}= 500 μm to 650 μm
1. n	24	48	6
2. s_{gave}	4.71 m^{-1}	4.97 m^{-1}	4.67 m^{-1}
3. $\sigma_{repeat(samp)}$	13 %	20 %	–
4. $\pm 2\sigma_{sg}$ limits	$\pm$1.2 m^{-1} ($\pm$ 25 %)	$\pm$2.0 m^{-1} ($\pm$ 40 %)	$\pm$1.7 m^{-1} ($\pm$ 37 %)
5. $u_{csg1ave}$ [a]	0.73 m^{-1} (14 %)	0.84 m^{-1} (17 %)	0.56 m^{-1} (12 %)
6. $u_{csg1ave}$ [b]	0.47 m^{-1} (10 %)	0.56 m^{-1} (11 %)	–
7. $u_{csg2ave}$ [c]	0.44 m^{-1} (9.5 %)	0.52 m^{-1} (11 %)	–
8. $u_{csg3ave}$ [d]	0.74 m^{-1} (16 %)	1.12 m^{-1} (23 %)	–

[a] Where u_{csg1} is determined in ASTM standard test method E 2246–02 [39]. For this calculation, the u_{samp} and u_{zcal} components in the u_{csg1} calculation in Eq. (SG16) are combined into one component. For this component, assuming a uniform (that is, rectangular) probability distribution, the limits are represented by a $\pm$20 nm variation in the z-value of the data points. Also, in ASTM standard test method E 2246–02, $u_{zres}=u_{xcal}=u_{xres}=0$ m^{-1}.

[b] Where u_{csg1} is determined in ASTM standard test method E 2246–05 [38] using Eq. (SG16).

[c] Where u_{csg2} is determined using Eq. (SG17).

[d] Where u_{csg3} is determined using Eq. (SG12).

Comments concerning the round robin data include the following:

a) *Plots*: In this round robin, random length cantilevers were measured. As such, there are at least two variables (orientation and length) as discussed below.

    i) *Orientation*: Figure SG9 is a plot of s_g versus orientation, which reveals no obvious orientation dependence.

    ii) *Length*: Figure SG10 is a plot of s_g versus design length, where the data indicate a decrease in the strain gradient for increasing length (for L_{des} = 400 μm to 600 μm) that levels off (from L_{des}=600 μm to 750 μm).

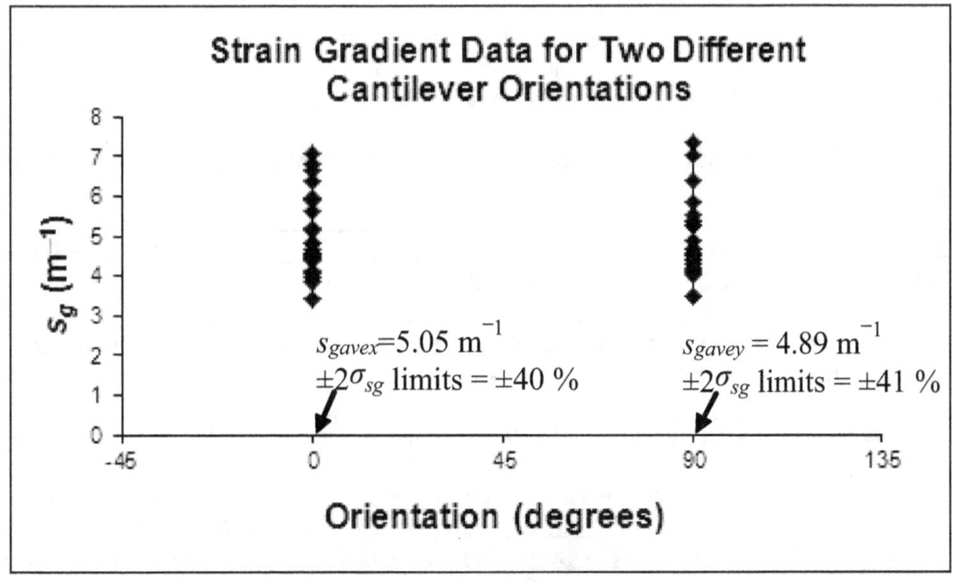

Figure SG9. A plot of s_g versus orientation [38].

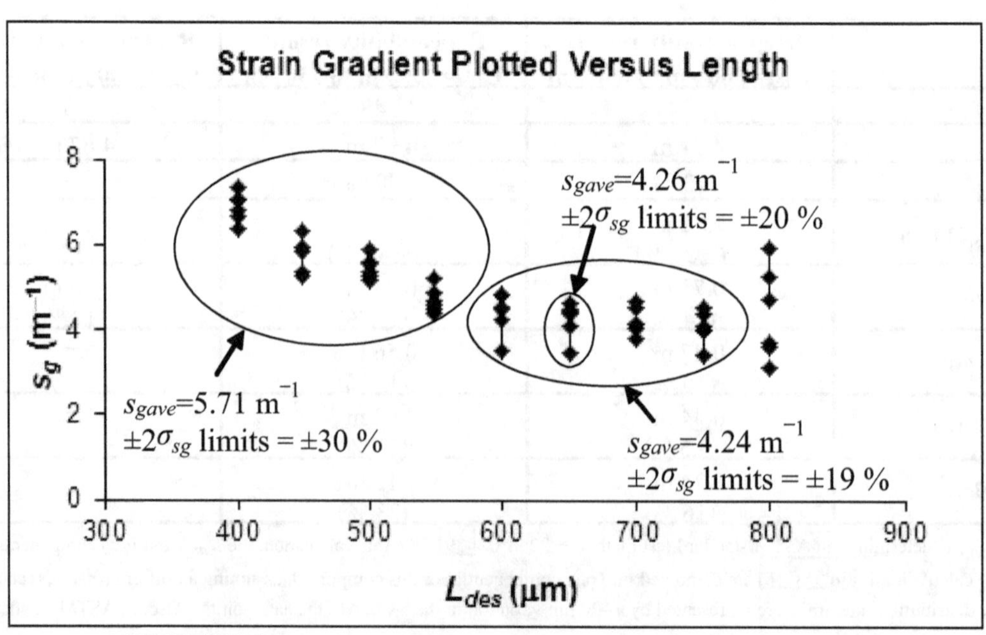

Figure SG10. A plot of s_g versus length for two different orientations.[39]

b) *Precision*: The *repeatability* and *reproducibility* precision data appear in Table SG8. In particular, for the ±2 σ_{sg} limits, the *repeatability* data (i.e., ±25 %) are tighter than the *reproducibility* data (i.e., ±37 %) for the same span of design lengths. This is due to the *repeatability* measurements being taken in the same laboratory using the same instrument by the same operator. It is interesting to note that the $u_{csglave}$ values for the *repeatability* measurements are slightly higher than the $u_{csglave}$ values for the *reproducibility* measurements.

c) *Bias*: No information can be presented on the bias of the procedure in ASTM standard test method E 2246 for measuring strain gradient because there is not a certified MEMS material for this purpose.

4.6 Using the MEMS 5-in-1 to Verify Strain Gradient Measurements

To compare your in-house strain gradient measurements with NIST measurements, you will need to fill out Data Analysis Sheet SG.3. (This data analysis sheet is accessible via the URL specified in the reference [13], a reproduction of which is given in Appendix 3.) After calibrating the instrument, locating the test structure, taking the measurements, and performing the calculations, the data on the completed form can be compared with the data on the SRM Certificate and the completed data analysis sheet supplied with the MEMS 5-in-1. Details of the procedure are given below.

Calibrate the instrument: Calibrate the instrument as specified in Sec. 4.2 for SRM measurements. Obtain the inputs for Table 1 in Data Analysis Sheet SG.3.

Locate the cantilever: In the third grouping of test structures, shown in Figs. SG1(a and b) on the MEMS 5-in-1 chips shown in Fig. 1 and Fig. 2 for SRM 2494 and SRM 2495, respectively, strain gradient measurements are made. Cantilever test structures are provided for this purpose, as shown in Fig. SG2(a) for SRM 2494 and as shown in Fig. SG3(a) for the poly2 cantilevers on SRM 2495. Data Analysis Sheet SG.3 requires measurements from one cantilever test structure. The specific test structure to be measured can be deduced from the data entered on the NIST-supplied Data Analysis Sheet SG.3 that accompanies the SRM.

For the strain gradient grouping of test structures for SRM 2494, as shown in Fig. SG1(a), the target test structure can be found as follows:
1. The input *design length* (i.e., input #4 on Data Analysis Sheet SG.3, a reproduction of which is given in Appendix 3) specifies the design length of the cantilever. The design length of the cantilever (in micrometers) is given at the top of each column of test structures in Fig. SG1(a) following the column number (i.e., 1 to 5), therefore *design length* can be used to locate the column in which the target test structure resides. Design lengths for the cantilever test structures are given in Table SG1.
2. The input *which cantilever* (i.e., input #6) specifies which cantilever in the column to measure (i.e., the "first," "second," "third," etc.) regardless of the orientation.

---

[39] Ibid.

3. The input *orientation* (i.e., input #8) can be used as a form of verification. The cantilevers are designed at both a $0°$ and a $180°$ orientation with the cantilevers having a $0°$ orientation being the first, second, and third cantilevers in each column and the cantilevers with a $180°$ orientation being the fourth, fifth, and sixth cantilevers in each column. Therefore, either $0°$ or $180°$ will be selected for *orientation*.

For the strain gradient grouping of test structures for SRM 2495, as shown in Fig. SG1(b), the target test structure can be found as follows:

1. The input *material* (i.e., input #3) is used to identify if a cantilever in a poly1 array is to be measured or if a cantilever in a poly2 array is to be measured. The two poly1 arrays in Fig. SG1(b) have a P1 designation and the one poly2 array has a P2 designation.
2. The input *orientation* (i.e., input #8) specifies the orientation of the cantilever array. The cantilevers in the lower left poly1 array have a $180°$ orientation and the cantilevers in the upper poly1 array have a $90°$ orientation. The poly2 array has a $180°$ orientation.
3. The input *design length* (i.e., input #4) specifies the design length of the cantilever. The design length of the cantilever (in micrometers) is given next to the second of three cantilevers of the same length, as can barely be seen in Fig. SG1(b). Therefore, *design length* can be used to locate a set of three possible target test structures. Design lengths for the cantilever test structures are given in Table SG1.
4. The input *which cantilever* (i.e., input #6) specifies which cantilever in the set of three possible target test structures of the same length in the array to measure (i.e., the "first," "second," "third," etc.). Since there are three instances of each test structure, the radio button corresponding to "first," "second," or "third" is used to identify the target test structure.

Take the measurements: Following the steps in ASTM standard test method E 2246 [3] for strain gradient measurements, the chip is oriented under the interferometric optics as shown in Fig. SG2(a) or Fig. SG3(a) [40] and one 3-D data set is obtained using the highest magnification objective that is available and feasible. (For the given design lengths, the magnification should be at least 10×.) The data are leveled and zeroed, and Traces a, b, c, d, and e are obtained. Traces a and e are used to calculate the misalignment angle, α. From Traces a and e, measurements of x_{upper} and n_t (typically from Edge 1) are entered into Data Analysis Sheet SG.3. The uncalibrated values for y_a and y_e are also recorded.

For SRM 2494, with Data Analysis Sheet SG.3, uncalibrated data points along the cantilever for (x_1, z_1), (x_2, z_2), and (x_3, z_3), are requested from Traces b, c, and d, shown in Fig. SG2(c).

For SRM 2495, with Data Analysis Sheet SG.3, there are data restrictions due to non-idealities in the geometry of the cantilever as discussed in Sec. 4.1 and Sec. 4.3. In particular, uncalibrated data points along the cantilever for (x_1, z_1), (x_2, z_2), and (x_3, z_3) are requested from Traces b, c, and d as shown in Fig. SG3(c), and these data should be taken along the cantilever such that the uncalibrated x value of each data point is greater than $x6_{ave}$ (assuming a $0°$ orientation of the cantilever).

Perform the calculations: Enter the data into Data Analysis Sheet SG.3 as follows:

1. Press one of the "Reset this form" buttons. (One of these buttons is located near the top of the data analysis sheet and the other is located near the middle of the data analysis sheet.)
2. Supply inputs to Table 1 through Table 3.
3. Press one of the "Calculate and Verify" buttons to obtain the results from the cantilever test structure. (One of these buttons is located near the top of the data analysis sheet and the other is located near the middle of the data analysis sheet.)
4. Verify the data by checking to see that all the pertinent boxes in the verification section at the bottom of the data analysis sheet say "ok". If one or more of the boxes say "wait," address the issue, if necessary, by modifying the inputs and recalculating.
5. Print out the completed data analysis sheet to compare both the inputs and outputs with those on the NIST-supplied data analysis sheet.

Compare the measurements: The MEMS 5-in-1 is accompanied by a Certificate. This Certificate specifies a strain gradient value, s_g, and the expanded uncertainty, U_{sg}, (with $k=2$) intending to approximate a 95 % level of confidence. It is your responsibility to determine an appropriate criterion for acceptance, such as given below:

---

[40] This orientation assumes that the pixel-to-pixel spacing in the x-direction of the interferometric microscope is smaller than or equal to the pixel-to-pixel spacing in the y-direction.

$$D_{sg} = \left| s_{g(customer)} - s_g \right| \leq \sqrt{U^2_{sg(customer)} - U^2_{sg}} \, , \qquad \text{(SG18)}$$

where D_{sg} is the absolute value of the difference between your strain gradient value, $s_{g(customer)}$, and the strain gradient value on the SRM Certificate, s_g, and where $U_{sg(customer)}$ is your expanded uncertainty value and U_{sg} is the expanded uncertainty on the SRM Certificate. If your measured value for strain gradient (as obtained in the newly filled out Data Analysis Sheet SG.3) satisfies your criterion for acceptance and there are no pertinent "wait" statements at the bottom of your Data Analysis Sheet SG.3, you can consider yourself to be appropriately measuring strain gradient according to the ASTM E 2246 strain gradient standard test method [3] according to your criterion for acceptance.

An effective strain gradient value is reported for SRM 2494, as shown in Fig. 1, due to non-idealities associated with the geometry and/or composition of the cantilever as discussed in Sec. 4.1 and Sec. 4.3. Most notably, the excessive curvature of the cantilevers on this chip makes it difficult to obtain a strain gradient correction term, $s_{gcorrection}$, as discussed in Sec. 4.3. When you use ASTM standard test method E 2246 with your own cantilever, you must be cognizant of the geometry and composition of your cantilever because this test method assumes an ideal geometry and composition, implying that you would be obtaining an "effective" strain gradient value if the geometry and/or composition of your cantilever deviates from the ideal.

Any questions concerning the measurements, analysis, or comparison can be directed to mems-support@nist.gov.

5   Grouping 4: Step Height

A step height measurement is defined here as the distance in the z-direction between an initial, flat, processed surface (or platform) and a final, flat, processed surface (or platform). These measurements can be used to determine thin film thickness values (see Sec. 8), which are an aid in the design and fabrication of MEMS devices [28-29].

This section on step height supplements SEMI standard test method MS2 [4], which more completely presents the purpose, scope, limitations, terminology, apparatus, and test structure design for MEMS as well as the calibration procedure, measurement procedure, calculations, precision and accuracy, etc. In this section, the NIST-developed step height test structures on SRM 2494 and SRM 2495 are described and illustrated in Sec. 5.1, the calibration procedure for step height measurements is described in Sec. 5.2, the step height measurement procedure in Sec. 5.3, the uncertainty analysis in Sec. 5.4, and the round robin results in Sec. 5.5. Section 5.6 describes how to use the MEMS 5-in-1 to verify step height measurements.

5.1 Step Height Test Structures

Step height measurements are taken in the fourth grouping of test structures, as shown in Fig. SH1(a) for SRM 2494 depicted in Fig. 1 in the Introduction and as shown in Fig. SH1(b) for SRM 2495 depicted in Fig. 2.

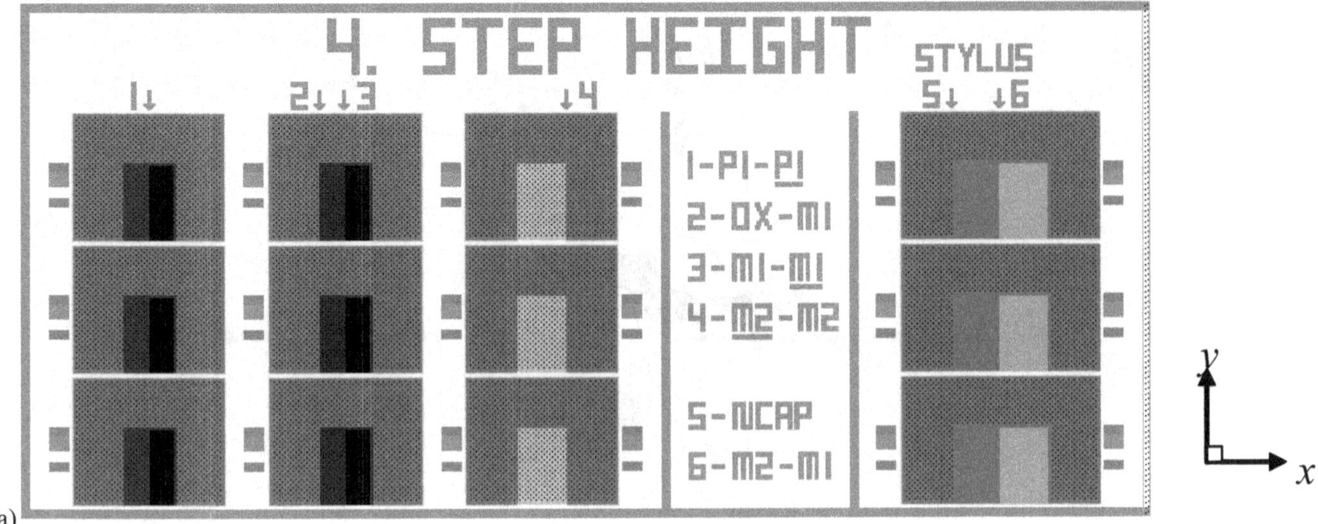

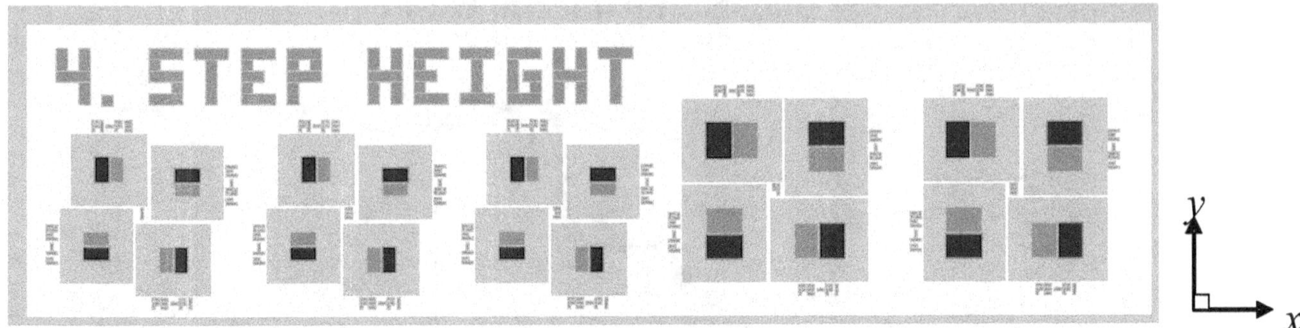

Figure SH1. The step height grouping of test structures on (a) SRM 2494, fabricated on a multi-user 1.5 μm CMOS process [8] followed by a bulk-micromachining etch, as depicted in Fig. 1 and (b) SRM 2495, fabricated using a polysilicon multi-user surface-micromachining MEMS process [9] with a backside etch, as depicted in Fig. 2.

A step height test structure from each of the groupings shown in Fig. SH1 is given in Fig. SH2(a) and Fig. SH3(a), respectively.

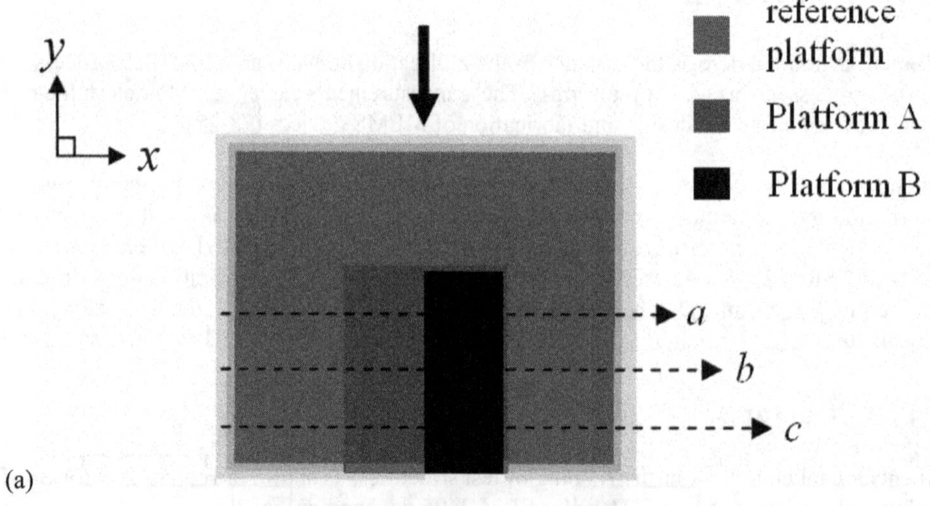

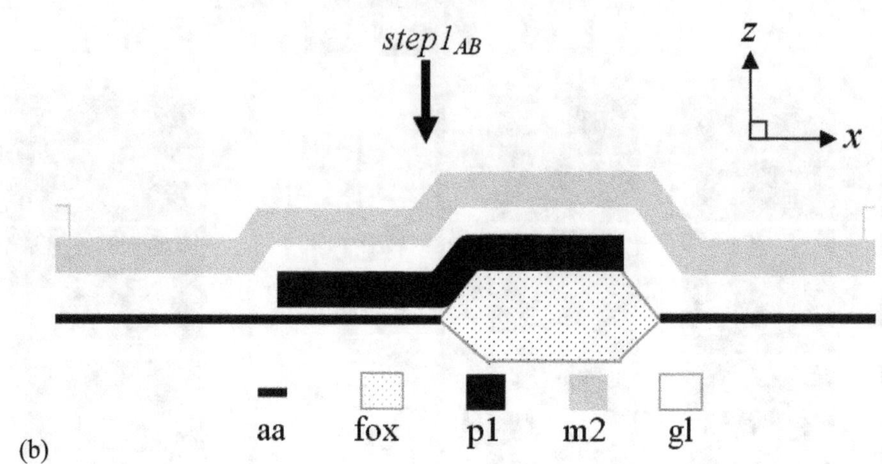

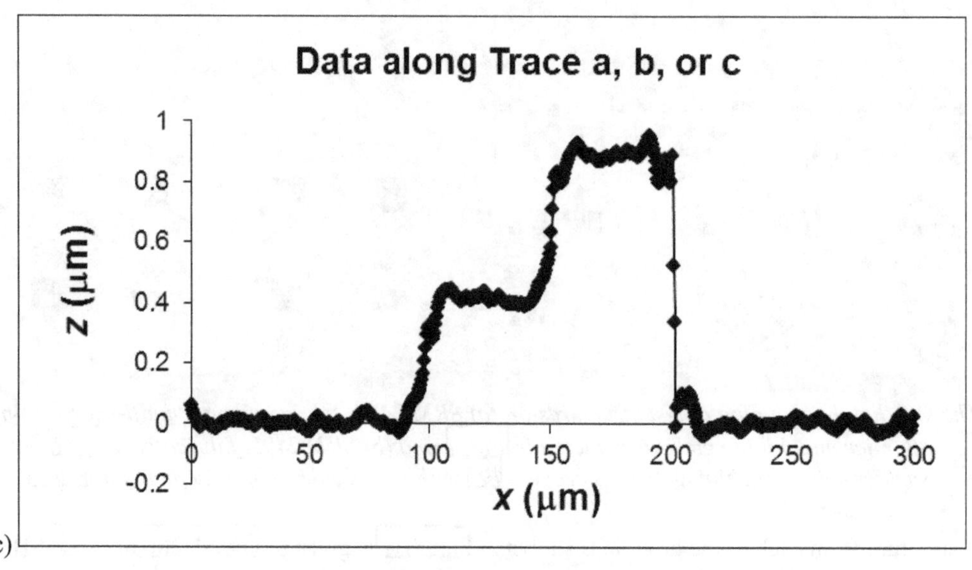

Figure SH2. *For a step height test structure on SRM 2494 as shown in Fig. 1, (a) a design rendition, (b) a cross section, and (c) an example of a 2-D data trace from (a).*

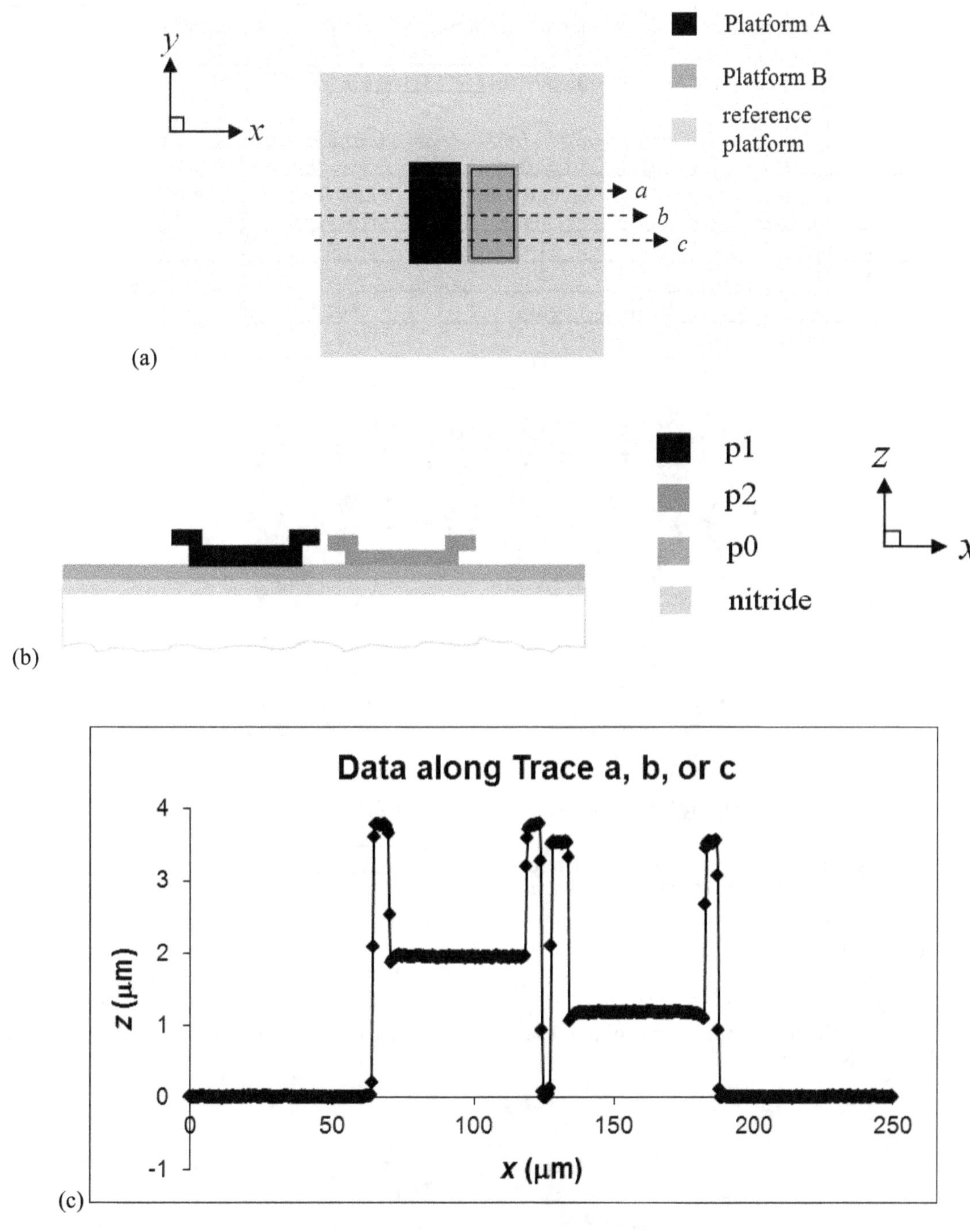

Figure SH3. For a step height test structure on SRM 2495 as shown in Fig. 2, (a) a design rendition, (b) a cross section, and (c) an example of a 2-D data trace from (a).

The step height test structures on SRM 2494 and SRM 2495 are described below.

For SRM 2494: In the grouping of step height test structures given in Fig. SH1(a) for SRM 2494, there are four distinct test structures (with three rows of each structure) for which six step height transitions are indicated by arrows above the topmost test structure in the row. We will only be concerned with those steps associated with an arrow. For the MEMS 5-in-1, only one of these six step heights is certified, in particular one of the three occurances of the specified step.

The six step height measurements in this grouping of test structures can be used to calculate the composite beam oxide thickness (for Young's modulus calculations).   It is important to keep in mind that the fourth test structure (associated with the fifth and sixth arrows) does not have a reflective top surface for each platform and as such is intended to be used with a stylus instrument. Consult Sec. 8 for details concerning this as well as the cross sections of each of the test structures given in Fig. SH1(a).

The approximate design dimensions of the four test structures in Fig. SH1(a) are given in Table SH1.

Table SH1. Design Dimensions (in Micrometers) as Depicted in Fig. SH4
For the Step Height Test Structures in Fig. SH1(a) For SRM 2494

	First Test Structure	Second Test Structure	Third Test Structure	Fourth Test Structure
w	≈150	≈150	≈150	≈150
l	≈50	≈50	≈100	≈100
r	≈94	≈94	≈94	≈94

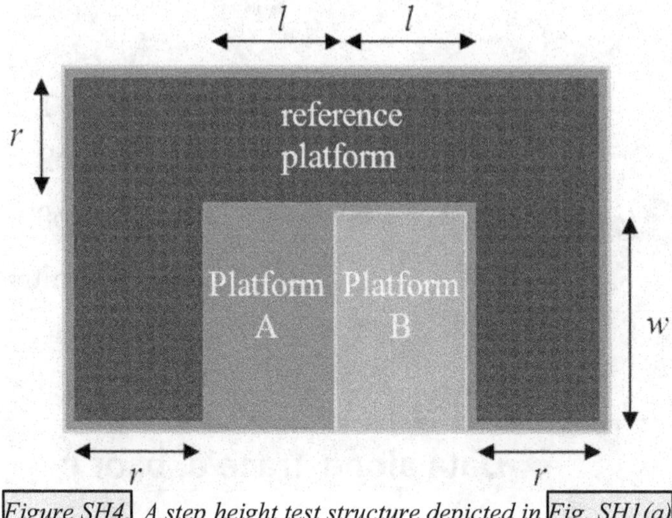

Figure SH4. A step height test structure depicted in Fig. SH1(a).

For SRM 2495: In the grouping of step height test structures given in Fig. SH1(b) for SRM 2495, the step height test structures are grouped in quads. The second quad (as indicated by the number " 2" in the center of the quad) is shown in Fig. SH5. There are four step height test structures within each quad, each with a different orientation. The upper left test structure has a 0° orientation, the bottom left has a 90° orientation, the bottom right has a 180° orientation and the upper right has a 270° orientation. Each of the test structures is a step from the top of the polysilicon layer called "poly1" or "p1" to the top of the polysilicon layer called "poly2" or "p2," or vice versa. In Fig. SH5 the "P1" and "P2" labels closest to the platform in the step can be used to determine which platform is made of poly1 and which is made of poly2. The design layer for the surrounding reference platform is called "poly0" or "p0."

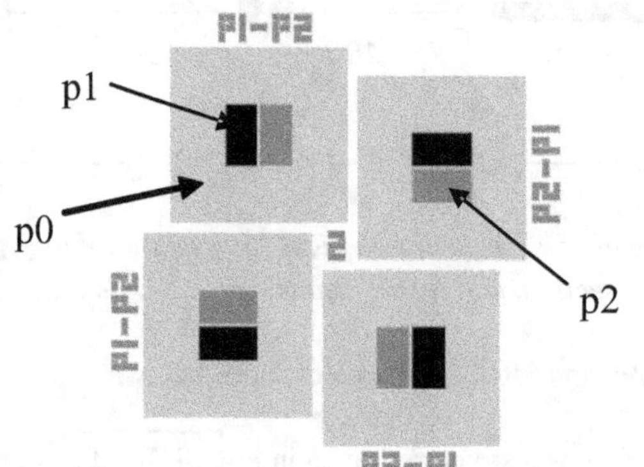

Figure SH5. Quad 2 in the step height grouping depicted in Fig. SH1(b).

There are two different sized quads in the step height grouping in Fig. SH1(b). The larger fourth and fifth quads are intended for use with a stylus instrument, although an optical instrument can also be used. The approximate design dimensions for the test structures in the two different sized quads are given in Table SH2, which refers to the labeling in Fig. SH6. As can be seen in Fig. SH6, the platform dimensions (including the anchor dimensions as given by the subscript "a") for Platform A and Platform B are the same.

Table SH2. *Design Dimensions (in Micrometers) as Depicted in Fig. SH6*
For the Step Height Test Structures in Fig. SH1(b) For SRM 2495

	For test structures in Quads 1, 2, and 3	For test structures in Quads 4 and 5
$l \times w$	60×110	110×160
$l_a \times w_a$	50×100	100×150
r	100	100
R_1	325	425
R_2	310	360

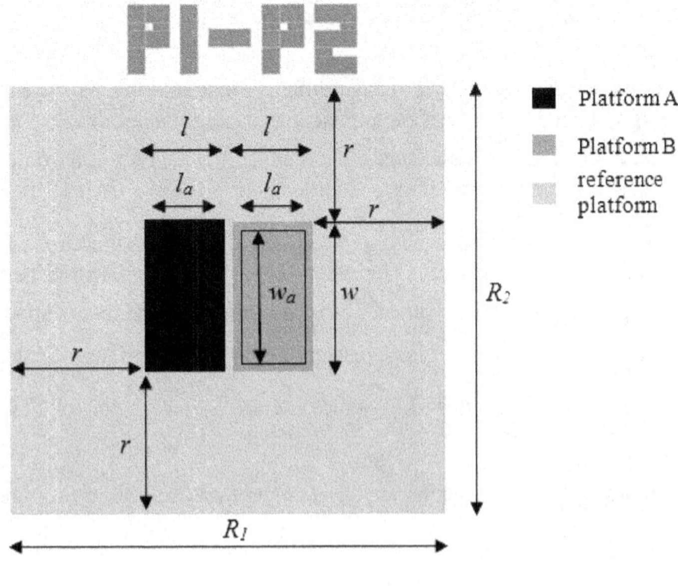

Figure SH6. A step height test structure depicted in Fig. SH1(b).

5.2 Calibration Procedures for Step Height Measurements

For step height measurements, the optical interferometer (or comparable instrument) is calibrated in the z-direction. The calibration procedure used with the MEMS 5-in-1 [4] is given below followed by the calibration procedure used when discussing earlier versions of the step height uncertainty equation presented in Sec. 5.4.

Calibration in the z-direction (as used with the MEMS 5-in-1) [4]:

1. Use the same "slope" value, if applicable, as obtained in Sec. 1.1.2.2. Verify that this slope value is adequate by using the instrument's prescribed calibration procedure to make sure that the difference between the instrument's step height measurements on a physical step height standard and the certified value of the physical step height standard is less than or equal to 1 %. If it is not, perform the steps in Sec. 1.1.2.2 to validate the step height measurements.
2. Before the data session:
   a. The height of the physical step height standard is recorded at six locations.
      i. If single-sided step height measurements are taken, the six measurements are taken with three measurements spread out evenly along each side of the physical step height standard. (The

measurements are taken within the specified certified area along the length and width of the step.[41])

    ii. If double-sided step height measurements are taken, the six measurements are spread out evenly along the certified area of the physical step height standard.

The mean value of the six measurements is called $\bar{z}_{before}$. The standard deviation of these six measurements is called σ_{before}.

    b. In addition, six step height measurements are taken at the same location on the physical step height standard. (This location is within the specified certified area along the length and width of the step.) The mean value of the six measurements taken at the same location is called $\bar{z}_{same1}$. The standard deviation of these six measurements is called σ_{same1}.

3. Similarly, after the data session:

    a. The height of the physical step height standard is recorded at six locations.

        i. If single-sided step height measurements are taken, the six measurements are taken with three measurements spread out evenly along each side of the physical step height standard. (The measurements are taken within the specified certified area along the length and width of the step.)

        ii. If double-sided step height measurements are taken, the six measurements are spread out evenly along the certified area of the physical step height standard.

The mean value of the six measurements is called $\bar{z}_{after}$. The standard deviation of these six measurements is called σ_{after}.

    b. In addition, six step height measurements are taken at the same location on the physical step height standard as before the data session. The mean value of the six measurements taken at the same location is called $\bar{z}_{same2}$. The standard deviation of these six measurements is called σ_{same2}.

[Note that if it can be demonstrated that a given instrument does not drift significantly during a data session, this second step can be skipped and it can be assumed that $\bar{z}_{before} = \bar{z}_{after}$, $\sigma_{before} = \sigma_{after}$, $\bar{z}_{same1} = \bar{z}_{same2}$, and $\sigma_{same1} = \sigma_{same2}$.

4. The average of the calibration measurements, $\bar{z}_{ave}$, is calculated using the following formula:

$$\bar{z}_{ave} = \frac{\bar{z}_{before} + \bar{z}_{after}}{2} \qquad . \tag{SH1}$$

5. The z-calibration factor for the instrument, cal_z, is determined using the following equation:

$$cal_z = \frac{cert}{\bar{z}_{ave}} \quad , \tag{SH2}$$

where $cert$ is the certified value of the physical step height standard. The z-data values obtained during the data session are multiplied by cal_z to obtain calibrated z-data values.

6. For uncertainty calculations, the quantities, σ_{cert} and z_{lin}, are recorded where σ_{cert} is the certified uncertainty of the physical step height standard and z_{lin} is the maximum relative deviation from linearity over the total scan range of the instrument (as quoted by the manufacturer or as determined in Sec. 1.1.2.2). Also, z_{drift}, σ_{6ave}, $\bar{z}_{6ave}$, σ_{6same}, and $\bar{z}_{6same}$ (as defined in the Definition of Symbols Section)[42] are determined as follows:

---

[41] The location of the certified area should be indicated in the Certificate accompanying the physical step height standard. Please note that the area close to the step transition is typically not included in the certified region.

[42] The values for z_{drift}, σ_{6ave}, $\bar{z}_{6ave}$, σ_{6same}, and $\bar{z}_{6same}$ as calculated here are uncalibrated to match the inputs to Data Analysis Sheet SH.1.a .

$$z_{drift} = \left| \bar{z}_{same1} - \bar{z}_{same2} \right| , \qquad \text{(SH3)}$$

$$\text{if } \sigma_{before} \geq \sigma_{after}, \text{ then } \sigma_{6ave} = \sigma_{before} \text{ and } \bar{z}_{6ave} = \bar{z}_{before}, \qquad \text{(SH4)}$$

$$\text{if } \sigma_{before} < \sigma_{after}, \text{ then } \sigma_{6ave} = \sigma_{after} \text{ and } \bar{z}_{6ave} = \bar{z}_{after}, \qquad \text{(SH5)}$$

$$\text{if } \sigma_{same1} \geq \sigma_{same2}, \text{ then } \sigma_{6same} = \sigma_{same1} \text{ and } \bar{z}_{6same} = \bar{z}_{same1}, \text{ and} \qquad \text{(SH6)}$$

$$\text{if } \sigma_{same1} < \sigma_{same2}, \text{ then } \sigma_{6same} = \sigma_{same2} \text{ and } \bar{z}_{6same} = \bar{z}_{same2}. \qquad \text{(SH7)}$$

Calibration in the z-direction (for discussing earlier versions of the step height uncertainty equations) [40,41]. The following measurements are taken on a calibrated double-sided physical step height standard:

1. Before the data session, six step height measurements are taken at the same location on the physical step height standard. This location is within the specified certified area along the length and width of the step [43] of the standard. The mean value of the six measurements taken at the same location is called $\bar{z}_{same1}$. The minimum of the six measurements is called z_{min1}. The maximum is called z_{max1}.

2. Similarly, after the data session, six step height measurements are taken on the physical step height standard at the same location as before the data session. The mean value of these six meaurements is called $\bar{z}_{same2}$. The minimum of the six measurements is called z_{min2}. The maximum is called z_{max2}. [Note that if it can be demonstrated that a given instrument does not drift significantly during a data session, this step can be skipped and it can be assumed that $\bar{z}_{same2} = \bar{z}_{same1}$, $z_{min2} = z_{min1}$, and $z_{max2} = z_{max1}$.]

3. The average of the calibration measurements, $\bar{z}_{same}$ is calculated using the following formula:

$$\bar{z}_{same} = \frac{\bar{z}_{same\,1} + \bar{z}_{same\,2}}{2}. \qquad \text{(SH8)}$$

4. The z-calibration factor for the instrument, cal_z, is determined using the following equation:

$$cal_z = \frac{cert}{\bar{z}_{same}} , \qquad \text{(SH9)}$$

where *cert* is the certified value of the physical step height standard. The z-data values obtained during the data session are multiplied by cal_z to obtain calibrated z-data values.

5. For uncertainty calculations, the quantities, σ_{cert} and z_{perc}, are recorded where σ_{cert} is the combined standard uncertainty of the calibrated physical step height standard and z_{perc} (also called z_{lin}) is the maximum relative deviation from linearity over the total scan range of the instrument as quoted by the manufacturer. Also, z_{drift}, $z_{repeat(shs)}$, and $\bar{z}_6$ (as defined in the Definition of Symbols Section) [44] are calculated using the following equations:

---

[43] The location of the certified area should be indicated in the Certificate accompanying the physical step height standard. Please note that the area close to the step transition is typically not included in the certified region.

[44] The values for z_{drift}, $z_{repeat(shs)}$, and $\bar{z}_6$ as calculated here are uncalibrated to match the inputs to Data Analysis Sheet SH.1.

$$z_{drift} = \left| \bar{z}_{same1} - \bar{z}_{same2} \right| , \qquad\qquad\qquad (SH10)$$

$$z_{repeat(shs)1} = z_{max1} - z_{min1} , \qquad\qquad\qquad (SH11)$$

$$z_{repeat(shs)2} = z_{max2} - z_{min2} , \qquad\qquad\qquad (SH12)$$

$$\text{if } z_{repeat(shs)1} \geq z_{repeat(shs)2} , \text{ then } z_{repeat(shs)} = z_{repeat(shs)1} \text{ and } \bar{z}_6 = \bar{z}_{same1} , \text{ and} \qquad (SH13)$$

$$\text{if } z_{repeat(shs)1} < z_{repeat(shs)2} , \text{ then } z_{repeat(shs)} = z_{repeat(shs)2} \text{ and } \bar{z}_6 = \bar{z}_{same2} . \qquad (SH14)$$

5.3 Step Height Measurement Procedure

For the MEMS 5-in-1, the step height measurements contributing to the SRM certification are taken from one step height test structure. Three 2-D data traces [such as, Trace a, Trace b, and Trace c, as shown in Fig. SH2(a) and Fig. SH3(a)] are taken along the top of the test structure, a cross section of which is given in Fig. SH2(b) and Fig. SH3(b), respectively. Example 2-D data traces are given in Fig. SH2(c) and Fig. SH3(c) respectively. All height measurements are with respect to the height of the surrounding or partially surrounding reference platform that is used to level and zero the data. For generic Test Structure N with platforms labelled X and Y, the individual platform height measurements from Trace a, Trace b, and Trace c (namely, $platNXa$, $platNXb$, $platNXc$, $platNYa$, $platNYb$, and $platNYc$) and the standard deviations from the two platforms associated with the step (namely, $s_{platNXa}$, $s_{platNXb}$, $s_{platNXc}$, $s_{platNYa}$, $s_{platNYb}$, and $s_{platNYc}$) are recorded,[45] being careful to extract these measurements from portions not close to the transitional edges. If the test structure in Fig. SH2(a) is called Test Structure 1, then for the step in Test Structure 1 from Platform A to Platform B, as pointed to by the arrow above the step, the platform height measurements from Trace a, Trace b, and Trace c would be $plat1Aa$, $plat1Ab$, $plat1Ac$, $plat1Ba$, $plat1Bb$, and $plat1Bc$ and the standard deviations would be $s_{plat1Aa}$, $s_{plat1Ab}$, $s_{plat1Ac}$, $s_{plat1Ba}$, $s_{plat1Bb}$, and $s_{plat1Bc}$.[46] Therefore, from the three profiles, twelve parameters (including both step heights and standard deviations) are obtained (six from Platform A and six from Platform B).

The step height from each profile (in general, $stepN_{XYt}$)[47] is given by:

$$stepN_{XYt} = \left(platNYt - platNXt \right) cal_z , \qquad\qquad (SH15)$$

where t is the data trace (a, b, c, etc.) being examined. For the step indicated by the arrow shown in Fig. SH2(a) from Platform A to Platform B, the equations are:

$$step1_{ABa} = \left(plat1Ba - plat1Aa \right) cal_z , \qquad\qquad (SH16)$$

$$step1_{ABb} = \left(plat1Bb - plat1Ab \right) cal_z , \text{ and} \qquad\qquad (SH17)$$

$$step1_{ABc} = \left(plat1Bc - plat1Ac \right) cal_z . \qquad\qquad (SH18)$$

The step height, $stepN_{XY}$, is the average of the values from the different surface profiles as given below:

---

[45] Consult the Definition of Symbols Section for the nomenclature used for $platNXt$ and $s_{platNXt}$.

[46] These are uncalibrated values to match the inputs to Data Analysis Sheet SH.1.a.

[47] Consult the Definition of Symbols Section.

$$stepN_{XY} = \frac{stepN_{XYa} + stepN_{XYb} + stepN_{XYc}}{3} \ . \tag{SH19}$$

For the step shown in Fig. SH2(a), the step height, $step1_{AB}$, is:

$$step\ 1_{AB} = \frac{step\ 1_{ABa} + step\ 1_{ABb} + step\ 1_{ABc}}{3} \ . \tag{SH20}$$

The calculation of the combined standard uncertainty, u_{cSH}, for $stepN_{XY}$ is described next.

5.4 Step Height Uncertainty Analysis

In this section, uncertainty equations are presented for use with step height. The first uncertainty equation (presented in Sec. 5.4.1) is used with the MEMS 5-in-1 [4]. An earlier equation [40] used in the round robin experiment is presented in Sec. 5.4.2.

5.4.1 Step Height Uncertainty Analysis for the MEMS 5-in-1

For the MEMS 5-in-1, the combined standard uncertainty, u_{cSH1a}, for step height measurements with eight uncertainty components is given by the following equation:

$$u_{cSH1a} = \sqrt{u_{Lstep}^2 + u_{Wstep}^2 + u_{cert}^2 + u_{cal}^2 + u_{repeat(shs)}^2 + u_{drift}^2 + u_{linear}^2 + u_{repeat(samp)}^2} \ , \tag{SH21}$$

where a number or a number and a letter following the subscript "SH" in "u_{cSH}" indicates the data analysis sheet that is used to obtain the combined standard uncertainty value. Therefore, u_{cSH1a} implies that Data Analysis Sheet SH.1.a [4,13] is used. In this equation,

- u_{Lstep} is the uncertainty of the measurement across the length of the step where the length is measured perpendicular to the edge of the step,[48]
- u_{Wstep} is the variation in measured step height values sampled across the width of the step,
- u_{cert} is the uncertainty of the value of the physical step height standard used for calibration,
- u_{cal} is the uncertainty of the measurements taken across the physical step height standard,
- $u_{repeat(shs)}$ is due to the *repeatability* of measurements on the physical step height standard,
- u_{drift} is the uncertainty due to the amount of drift during the data session,
- u_{linear} is the uncertainty of a measurement due to the deviation from height linearity of the data scan, and
- $u_{repeat(samp)}$ is the uncertainty of step height *repeatability* measurements taken on test structures processed similarly to the one being measured.

Table SH3 provides the equations for the uncertainty components. In determining the combined standard uncertainty, a Type B evaluation [19-21] (i.e., one that uses means other than the statistical Type A analysis) is used for each source of uncertainty, except where noted. This table can be referenced as each component is discussed.

The first uncertainty component in Eq. (SH21) and listed in Table SH3 is u_{Lstep}. This uncertainty is due to platforms that are not level with respect to the reference platform, as seen in a surface profile. [Fig. SH2(c) is an example of such a profile.] To estimate this component of uncertainty, the standard deviations along each platform in each surface profile are obtained (for example, $s_{plat1Xa}$, $s_{plat1Xb}$, and $s_{plat1Xc}$ for Platform X and $s_{plat1Ya}$, $s_{plat1Yb}$, and $s_{plat1Yc}$ for Platform Y). For each platform, the standard deviations from the different profiles are averaged together then calibrated as given in the following equations:

---

[48] Stated differently, u_{Lstep} is the uncertainty in the step height due to the variation in the topography along each platform (that is not due to roughness).

$$s_{platNXave} = \left(\frac{s_{platNXa} + s_{platNXb} + s_{platNXc}}{3} \right) cal_z \quad \text{and} \tag{SH22}$$

$$s_{platNYave} = \left(\frac{s_{platNYa} + s_{platNYb} + s_{platNYc}}{3} \right) cal_z . \tag{SH23}$$

The equation for u_{Lstep} would then be obtained by adding $s_{platNXave}$ and $s_{platNYave}$ in quadrature. However, the fine scale roughness (hidden within the above standard deviations) should be discounted from the uncertainty calculation of u_{Lstep} because the roughness should not effect the step height value. For this analysis, the roughness is assumed to be the smallest of all the standard deviations obtained for a given surface material. In other words, the roughnesses $s_{roughNX}$ and $s_{roughNY}$ are the uncalibrated surface roughnesses of *platNX* and *platNY*, respectively, and are equated with the smallest of all the values obtained for $s_{platNXt}$ and $s_{platNYt}$, respectively, as given below:

$$s_{roughNX} = MIN(s_{platNXa}, s_{platNXb}, s_{platNXc}) \quad \text{and} \tag{SH24}$$

$$s_{roughNY} = MIN(s_{platNYa}, s_{platNYb}, s_{platNYc}) . \tag{SH25}$$

However, if the surfaces of *platNX*, *platNY*, and *platNr* all have identical compositions, then $s_{roughNX}$ equals $s_{roughNY}$, which equals the smallest of all the values obtained for $s_{platNXt}$, $s_{platNYt}$, and $s_{platNrDt}$.[49] The surface roughness of each platform is then subtracted quadratically in the calculation of u_{Lstep} to obtain the equation given in the fourth column of Table SH3,[50] which assumes a Gaussian distribution.

Table SH3. Step Height Uncertainty Equations for the MEMS 5-in-1 [4.10][a]

	stepN$_{XYmin}$ **and** stepN$_{XYmax}$	**G or U**[b] **/A or B**[c]	**equation**
1. u$_{Lstep}$	–	G / B	$u_{Lstep} = \sqrt{[s^2_{platNXave} - (s_{roughNX}cal_z)^2] + [s^2_{platNYave} - (s_{roughNY}cal_z)^2]}$
2. u$_{Wstep}$	–	G / A	$u_{Wstep} = \sigma_{Wstep} = STDEV(stepN_{XYa}, stepN_{XYb}, stepN_{XYc})$
3. u$_{cert}$	–	G / B	$u_{cert} = \dfrac{\sigma_{cert}}{cert} \lvert stepN_{XY} \rvert$
4. u$_{cal}$	–	G / A	$u_{cal} = \dfrac{\sigma_{6ave}}{\overline{z}_{6ave}} \lvert stepN_{XY} \rvert$
5. u$_{repeat(shs)}$	$stepN_{XY\,min} = \lvert stepN_{XY} \rvert - 3\lvert stepN_{XY} \rvert \dfrac{\sigma_{6same}}{\overline{z}_{6same}}$ $stepN_{XY\,max} = \lvert stepN_{XY} \rvert + 3\lvert stepN_{XY} \rvert \dfrac{\sigma_{6same}}{\overline{z}_{6same}}$	G / A	$u_{repeat(shs)} = \dfrac{stepN_{XY\,max} - stepN_{XY\,min}}{6} = \dfrac{\sigma_{6same}}{\overline{z}_{6same}} \lvert stepN_{XY} \rvert$
6. u$_{drift}$	$stepN_{XY\,min} = \lvert stepN_{XY} \rvert - \lvert stepN_{XY} \rvert \dfrac{z_{drift}cal_z}{2cert}$	U / B	$u_{drift} = \dfrac{stepN_{XY\,max} - stepN_{XY\,min}}{2\sqrt{3}}$

---

[49] Consult the Definition of Symbols Section as needed.

[50] This equation is different than the one presented in [10].

	$stepN_{XY\,max} = \left|stepN_{XY}\right| + \left|stepN_{XY}\right|\dfrac{z_{drift}cal_z}{2cert}$		$= \dfrac{z_{drift}cal_z}{2\sqrt{3}cert}\left|stepN_{XY}\right|$
7. u_{linear}	$stepN_{XY\,min} = \left|stepN_{XY}\right| - \left|stepN_{XY}\right|z_{lin}$ $stepN_{XY\,max} = \left|stepN_{XY}\right| + \left|stepN_{XY}\right|z_{lin}$	U / B	$u_{linear} = \dfrac{stepN_{XY\,max} - stepN_{XY\,min}}{2\sqrt{3}}$ $= \dfrac{z_{lin}}{\sqrt{3}}\left|stepN_{XY}\right|$
8. $u_{repeat(samp)}$	–	G / A	$u_{repeat(samp)} = \sigma_{repeat(samp)}\left|stepN_{XY}\right|$

[a] Refer to the Definition of Symbols Section as needed

[b] "G" indicates a Gaussian distribution and "U" indicates a uniform distribution

[c] Type A or Type B analysis

The uncertainty equation for u_{Wstep} is determined from σ_{Wstep}, the calibrated one sigma standard deviation of the step height measurements $stepN_{XYa}$, $stepN_{XYb}$, and $stepN_{XYc}$, as given in the fourth column of Table SH3. This is a statistical Type A component.

The uncertainty equation for u_{cert} is determined from $cert$ (the certified value of the double-sided physical step height standard used for calibration) and σ_{cert} (the certified one sigma uncertainty of the calibrated physical step height standard) as given in Table SH3. The uncertainty of the measured step height is assumed to scale linearly with height. Therefore, u_{cert} is calculated using the equation given in Table SH3.

The uncertainty equation for u_{cal} is determined from σ_{6ave} [the maximum of two uncalibrated values (σ_{before} and σ_{after})] and $\overline{z}_{6ave}$ (the uncalibrated average of the six calibration measurements from which σ_{6ave} is found) as given in Table SH3. The uncertainty of the measured step height is assumed to scale linearly with height. Therefore, u_{cal} is calculated using the equation given in Table SH3.

The uncertainty equation for $u_{repeat(shs)}$ is determined from the minimum and maximum step height values (namely, $stepN_{XYmin}$ and $stepN_{XYmax}$, respectively) as given in Table SH3 where σ_{6same} is the maximum of two uncalibrated values (σ_{same1} and σ_{same2}) and where $\overline{z}_{6same}$ is the uncalibrated average of the six calibration measurements from which σ_{6same} is found. The uncertainty of the measured step height is assumed to scale linearly with height. Assuming a Gaussian distribution (and assuming u_{Lstep}, u_{Wstep}, u_{cert}, u_{cal}, u_{drift}, u_{linear}, and $u_{repeat(samp)}$ equal zero), the value for $stepN_{XY}$ lies between $stepN_{XYmin}$ and $stepN_{XYmax}$. Therefore, $u_{repeat(shs)}$ is calculated using the equation given in Table SH3.[51]

In the same way, u_{drift} is calculated (however a uniform distribution is assumed), resulting in the equation listed in Table SH3, where z_{drift} is the uncalibrated positive difference between the averages of the six calibration measurements taken before and after the data session (at the same location on the physical step height standard used for calibration).

The uncertainty equation for u_{linear} is calculated from the minimum and maximum step height values (namely, $stepN_{XYmin}$ and $stepN_{XYmax}$, respectively) as given in Table SH3, where z_{lin} is the maximum relative deviation from linearity over the instrument's total scan range, as quoted by the instrument manufacturer or as determined in Sec. 1.1.2.2. The uncertainty of the measured step height is assumed to scale linearly with height. Assuming a uniform distribution, u_{linear} can be calculated using the equation given in Table SH3.

The last uncertainty component in Eq. (SH21) is $u_{repeat(samp)}$, the uncertainty of step height _repeatability_ measurements taken on test structures processed similarly to the one being measured. For example, Table 3 in Sec. 1.13 specifies that the relative

---

[51] This equation is different than the one presented in [10].

standard deviation $\sigma_{repeat(samp)}$ is 3.95 % for the measurement *repeatability* for step height test structures, fabricated by a bulk micromachining process similar to that used to fabricate SRM 2494. Therefore, $u_{repeat(samp)}$ is calculated using the equation given in Table SH3.

The expanded uncertainty for step height, U_{SH}, is calculated using the following equation:

$$U_{SH} = ku_{cSH1a} = 2u_{cSH1a} \, ,$$

$$(SH26)$$

where the k value of 2 approximates a 95 % level of confidence.

Reporting results [9-21]. Since it can be assumed that the estimated values of the uncertainty components are either approximately uniformly or Gaussianly distributed with approximate combined standard uncertainty u_{cSH1a}, the step height is believed to lie in the interval $stepN_{XY} \pm u_{cSH1a}$ (expansion factor $k=1$) representing a level of confidence of approximately 68 %.

5.4.2 Previous Step Height Uncertainty Analysis

In this section, an earlier uncertainty equation is presented that was used in the round robin experiment [10]. The equation includes six sources of uncertainty with all other sources of uncertainty considered negligible. This step height uncertainty equation (as calculated in SEMI standard test method MS2-1109 [40]) with six sources of uncertainty is as follows:

$$u_{cSH1} = \sqrt{u_{Lstep}^2 + u_{Wstep}^2 + u_{cert}^2 + u_{repeat(shs)}^2 + u_{drift}^2 + u_{linear}^2} \, ,$$

$$(SH27)$$

where the number following the subscript "SH" in "u_{cSH}" indicates the data analysis sheet that is used to obtain the uncertainty value. Therefore, u_{cSH1} implies that Data Analysis Sheet SH.1 [13] is used. The above equation is basically Eq. (SH21) without the components u_{cal} and $u_{repeat(samp)}$. Also, the calculations of u_{Lstep} and $u_{repeat(shs)}$ in Eq. (SH27) are slightly different from the calculations of u_{Lstep} and $u_{repeat(shs)}$ in Eq. (SH21).

The uncertainty for u_{Lstep} in Eq. (SH27) is calculated using the following equation [10]:

$$u_{Lstep} = \sqrt{\left[s_{platNXave} - (s_{roughNX}cal_z)\right]^2 + \left[s_{platNYave} - (s_{roughNY}cal_z)\right]^2} \, .$$

$$(SH28)$$

The uncertainty equation for $u_{repeat(shs)}$ in Eq. (SH27) is determined from the minimum and maximum step height values (namely, $stepN_{XYmin}$ and $stepN_{XYmax}$, respectively) as given below:

$$stepN_{XY\,min} = \left|stepN_{XY}\right| - \left|stepN_{XY}\right| \frac{z_{repeat(shs)}}{2\bar{z}_6} \, ,$$

$$(SH29)$$

$$stepN_{XY\,max} = \left|stepN_{XY}\right| + \left|stepN_{XY}\right| \frac{z_{repeat(shs)}}{2\bar{z}_6} \, ,$$

$$(SH30)$$

where $z_{repeat(shs)}$ is the maximum of two uncalibrated values; one of which is the positive difference between the minimum and maximum values of the six calibration measurements taken at a single location on the calibration step before the data session and the other is the positive difference between the minimum and maximum values of the six calibration measurements taken at this same location after the data session and where $\bar{z}_6$ is the uncalibrated average of the six calibration measurements from which $z_{repeat(shs)}$ is found. The uncertainty of the measured step height is assumed to scale linearly with height. Assuming a uniform distribution (and assuming u_{Lstep}, u_{Wstep}, u_{cert}, u_{drift}, and u_{linear} equal zero), the value for $stepN_{XY}$ lies between $stepN_{XYmin}$ and $stepN_{XYmax}$. Therefore, $u_{repeat(shs)}$ is calculated using the following equation:

$$u_{repeat\,(shs)} = \frac{stepN_{XY\,max} - stepN_{XY\,min}}{2\sqrt{3}} = \frac{z_{repeat\,(shs)}}{2\sqrt{3}\,\bar{z}_6}\left|stepN_{XY}\right|$$

(SH31)

Table SH4 gives example values for each of the uncertainty components in Eq. (SH27) as well as the value for the combined standard uncertainty, u_{cSH1}.

Table SH4. *Example Step Height Uncertainty Values*
From a Round Robin Bulk-Micromachined CMOS Chip

	source of uncertainty or descriptor	uncertainty values
1. u_{Lstep}	variations across the length of the step	**0.011 µm** (using $s_{platNXave}$=0.0118 µm, $s_{platNYave}$=0.0102 µm, and $s_{roughNX}$=$s_{roughNY}$=0.0036 µm)
2. u_{Wstep}	variations across the width of the step	**0.0073 µm** (using $stepN_{XYa}$=0.4928 µm, $stepN_{XYb}$=0.4814 µm, and $stepN_{XYc}$=0.4949 µm)
3. u_{cert}	certified value of physical step height standard used for calibration	**0.0041 µm** (using $cert$=9.887 µm, σ_{cert}=0.083 µm, and $stepN_{XY}$=0.490 µm)
4. $u_{repeat(shs)}$	*repeatability* of measurement on physical step height standard	**0.00034 µm** (using $z_{repeat(shs)}$ =0.024 µm, $\bar{z}_6$ =9.876 µm, and cal_z=1.00031)
5. u_{drift}	drift during data session	**0.00023 µm** (using z_{drift}=0.016 µm and $cert$=9.887 µm)
6. u_{linear}	height linearity of data scan	**0.0028 µm** (using z_{perc}=1.0 %)
u_{cSH1} [a]	uncertainty of step height measurement	**0.014 µm** $=\sqrt{u_{Lstep}^2 + u_{Wstep}^2 + u_{cert}^2 + u_{repeat(shs)}^2 + u_{drift}^2 + u_{linear}^2}$

[a] This u_{cSH1} uncertainty (times 3) is associated with the first *repeatability* data point plotted in Fig. SH8 and Fig. SH9 (for TS1 in Quad 1).

5.5 Step Height Round Robin Results

The round robin *repeatability* and *reproducibility* results are given in this section for step height measurements. The *repeatability* data were taken in one laboratory using a stroboscopic interferometer operated in the static mode (see Sec. 1.1.2). A bulk-micromachined CMOS test chip with step height test structures arranged in quads, as shown in Fig. SH7 and processed in a similar way to the MEMS 5-in-1 chip, was used in the round robin. Four step height measurements were taken from each of the four test structures in each of three quads. Therefore, 48 step height measurements were obtained where each step height

105

measurement is the average of three measurements taken at the different positions (a, b, c) somewhat evenly spaced along the width of the step as specified in Sec. 5.3.

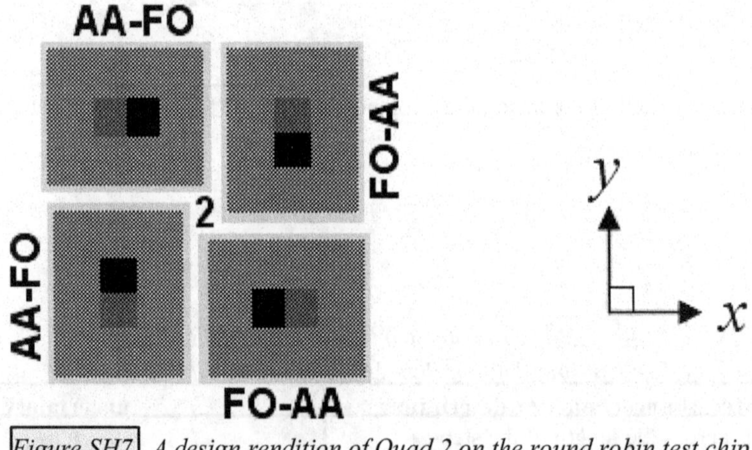

Figure SH7. A design rendition of Quad 2 on the round robin test chip

For the *reproducibility* data, seven participants were identified and each participant was supplied with a round robin test chip. The participant was asked to obtain the step height from any two test structures in the first of the three quads of step height test structures. Following SEMI standard test method MS2-1109 [40] for step height measurements, the raw, uncalibrated measurements were recorded on Data Analysis Sheet SH.1 [13].

Table SH5 presents the step height *repeatability* and *reproducibility* results. In this table, n indicates the number of calculated step height values. The averages (namely $|stepN_{ABave}|$) of the *repeatability* and *reproducibility* measurement results are listed next followed by $\sigma_{repeat(samp)}$, which is the relative standard deviation of the *repeatability* step height measurements. Then, $2\sigma_{stepNAB}$ is given, which is 2.0 times the standard deviation of the *repeatability* or *reproducibility* measurements. Below this, the average of the *repeatability* or *reproducibility* combined standard uncertainty values (u_{cSHave}) for different calculations are presented. (It is interesting in comparing the $2\sigma_{stepNAB}$ values for $|stepN_{AB}|$ that the *repeatability* value is larger than the *reproducibility* value. The reason for this anomaly is a challenge to explain.)

Table SH5. Step Height Measurement Results

	Repeatability results	Reproducibility results
n	48	14
$\mathbf{|stepN_{ABave}|}$	0.477 μm	0.481 μm
$\sigma_{repeat(samp)}$	3.95 %	–
$2\sigma_{stepNAB}$	7.9 %	6.2 %
$\mathbf{u}_{cSH1ave}$ [a]	0.014 μm (3.0 %)	0.014 μm (3.0 %)
$\mathbf{u}_{cSH1aave}$ [b]	0.024 μm (5.0 %)	–

[a] As determined using Eq. (SH27)

[b] As determined using Eq. (SH21)

Figs. SH8 and SH9 are plots of the step height round robin results. In these figures, the *repeatability* data are plotted first, after which the results from the seven participants [52] are plotted. At the top of these figures, $|stepN_{ABave}|$ and $3u_{cSHave}$ (as obtained or derived from Table SH5) are specified for the *repeatability* data. The values for $|stepN_{ABave}| \pm 3u_{cSHave}$ are also plotted in these

---

[52] Participant #2 provided stylus profilometer results using average roughness values instead of standard deviation values (because that instrument did not provide standard deviation values). Therefore, in the analysis for that laboratory average roughness values were inserted into the data sheet for analysis as opposed to standard deviation values.

figures with both the *repeatability* and *reproducibility* data. [53] As an observation, all of the *reproducibility* results fall comfortably between the *repeatability* bounds of $|stepN_{ABave}| \pm 3u_{cSHave}$.

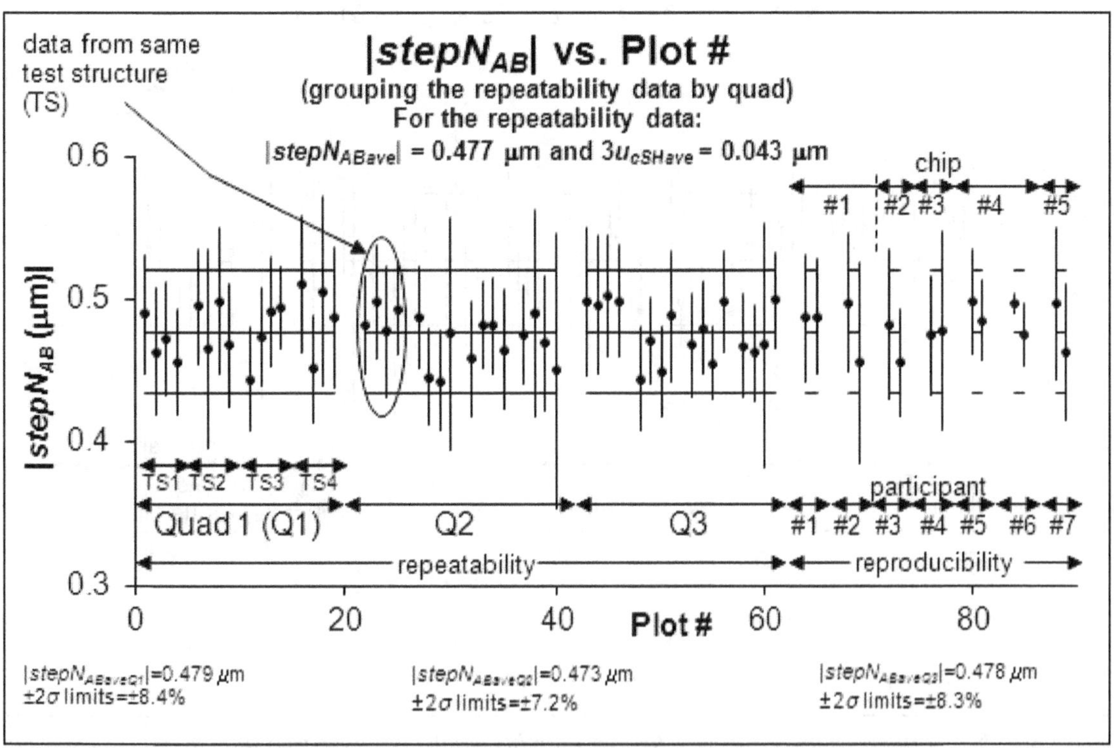

Figure SH8. Step height round robin results with the repeatability results grouped according to quad. [54]

Table SH6. Step Height Repeatability Data Grouped by Quad

	Q1	Q2	Q3
n	16	16	16
$|stepN_{ABave}|$	0.479 µm	0.473 µm	0.478 µm
$\sigma_{repeat(samp)}$	4.2 %	3.6 %	4.2 %
$\pm2\sigma$ limits for $|stepN_{AB}|$	±8.4 %	±7.2 %	±8.3 %
$u_{cSH1ave}$ [a]	0.015 µm (3.1 %)	0.015 µm (3.2 %)	0.013 µm (2.8 %)
$u_{cSH1aave}$ [b]	0.025 µm (5.2 %)	0.023 µm (4.8 %)	0.024 µm (5.0 %)

[a] As determined using Eq. (SH27)

[b] As determined using Eq. (SH21)

Figure SH8 groups the *repeatability* results by quad number. The results from Quad 1 are plotted first, followed by the results from Quad 2, then the results from Quad 3. Within the results of each quad, the results are grouped according to test structure number[55] with the results from Test Structure 1 plotted first, followed by the results from Test Structure 2, etc. For each quad, the average step height value and the ±2 σ limits for this value are given at the bottom of Fig. SH8 and also in Table SH6. The results among the quads are comparable implying there are no discernable variations in the step height value between neighboring quads.

---

[53] Table SH3 specifies the value of each of the uncertainty components comprising the $3u_{SH1}$ uncertainty bars for the first *repeatability* data point plotted in Fig. SH8 and in Fig. SH9 (for TS1 in Quad 1).

[54] *Republished with permission from Semiconductor Equipment and Materials International, Inc. (SEMI) © 2011.*

[55] The upper left hand step height test structure in a quad, such as shown in Fig. SH1, is called Test Structure 1 and it has a 0° orientation. Test Structure 2 (the upper right test structure) has a 270° orientation, Test Structure 3 (the bottom right test structure) has a 180° orientation, and Test Structure 4 (the bottom left test structure) has a 90° orientation.

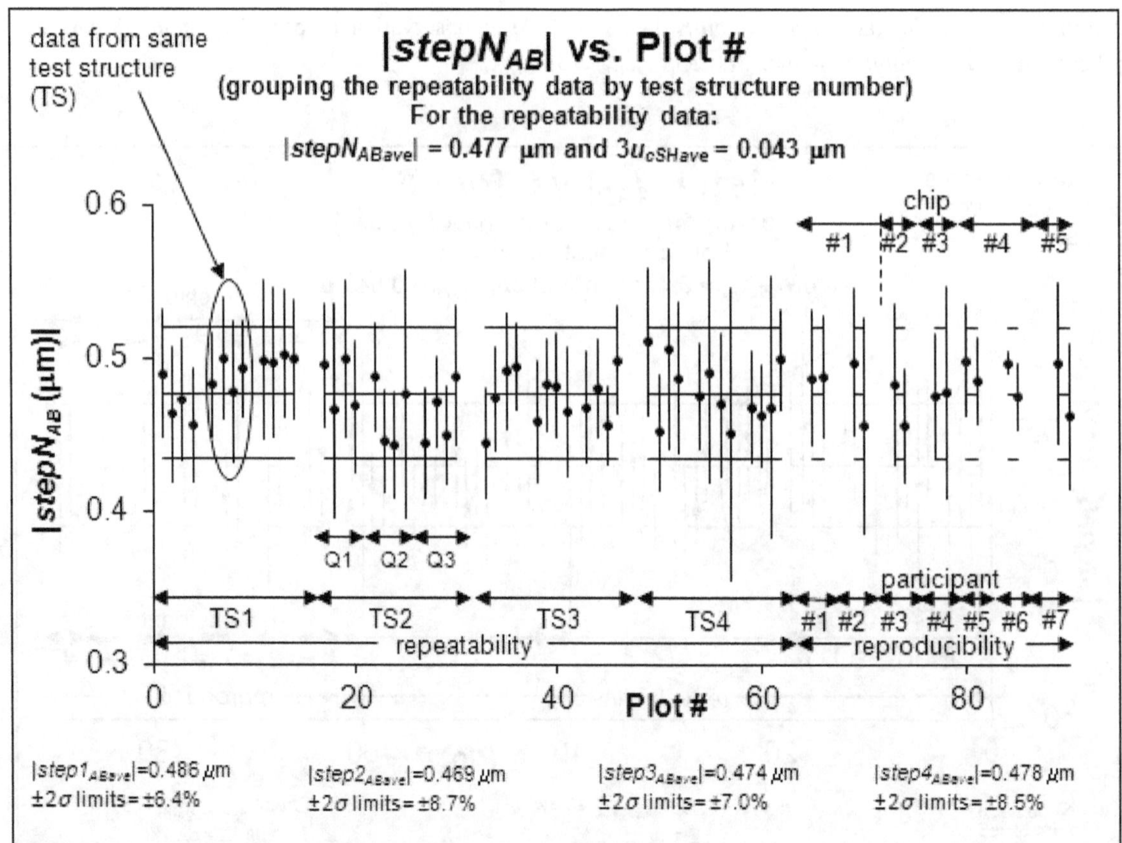

Figure SH9. *Step height round robin results with the repeatability results grouped according to test structure number*[56]

Table SH7. *Step Height Repeatability Data Grouped by Test Structure*

	TS1	TS2	TS3	TS4
n	12	12	12	12
|stepN_ABave|	0.486 µm	0.469 µm	0.474 µm	0.478 µm
$\sigma_{repeat(samp)}$	3.2 %	4.4 %	3.5 %	4.3 %
±2σ limits for |stepN_AB|	±6.4 %	±8.7 %	±7.0 %	±8.5 %
$u_{cSH1ave}$ [a]	0.014 µm (2.8 %)	0.015 µm (3.1 %)	0.011 µm (2.4 %)	0.018 µm (3.7 %)
$u_{cSH1aave}$ [b]	0.021 µm (4.3 %)	0.025 µm (5.4 %)	0.020 µm (4.2 %)	0.027 µm (5.6 %)

[a] As determined using Eq. (SH27)

[b] As determined using Eq. (SH21)

Figure SH9 groups the *repeatability* results by test structure number. The results from Test Structure 1 (TS1) are plotted first, followed by the results from Test Structure 2 (TS2), followed by the results from Test Structure 3 (TS3), then Test Structure 4 (TS4). Within the results for each test structure, the results are grouped according to quad with the results from Quad 1 plotted first, followed by the results from Quad 2, then the results from Quad 3. As in Fig. SH8, for each test structure the average step height value and the ±2 σ limits for this value are given at the bottom of Fig. SH9 and also in Table SH7. It is interesting that TS1 and TS3 (which are rotated ±90° with respect to TS2 and TS4) have comparable ±2 σ limits as do TS2 and TS4; however the ±2σ limits for TS1 and TS3 are slightly less than the ±2 σ limits for TS2 and TS4 when they should be comparable. In addition, there are more variations in the average step height value between rotated test structures (as shown in Fig. SH9 and Table SH7) than variations between quads (as shown in Fig. SH8 and Table SH6).

The platform surfaces involved in the measured steps are not ideal surfaces. Oftentimes they are tilted (even though the data are leveled with respect to the reference platform) and the data jagged. The standard deviations obtained from these surfaces is affected by the precise selection of the analysis regions (including the number of data points within these regions). A more comprehensive determination of the length and width variations may be necessary when dealing with tilt. *Repeatability* might also be improved by calculating the step height from fitted straight lines as described for NIST step height calibrations [42] and outlined in ASTM E 2530 [43]. As given in [42], "For step height measurements, one of several algorithms may be used. For single-sided steps, a straight line is fitted by the method of least squares to each side of the step transition, and the height is calculated from the relative position of these two lines extrapolated to the step edge."

Calibration of the interferometer or comparable instrument in the out-of-plane z-direction is considered mandatory for step height measurements. Without this calibration, a bias to the measurements is expected with the direction and degree of the bias being different for each magnification.

5.6 Using the MEMS 5-in-1 to Verify Step Height Measurements

To compare your in-house step height measurements with NIST measurements, you will need to fill out Data Analysis Sheet SH.1.a. (This data analysis sheet is accessible via the URL specified in the reference [1], a reproduction of which is given in Appendix 4.) After calibrating the instrument, locating the test structure, taking the measurements, and performing the calculations, the data on your completed form can be compared with the data on the SRM Certificate and the completed data analysis sheet supplied with the MEMS 5-in-1. Details of the procedure are given below.

Calibrate the instrument: Calibrate the instrument as given in Sec. 5.2. Obtain the inputs for Table 1 in Data Analysis Sheet SH.1.a.

Locate the step height test structure: In the fourth grouping of test structures on the MEMS 5-in-1 chips shown in Fig. 1 and Fig. 2 for SRM 2494 and SRM 2495, respectively, step height measurements are made. Step height test structures are provided for the purpose, such as shown in Fig. SH1(a) for SRM 2494 and as shown in Fig. SH1(b) for SRM 2495. Specifications for the step height test structures on the chips shown in Fig. 1 and Fig. 2 are given in Table SH1 and Table SH2, respectively.

Data Analysis Sheet SH.1.a requires measurements from one step height test structure on the MEMS 5-in-1 chip. The step height test structure to be measured can be deduced from the data entered on the NIST version of the Data Analysis Sheet SH.1.a, which accompanies the SRM.

For the step height grouping of test structures for SRM 2494, as shown in Fig. SH1(a), the target test structure can be found as follows:
1. The input *which* (i.e., input #4 on Data Analysis Sheet SH.1.a) specifies which of the six step height measurements to take in the fourth grouping of test structures. Using the arrows as a guide in Fig. SH1(a), the first measurement is taken from the first step height test structure, the second and third measurements are taken from the second step height test structure, the fourth measurement is taken from the third step height test structure, and the fifth and sixth measurements are taken from the fourth step height test structure. (Measurements from the fourth step height test structure should be taken with a stylus instrument or instrument not affected by the reflectivity of the sample surface, unless the chip is covered with a smooth reflective material before measurement.)
2. The input *which2* (i.e., input #5) specifies which iteration of the test structure in the set of three possible target test structures of the same design where "first" corresponds to the topmost test structure in the column.

For the step height grouping of test structures for SRM 2495, as shown in Fig. SH1(b), the target test structure can be found as follows:
1. The input *which* (i.e., input #4) indicates which of the five quads depicted in Fig. SH1(b) contains the target step height test structure.
2. The input *orient* (i.e., input #6) is used to locate the target test structure within the selected quad. There are four step height test structures in each quad with each test structure having a different orientation ($0°$, $90°$, $180°$, or $270°$). The upper left hand step height test structure has a $0°$ orientation, the bottom left test structure has a $90°$ orientation, the bottom right test structure has a $180°$ orientation, and the upper right test structure has a $270°$ orientation.

Take the measurements: Following the steps in SEMI standard test method MS2 [4] for measuring step heights, the step height measurements are obtained using the highest magnification objective that is available and feasible (e.g., a $20\times$ objective). The data are leveled and zeroed with respect to the top of the surrounding or partially surrounding reference platform. Three 2-D data traces (Trace a, Trace b, and Trace c, as shown in Fig. SH2 and Fig. SH3) are used to obtain the step height measurement. For the step height test structure in Fig. SH2, given the step of interest pointed to by the arrow in this figure, the platform height measurements for Platforms A and B and standard deviations along Traces a, b, and c are recorded. For the test structure shown in Fig. SH3, the measurements and standard deviations are obtained from both central platforms. Therefore,   twelve measurements are obtained, six from the first platform and six from the second platform. For example, the four quantities calculated from Trace a are *platNXa*, *platNYa*, *s*$_{platNXa}$, and *s*$_{platNYa}$. Analogous quantities are calculated from Traces b and c.

Perform the calculations: Enter the data into Data Analysis Sheet SH.1.a as follows:
1. Press the "Reset this form" button. (One of these buttons is located near the top of the data analysis sheet and the other is located near the middle of the data analysis sheet.)
2. Supply the inputs to Table 1 and Table 2.
3. Press the "Calculate and Verify" button to obtain the results for the step height test structure. (One of these buttons is located near the top of the data analysis sheet and the other is located near the middle of the data analysis sheet.)
4. Verify the data by checking to see that all the pertinent boxes in the verification section at the bottom of the data analysis sheet say "ok". If one or more of the boxes say "wait," address the issue, if necessary, by modifying the inputs and recalculating.
5. Print out the completed data analysis sheet to compare both the inputs and outputs with those on the NIST-supplied data analysis sheet.

Compare the measurements: The Certificate accompanying the MEMS 5-in-1 specifies a step height value (for example, *step1*$_{AB}$) and the expanded uncertainty, U_{SH}, (with k=2) intended to approximate a 95 % level of confidence. It is your responsibility to determine an appropriate criterion for acceptance, such as given below:

$$D_{SH} = \left| step1_{AB(customer)} - step1_{AB} \right| \leq \sqrt{U_{SH(customer)}^2 - U_{SH}^2} \, , \tag{SH32}$$

where D_{SH} is the absolute value of the difference between your step height value, e.g., *step1*$_{AB(customer)}$, and the step height value on the SRM Certificate, *step1*$_{AB}$, and where $U_{SH(customer)}$ is your expanded uncertainty value and U_{SH} is the expanded uncertainty on the SRM Certificate. If your measured value for step height (as obtained in the newly filled out Data Analysis Sheet SH.1.a) satisfies your criterion for acceptance and there are no pertinent   "wait" statements at the bottom of your Data Analysis Sheet SH.1.a, you can consider yourself to be appropriately measuring step height according to the SEMI MS2 step height standard test method [4] according to your criterion for acceptance.

Any questions concerning the measurements, analysis, or comparison can be directed to mems-support@nist.gov.

6   Grouping 5: In-Plane Length

An in-plane length (or deflection) measurement is defined as the experimental determination of the straight-line distance between two transitional edges in a MEMS device. A transitional edge is an edge of a MEMS structure that is characterized by an abrupt change in surface slope. Many times, more precise in-plane length values can be obtained by using the design dimensions as opposed to using measurements taken with an optical interferometric microscope (which typically provides more precise measurements than an optical microscope [14]). Therefore, ASTM standard test method E 2244 [5] on in-plane length measurements is used when measuring in-plane deflections and when measuring lengths in fabrication processes that are being developed.

This section on in-plane length is not meant to replace but to supplement the ASTM standard test method E 2244 [5], which more completely presents the scope, significance, terminology, apparatus, and test structure design as well as the calibration procedure, measurement procedure, calculations, precision and bias data, etc. In this section, the NIST-developed in-plane length test structures on SRM 2494 and SRM 2495, as shown in Fig. 1 and Fig. 2, respectively, in the Introduction are given in Sec. 6.1. Sec. 6.2 discusses the calibration procedure for the in-plane length measurements, and Sec. 6.3 discusses the in-plane length measurement procedure. Following this, the uncertainty analysis is presented in Sec. 6.4, the round robin results are presented in Sec. 6.5, and Sec. 6.6 describes how to use the MEMS 5-in-1 to verify in-plane length measurements.

6.1 In-Plane Length Test Structures

In-plane length measurements are taken in the fifth grouping of test structures, as shown in Fig. L1(a) for SRM 2494 depicted in Fig. 1 and as shown in Fig. L1(b) for SRM 2495 depicted in Fig. 2.

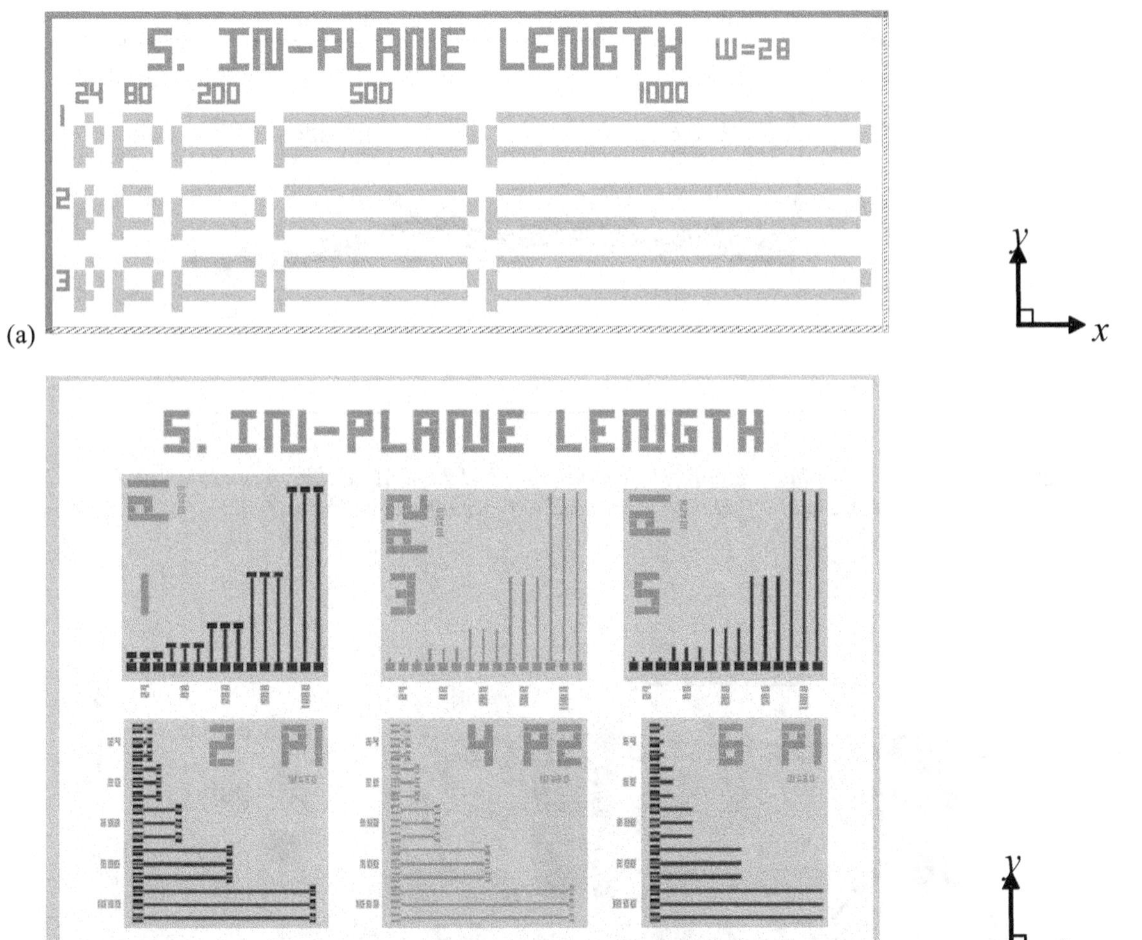

Figure L1. The in-plane length grouping of test structures (a) on SRM 2494, fabricated on a multi-user 1.5 µm CMOS process [8] followed by a bulk-micromachining etch, as depicted in Fig. 1 and (b) on SRM 2495, fabricated using a polysilicon multi-user surface-micromachining MEMS process [9] with a backside etch, as depicted in Fig. 2.

An in-plane length test structure from the grouping shown in Figs. L1(a and b) is given in Fig. L2(a) and Fig. L3(a), respectively, with an applicable data trace taken from these test structures given in Fig. L2(b) and Fig. L3(b) respectively.

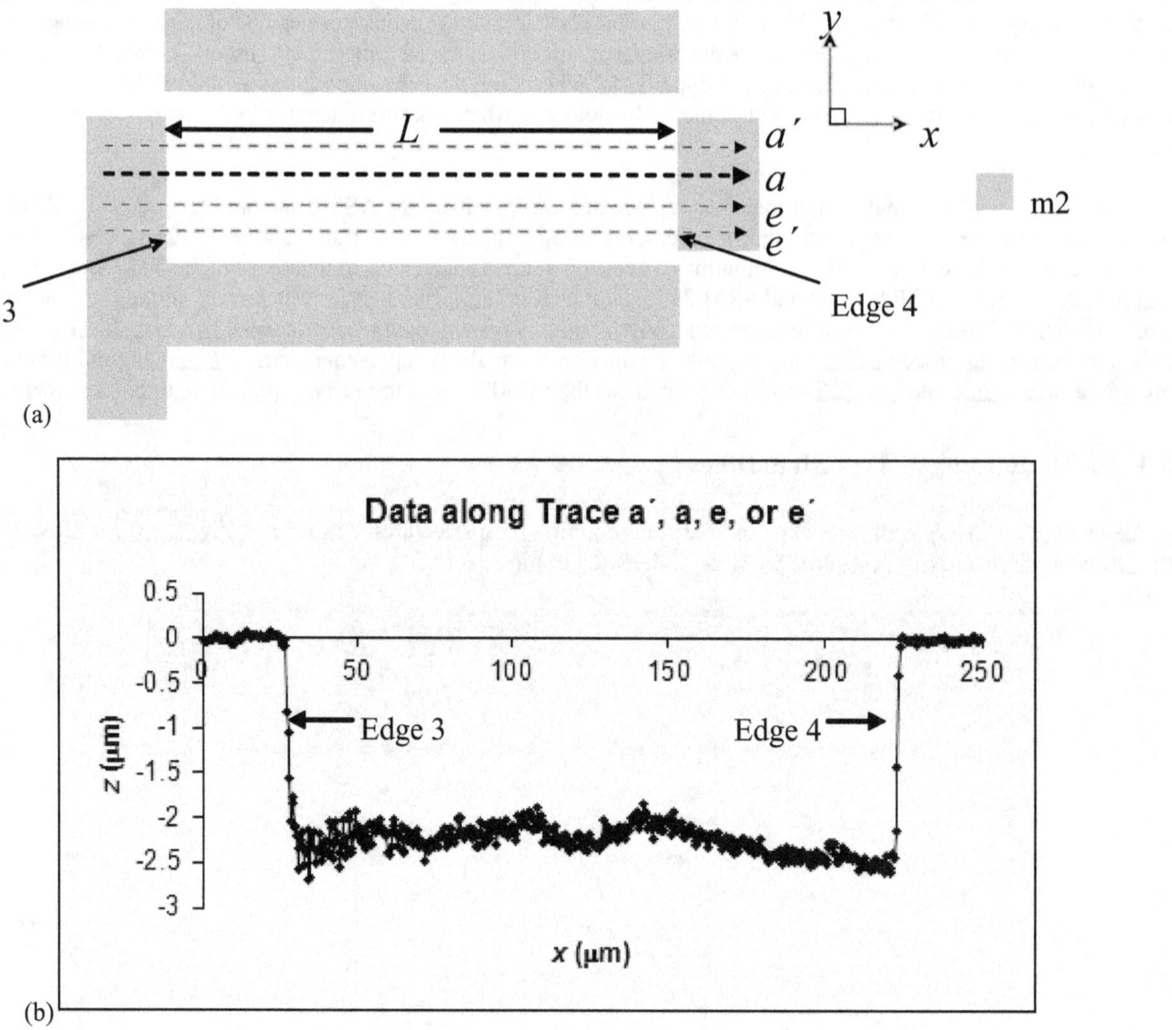

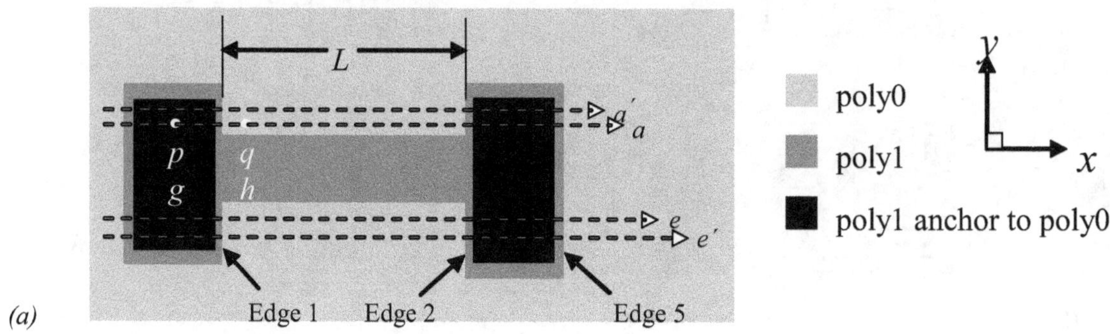

Figure L2. For an in-plane length test structure on SRM 2494, (a) a design rendition and (b) an example of a 2-D data trace used to determine L, as shown in (a).

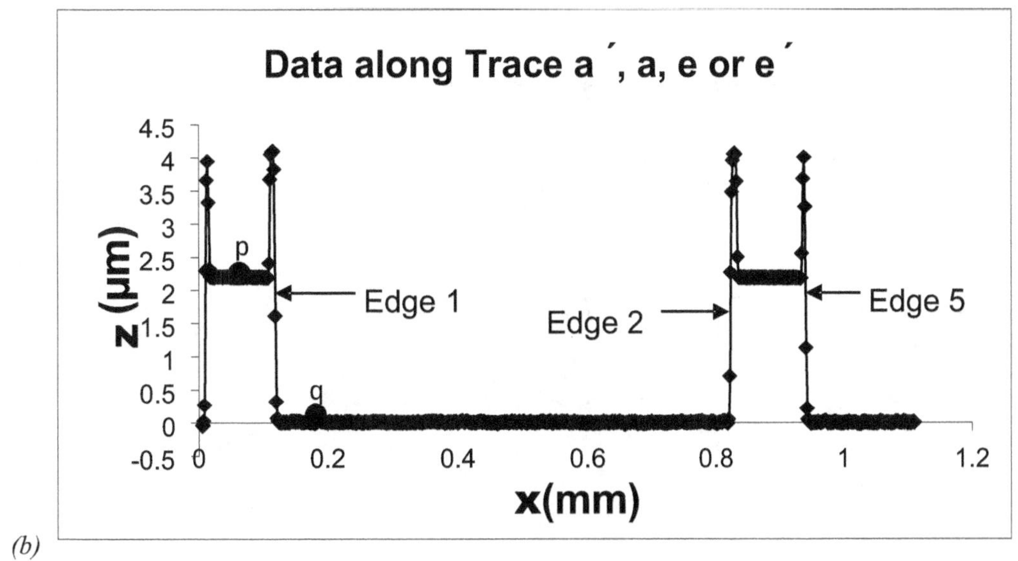

(b)

Figure L3. For a poly1 in-plane length test structure on SRM 2495, (a) a design rendition and (b) an example of a 2-D data trace used to determine L, as shown in (a).

The in-plane length measurements are made with an interferometric microscope. Many interferometric microscopes are purchased with five magnifications (5 ×, 10×, 20×, 40×, and 80 ×). Therefore, for each of these magnifications, an in-plane length test structure is provided in both of the in-plane length groupings of test structures in Figs. L1(a and b). The design length, L_{des}, for the test structure designed for the specified magnification is given in Table L1. This table also includes a maximum field of view for each magnification for a representative c-mount camera. [57] In most cases, L_{des} is at least 70 µm less than this value.

Table L1. Design Lengths on SRM 2494 and SRM 2495 for the Given Magnifications

Magnification	Calibrated Maximum Field of View (in the x-direction)	L_{des}
5×	1165.00 µm	1000 µm
10×	599.998 µm	500 µm
20×	287.00 µm	200 µm
40×	150.000 µm	80 µm
80×	75.0000 µm	24 µm

For SRM 2494: As can be seen in Fig. L1(a), the design length, L_{des}, for the in-plane length test structures is specified at the top of each column of test structures. This is the design length specified in Table L1. The design width is 28 µm. There are three occurrences of each in-plane length test structure, one of which is shown in Fig. L2(a) and Fig. L4. For each test structure, the following three types of in-plane length measurement can be obtained for the given design length:
1. An outside edge-to-outside edge length measurement, as given by L_{oo} in Fig. L4, where Edge 1 and Edge 2 are considered outside edges,
2. An inside edge-to-inside edge length measurement, as given by L_{ii} in Fig. L4, where Edge 3 and Edge 4 are considered inside edges, and
3. An inside edge-to-outside edge length measurement, as given by L_{io} in Fig. L4, where Edge 5 is considered an inside edge and Edge 6 is considered an outside edge. (We will consider this measurement comparable to an outside edge-to-inside edge length measurement, L_{oi}.)

---

[57] For this interferometric microscope, the resolution in the *x*-direction is better than the resolution in the *y*-direction.

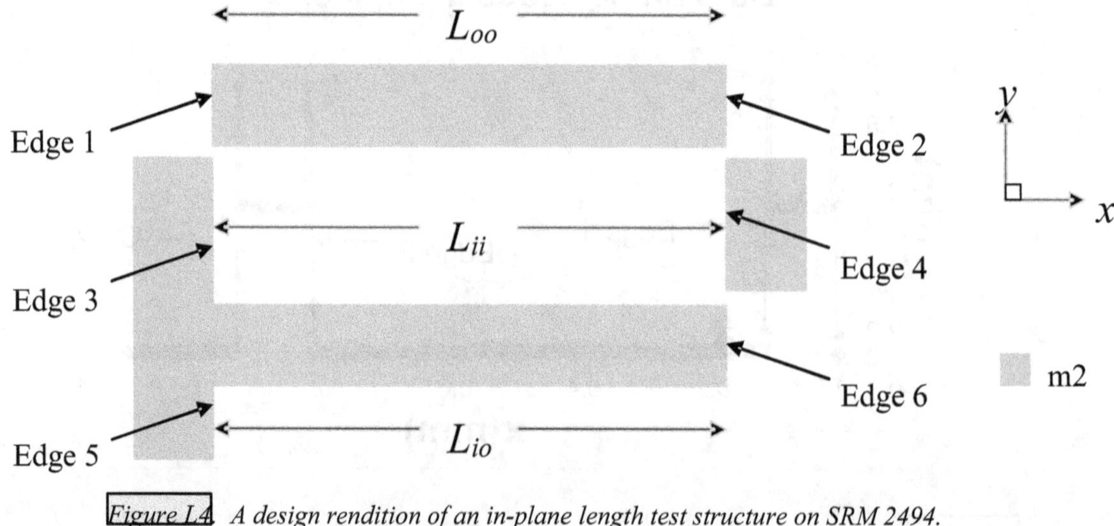

Figure L4. A design rendition of an in-plane length test structure on SRM 2494.

The design layer for the test structure given in Fig. L4 is metal2 (or m2), which is the topmost metal layer in the process. This m2 layer is encompassed in oxide. The 2-D data trace given in Fig. L2(b) shows the existence of two transitional edges assumed to be Edge 3 and Edge 4 in Fig. L2(a). In fact, these edges may be due to the oxide layer as it traverses over the metal2. This would cause an offset, L_{offset}, to the measured m2-to-m2 in-plane length value, L, given in Fig. L2(a) as discussed in Sec. 6.3.

For SRM 2495: As seen in Fig. L1(b), the in-plane length grouping of test structures on SRM 2495 includes both fixed-fixed beams and cantilevers [comprised of either poly1 (or p1) or poly2 (or p2)] as in-plane length test structures. For the measurements supplied on the SRM 2495 Certificates, we will only be concerned with the fixed-fixed beam arrays. There are two p1 fixed-fixed beam arrays and one p2 fixed-fixed beam array. The two p1 arrays of fixed-fixed beams are located on the left hand side of the in-plane length grouping of test structures in Fig. L1(b) and the one p2 array is the middle array in the bottom row. The beams in the bottom p1 array and the p2 array have a 0° orientation and the beams in the top p1 array have a 90° orientation. The fixed-fixed beams in these three arrays are designed with the lengths, L_{des}, specified in Table L1, with three occurrences of each beam. This design length is specified (in micrometers) next to the second occurrence of each test structure. The design width of all these beams is 20 μm. An example poly1 fixed-fixed beam test structure is shown in Fig. L3(a). The in-plane length measurement, L, is taken between the edges of the anchor lips of the fixed-fixed beam [i.e., between Edges 1 and 2 in Fig. L3(a)]. An example 2-D data trace is given in Fig. L3(b).

6.2 Calibration Procedure for In-Plane Length Measurements

For SRM in-plane length measurements, the interferometric microscope is calibrated in the z-direction to obtain the z-calibration factor, cal_z, as specified in Sec. 5.2 for step height calibrations. The interferometric microscope is also calibrated in the x- and y-directions as given below. These calibration procedures are the same as those for residual strain and strain gradient measurements, as indicated in Sec. 3.2 and Sec. 4.2 respectively.

To calibrate the interferometric microscope in the x- and y-directions [2-3,5], a 10 μm grid (or finer grid) ruler is used with each combination of lenses. The following calibration is performed on a yearly basis, or after the instrument has been serviced:

1. The ruler is oriented in the x-direction and $ruler_x$ is recorded as the maximum field of view in the x-direction for the given combination of lenses (as measured on the screen of the interferometric microscope). The value for σ_{xcal} is estimated, where σ_{xcal} is the standard deviation in a ruler measurement in the interferometric microscope's x-direction for the given combination of lenses.

2. The x-calibration factor, cal_x, is calculated using the following equation:

$$cal_x = \frac{ruler_x}{scope_x}, \qquad (\text{L1})$$

where $scope_x$ is the interferometric microscope's maximum field of view in the x-direction for the given combination of lenses (or typically, the maximum x-value obtained from an extracted 2-D data trace).

3. The above two steps are repeated in the y-direction to obtain cal_y.
4. The x- and y-data values obtained during the data session are multiplied by the appropriate calibration factor to obtain calibrated x- and y-data values.

6.3 In-Plane Length Measurement Procedure

In-plane length measurements are taken between two transitional edges, such as Edges 1 and 2 in Fig. L3(a). A transitional edge is an edge of a MEMS structure that is characterized by an abrupt change in surface slope, as seen in Fig. L3(b) for say, Edges 1 and 2. From each of four 2-D data traces, an x-value is obtained at each transitional edge defining the in-plane length measurement. The two outermost data traces [e.g., Traces a′ and e′ in Fig. L3(a)] are typically used for alignment purposes. From these outermost data traces, the y-values are also obtained. The in-plane length is calculated using the acquired x- and y-values.

To obtain an in-plane length measurement, consult the standard test method [5]. Briefly, the following steps are taken for SRM 2495, and a slightly modified procedure [5] is followed for SRM 2494:

1. If the transitional edges that define the in-plane length measurement face each other [such as Edge 1 and Edge 2 in Fig. L3(a)], Traces a′, a, e, and e′, also shown in Fig. L3(a), are extracted from a 3-D data set. The uncalibrated values from Edge 1 and Edge 2 (namely, for $x1_{uppert}$ and $x2_{uppert}$ to be described in the next step) are obtained from each 2-D data trace [one of which is shown in Fig. L3(b)]. The trailing subscript "t" indicates the data trace ($a′$, a, e, or $e′$) being examined. The y values associated with Traces a′ and e′ are also recorded.

2. To find x_{upper}: The upper transitional x-data value, x_{upper}, is found as follows. The x values are examined between Point p and Point q in Figs. L3(a and b). The x-value that most appropriately locates the upper corner of the transitional edge is called x_{upper}, or $x1_{upper}$ in this case since it is associated with Edge 1. At times it is easy to identify one point that accurately locates the upper corner of the transitional edge. In this case, the maximum uncertainty associated with the identification of this point is $\pm n_t x_{res} cal_x$, where $n_t = 1$ and x_{res} is the uncalibrated resolution of the interferometric microscope in the x-direction. This value of $n1_t = 1$ is also recorded. For a less obvious point that locates the upper corner of the transitional edge, the value for $n1_t$ would be larger than one. If $n1_t$ is larger than four, it is recommended that another 2-D data trace be obtained or another 3-D data set.

3. If an in-plane length measurement is determined between transitional edges that face each other, such as Edge 1 and Edge 2 in Figs. L3(a and b), then the measured in-plane length for each 2-D data trace, L_{meast}, is calculated using the following equation:

$$L_{meast} = (x2_{uppert} - x1_{uppert}) cal_x \; , \tag{L2}$$

where a trailing subscript "t" indicates the data trace ($a′$, a, e, or $e′$) being examined. The measured length, L_{meas}, is calculated as follows:

$$L_{meas} = \frac{L_{measa′} + L_{measa} + L_{mease} + L_{mease′}}{4} \; . \tag{L3}$$

To account for misalignment, the aligned length, L_{align}, is calculated using the following equation:

$$L_{align} = L_{meas} \cos\alpha \; , \tag{L4}$$

where the misalignment angle, α, is shown in Fig. L5(a). This misalignment angle is typically determined using the two outermost data traces [$a′$ and $e′$ in this case, as seen in Fig. L3(a)] and is calculated to be either α_1 or α_2 using either $\Delta x1$ or $\Delta x2$, respectively [as seen in Fig. L5(b)]. The equations for $\Delta x1$ and $\Delta x2$ are:

$$\Delta x1 = x1_{uppera′} - x1_{uppere′} \; , \text{and} \tag{L5}$$

$$\Delta x2 = x2_{uppera'} - x2_{uppere'} \quad . \tag{L6}$$

The equation for α is as follows:

$$\alpha = \tan^{-1}\left[\frac{\Delta x\ cal_x}{\Delta y\ cal_y}\right] \quad , \text{ where} \tag{L7}$$

$$\Delta y = y_{a'} - y_{e'} \quad , \tag{L8}$$

and where ya' and ye' are the uncalibrated y-values associated with Traces a′ and e′, respectively. In addition,

$$\text{if } nl_{a'} + nl_{e'} \le n2_{a'} + n2_{e'}, \text{ then } \alpha = \alpha_1 \text{ and } \Delta x = \Delta x1, \tag{L9}$$

$$\text{and if } nl_{a'} + nl_{e'} > n2_{a'} + n2_{e'}, \text{ then } \alpha = \alpha_2 \text{ and } \Delta x = \Delta x2. \tag{L10}$$

The effect of the misalignment angle, α, is expected to be much smaller (almost negligible) for the shorter length measurements (≤ 200 µm at a magnification of 20×) because for these measurements it is easier to visually align the sample within the field of view of the interferometric microscope before taking a 3-D data set than it is for the longer length measurements (≥ 500 µm at a magnification of 10×).

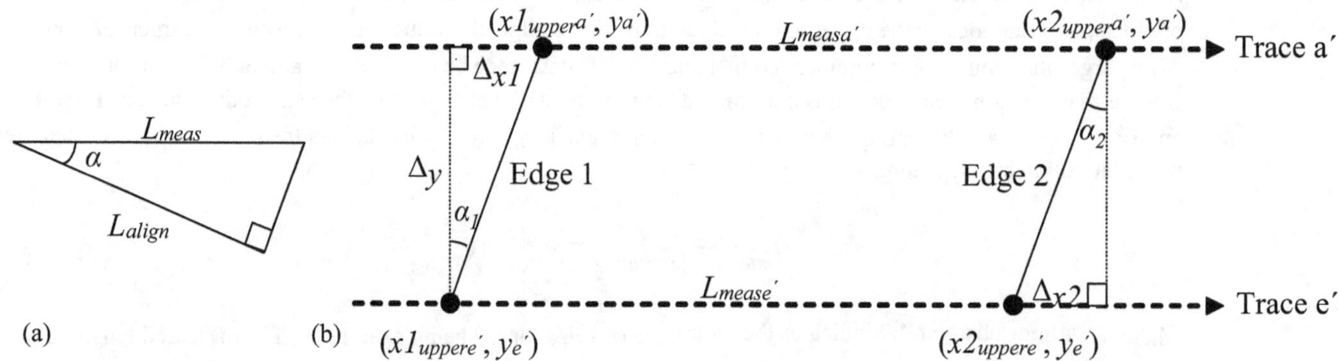

Figure L5. Drawings depicting (a) the misalignment angle, α, and (b) the misalignment between the 2-D data traces a′ and e′ and Edges 1 and 2. In drawing (b) it is assumed the x- and y-values are calibrated.

The equation for the in-plane length, L, is a follows:

$$L = L_{align} + L_{offset} \quad , \tag{L11}$$

where L_{offset} is the in-plane length correction term for the given type of in-plane length measurement [for example, for an inside edge-to-inside edge length measurement, L_{ii} as seen in Fig. (L4)] on similar structures, when using similar calculations, and for a given magnification of a given interferometric microscope. As specified in Sec. 6.1 for SRM 2494, Edges 3 and 4 in Fig. L2(b) may be due to the oxide covering the m2 layer in Fig. L2(a), in which case L_{offset} is used as a correction term to provide a measurement between the two m2 lines in Fig. L2(a). In addition, the interferometric software may treat data at transitional edges differently (for example, if neighboring

116

data points are averaged together) which could add to the offset of an in-plane length measurement. The determination of L_{offset} is discussed in the next step.

4. To determine L_{offset} (mentioned above) for a given magnification of a given interferometric microscope for a given structure in a given process, at least twelve 3-D data sets are obtained and L_{align} is calculated as given above. The average of these twelve or more measurements, $L_{alignave}$, is subtracted from L_{des} to find L_{offset}, where L_{des} is the design length. An alternate measurement of L_{offset} is determined from four measurements of L_{align} for both L_{ii} and L_{oo} in Fig. L4 using four 3-D data sets. The average of these values (namely, $L_{iialignave}$ and $L_{ooalignave}$, respectively) is determined. Then L_{offset} is calculated as $(L_{ooalignave} - L_{iialignave})/2$.

5. If the transitional edges that define the in-plane length measurement are oriented in the same direction and have similar slopes and magnitudes [such as Edge 1 and Edge 5 in Figs. L3(a and b)], Traces a′, a, e, and e′ are extracted from a 3-D data set as specified in step 1. However, for each data trace, a measured x-value is obtained at the upper portion of each transitional edge (x_{upper}) as specified in step 2 or a measured x-value is obtained at the lower portion of each transitional edge (x_{lower}) as also specified in step 2 but replacing "upper" with "lower." The upper values are used unless the lower values are easier to locate. (In other words, the upper values are used unless $n1_t + n5_t$ for the lower values are typically smaller than those for the upper values.) To find the in-plane length, the equations are similar to those used above when the in-plane length measurement is taken between transitional edges that face each other; however, when and if appropriate the lower values replace the upper values and the pertinent edges are referenced. Due to the similarities of the edges involved when the transitional edges that define the in-plane length measurement face the same direction, a length correction term, L_{offset}, is not needed and the in-plane length, L, is equated with L_{align}, as given in Eq. (L4).

6.4 In-Plane Length Uncertainty Analysis

In this section, three uncertainty equations are presented for use with in-plane length. The first one (presented in Sec. 6.4.1) is used for the MEMS 5-in-1. The other two (presented in Sec. 6.4.2) were used in the round robin experiment and other previous work.

6.4.1 In-Plane Length Uncertainty Analysis for the MEMS 5-in-1

This section presents the six-term combined standard uncertainty equation [5] used with the MEMS 5-in-1. The six sources of uncertainty are a) the uncertainty of the in-plane length measurement (u_L), b) the uncertainty due to four measurements taken on the test structure at different locations ($u_{repeat(L)}$), c) the uncertainty of the calibration of the interferometric microscope in the x-direction (u_{xcal}), d) the alignment uncertainty (u_{align}), e) the uncertainty of the value for L_{offset} (u_{offset}), and f) the uncertainty due to the *repeatability* of similar measurements taken on test structures processed similarly to the one being measured ($u_{repeat(samp)}$). The combined standard uncertainty equation [5] can be written as follows:

$$u_{cL0} = \sqrt{u_L^2 + u_{repeat(L)}^2 + u_{xcal}^2 + u_{align}^2 + u_{offset}^2 + u_{repeat(samp)}^2} \; , \tag{L12}$$

where a number (or a number and a letter) following the subscript "L" in "u_{cL}" indicates the data analysis sheet that is used to obtain the combined standard uncertainty value. Therefore, u_{cL0} implies that Data Analysis Sheet L.0 [5,13] is used. In determining the combined standard uncertainty, a statistical Type A analysis is used to obtain $u_{repeat(L)}$ and $u_{repeat(samp)}$. The other sources of uncertainty are obtained using a Type B evaluation [19-21] (i.e., one that uses means other than the statistical Type A analysis).

Table L2 provides the equations for the uncertainty components. The first uncertainty component in Eq. (L12) and listed in Table L2 is u_L. The value for u_L is determined from the minimum and maximum length values (namely, L_{minL} and L_{maxL}) as calculated using the equations given in Table L2. Assuming a 99.7 % nominal probability of coverage, the range of values for L due to this component is $L_{maxL} - L_{minL}$. Further, if a Gaussian probability distribution is assumed, u_L is calculated using the formula given in the last column of Table L2, which is repeated below:

$$u_L = \frac{L_{maxL} - L_{minL}}{6} \qquad\qquad (L13)$$

It is assumed that the uncertainty associated with the pixel resolution in the x-direction is incorporated within this uncertainty component.

The second uncertainty component in Eq. (L12) and listed in Table L2 is $u_{repeat(L)}$. This uncertainty component is determined from $\sigma_{repeat(L)}$, the calibrated one sigma standard deviation of the in-plane length measurements $L_{measa'}$, L_{measa}, L_{mease}, and $L_{mease'}$. The equation used to calculate $u_{repeat(L)}$ is given in the last column of Table L2. This is a statistical Type A component.

Table L2. In-Plane Length Uncertainty Equations for the MEMS 5-in-1 [5]

	L_{min}	L_{max}	G or U[a] / A or B[b]	equation
1. u_L	$L_{minL} = L_{measmin}\cos\alpha + L_{offset}$ where $L_{measmin} = (L_{measmina'} + L_{measmina} + L_{measmine} + L_{measmine'})/4$ $L_{measmint} = L_{meast} - (n1_t + n2_t)x_{res}cal_x$	$L_{maxL} = L_{measmax}\cos\alpha + L_{offset}$ where $L_{measmax} = (L_{measmaxa'} + L_{measmaxa} + L_{measmaxe} + L_{measmaxe'})/4$ $L_{measmaxt} = L_{meast} + (n1_t + n2_t)x_{res}cal_x$	G / B	$u_L = \dfrac{L_{maxL} - L_{minL}}{6}$
2. $u_{repeat(L)}$	–	–	G / A	$u_{repeat(L)} = \sigma_{repeat(L)}\cos\alpha$ $= STDEV(L_{measa'}, L_{measa}, L_{mease}, L_{mease'})\cos\alpha$
3. u_{xcal}	–	–	G / B	$u_{xcal} \approx \left(\dfrac{\sigma_{xcal}}{ruler_x}\right)L_{meas}\cos\alpha$
4. u_{align}	$L_{minalign} = L_{meas}\cos\alpha_{min} + L_{offset}$ using Eq. (L14) for α_{min}	$L_{maxalign} = L_{meas}\cos\alpha_{max} + L_{offset}$ using Eq. (L15) for α_{max}	U / B	$u_{align} = \left|\dfrac{L_{maxalign} - L_{minalign}}{2\sqrt{3}}\right|$
5. u_{offset}	–	–	G / B	$u_{offset} = \dfrac{|L_{offset}|}{3}$
6. $u_{repeat(samp)}$	–	–	G / A	$u_{repeat(samp)} = \sigma_{repeat(samp)'}$

[a] "G" indicates a Gaussian distribution and "U" indicates a uniform distribution

[b] Type A or Type B analysis

The third uncertainty component in Eq. (L12) and listed in Table L2 is u_{xcal}. This uncertainty component is determined from the estimated value of σ_{xcal} obtained in Sec. 6.2 and is assumed to scale linearly with the aligned length (i.e., $L_{meas}\cos\alpha$) as given in Table L2. This uncertainty component assumes that the uncertainty of the calibration is due solely to the uncertainty of the calibration in the x-direction (in other words, the effect of σ_{ycal} on the misalignment angle, α, is considered negligible). Similarly, it is assumed that the effect of σ_{xcal} on the misalignment angle, α, is considered negligible.

The fourth uncertainty component in Eq. (L12) and listed in Table L2 is u_{align}. This uncertainty component is determined from the minimum and maximum length values (namely, $L_{minalign}$ and $L_{maxalign}$) given in this table resulting from calculated minimum and maximum values for the misalignment angle, α. These equations for α_{min} and α_{max} are given below:

$$\alpha_{min} = \tan^{-1}\left[\frac{\Delta x\ cal_x}{\Delta y\ cal_y} - \frac{2x_{res}\ cal_x}{\Delta y\ cal_y}\right] , \tag{L14}$$

$$\alpha_{max} = \tan^{-1}\left[\frac{\Delta x\ cal_x}{\Delta y\ cal_y} + \frac{2x_{res}\ cal_x}{\Delta y\ cal_y}\right] , \tag{L15}$$

where Δy is given in Eq. (L8) and Δx is equated with either $\Delta x1$ or $\Delta x2$ in Eq. (L5) or Eq. (L6), respectively, as determined in Eq. (L9) or Eq. (L10). A uniform probability distribution is assumed in the calculation of u_{align}.

The fifth uncertainty component in Eq. (L12) and listed in Table L2 is u_{offset}, which is assumed to be equivalent to $|L_{offset}|$, as determined in Sec. 6.3, divided by three. Besides the uncertainty of the value of L_{offset}, u_{offset} is also assumed to incorporate geometrical uncertainties along the applicable edges. For in-plane length measurements when the transitional edges face the same direction, it is assumed that $L_{offset} = u_{offset} = 0$.

The sixth uncertainty component in Eq. (L12) and listed in Table L2 is $u_{repeat(samp)}$, which is equated with $\sigma_{repeat(samp)}{}'$, the *repeatability* standard deviation for in-plane length test structures. This value of $\sigma_{repeat(samp)}{}'$ is found from at least twelve 3-D data sets from which twelve values of L_{align} are calculated as given in Sec. 6.3. The standard deviation of the twelve or more measurements of L_{align} is equated with $\sigma_{repeat(samp)}{}'$. Table 3 in Sec. 1.13 specifies that $\sigma_{repeat(samp)}{}'$ for test structures with edges that face each other, with an approximate design length of 200 μm, and fabricated by a bulk micromachining process similar to that used to fabricate SRM 2494 is 1.1565 μm when using a magnification of 25$\times$.

The expanded uncertainty for in-plane length, U_L, is calculated using the following equation:

$$U_L = ku_{cL0} = 2u_{cL0} , \tag{L16}$$

where the k value of 2 yields an interval that approximates a 95 % confidence interval.

Reporting results [9-21]. Since it can be assumed that the estimated values of the uncertainty components are either approximately uniformly or Gaussianly distributed with approximate combined standard uncertainty u_{cL0}, the in-plane length is believed to lie in the interval $L \pm u_{cL0}$ (expansion factor $k=1$) representing a level of confidence of approximately 68 %.

6.4.2 Previous In-Plane Length Uncertainty Analyses

Two previous versions of the combined standard uncertainty equation exist. For these versions, the in-plane length, L, is calculated using the following equation:

$$L = L_{meas} = \frac{L_{minL} + L_{maxL}}{2} , \tag{L17}$$

where L_{minL} and L_{maxL} are specified in a following paragraph. In addition, alignment is ensured (such that $\alpha = 0$) and L is determined from one data trace.

The first combined standard uncertainty equation is used with the round robin results that are presented in the next section. The following three term uncertainty equation is used [44]:

$$u_{cL1} = \sqrt{u_L^2 + u_{xcal}^2 + u_{xres}^2} \ .$$

(L18)

The first uncertainty component in the above equation is u_L, which is determined from the minimum and maximum length values (namely, L_{minL} and L_{maxL}) as calculated from one data trace using the following equations for transitional edges that face each other:[58]

$$L_{minL} = \left(x2_{lower} - x1_{lower} \right) cal_x , \text{ and}$$

(L19)

$$L_{maxL} = \left(x2_{upper} - x1_{upper} \right) cal_x .$$

(L20)

Assuming a 99.7 % nominal probability of coverage, the range of values for L due to this component is $L_{maxL} - L_{minL}$. Further, if a Gaussian probability distribution is assumed, u_L is calculated using Eq. (L13). For transitional edges that are oriented in the same direction, the following equations are used:

$$L_{minL} = L - 2x_{res} cal_x , \text{ and}$$

(L21)

$$L_{maxL} = L + 2x_{res} cal_x ,$$

(L22)

such that u_L in Eq. (L13) can be simplified to be:

$$u_L = \frac{2 x_{res} \, cal_x}{3} .$$

(L23)

This is a Type B component.

The second uncertainty component in Eq. (L18) is u_{xcal}, which is calculated using the equation given in Table L2 with $\alpha = 0$ and L_{meas} determined using Eqs. (L17), (L19), and (L20) or Eqs. (L17), (L21), and (L22), as appropriate.

The third uncertainty component in Eq. (L18) is u_{xres}. For this uncertainty component, L_{minres} and L_{maxres} are determined from one 2-D data trace using the following equations:

$$L_{minres} = L_{meas} - x_{res} cal_x ,$$

(L24)

$$L_{maxres} = L_{meas} + x_{res} cal_x ,$$

(L25)

where x_{res} is the uncalibrated pixel resolution of the interferometric microscope in the x-direction. Assuming a uniform distribution for this uncertainty component of width $2x_{res}$, u_{xres} is calculated using the equation given below:

---

[58] Consult ASTM E 2244–05 [44] for slight differences from the MEMS 5-in-1 approach concerning the determination of the upper and lower x-values used to obtain L_{minL} and L_{maxL}.

$$u_{xres} = \frac{L_{maxres} - L_{minres}}{2\sqrt{3}} = \frac{x_{res}cal_x}{\sqrt{3}}, \tag{L26}$$

where u_{xres} is a Type B component.

Eq. (L18) is used in Data Analysis Sheet L.1 [13] with Table L3 giving example values for each of these uncertainty components as well as the combined standard uncertainty value, u_{cL1}. The original uncertainty equation [45] consists of only the first term in Eq. (L18).

Table L3. Example In-Plane Length Uncertainty Values
From a Round Robin Surface-Micromachined Chip Using Eq. (L18)

	source of uncertainty or descriptor	uncertainty values
1. u$_L$	in-plane length measurement	**0.911 μm**[a] (using $x1_{upper}$ = 18.872 μm, $x1_{lower}$ = 21.9532 μm, $x2_{lower}$ = 215.68 μm, $x2_{upper}$ = 217.991 μm, and cal_x =1.01385)
2. u$_{xcal}$	calibration in x-direction	**0.463 μm** (using σ_{xcal} = 0.667 μm, $ruler_x$ /cal_x = $inter_x$ = 283.08 μm,[b] and L =199.14 μm)
3. u$_{xres}$	interferometric resolution in the x-direction	**0.225 μm** (using x_{res} = 0.3851 μm)
u$_{cL1}$	combined standard uncertainty for in-plane length	**1.05 μm**[c] $= \sqrt{u_L{}^2 + u_{xcal}{}^2 + u_{xres}{}^2}$

[a] This u_l uncertainty value is one of the 48 *repeatability* values which comprise the average value, u_{Lave}, as given in Table L4 and plotted in Fig. L9 for L_{des}=200 μm.

[b] $Inter_x$ is also called $scope_x$.

[c] For transitional edges that face each other, such as Edges 1 and 2 in Figs. L3(a and b) at a magnification of 20X [11,44]

6.5 In-Plane Length Round Robin Results

The round robin *repeatability* and *reproducibility* results are given in this section for in-plane length measurements. The *repeatability* data were taken in one laboratory using an optical interferometric microscope (see Sec. 1.1.2). Unlike the MEMS 5-in-1 chip shown in Fig. 2, a similarly processed surface-micromachined test chip (from run #46 [9] and without the backside etch) was fabricated on which in-plane length measurements were taken from each of the fixed-fixed beams depicted in Fig. L6. The design length (in micrometers) of each of these beams is specified to the left of these test structures. This design length corresponds to the design length of L_A in Fig. L7. The measurement of L_A [or L in Fig. L3(a)] is obtained from the data along Edges 1 and 2 using Trace a or Trace e. Example 2-D data from Trace a or Trace e are given in Fig. L3(b). Trace a or Trace e can also be used to obtain the in-plane length measurement, L_B, as shown in Fig. L7 from data along Edges 1 and 5. The design lengths for the measurements of L_B on the test structures shown in Fig. L6 are 1035 μm, 535 μm, 235 μm, 115 μm, and 60 μm (35 μm more than the corresonding design values for L_A). For the *repeatability* data, 24 measurements of L_A and 24 measurements of L_B were taken on each of the five test structures in Fig. L6 and the same number of measurements were taken on a similiarly designed array of test structures with a 90° orientation with respect to the orientation shown in Fig. L6.

121

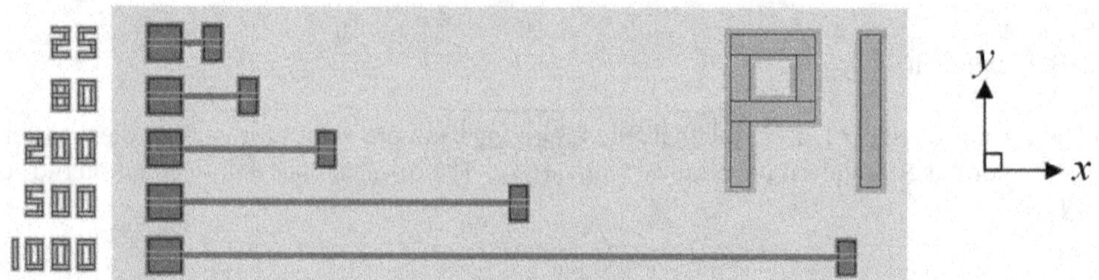

Figure L6. Poly1 fixed-fixed beam test structures for in-plane length measurements on the round robin test chip.

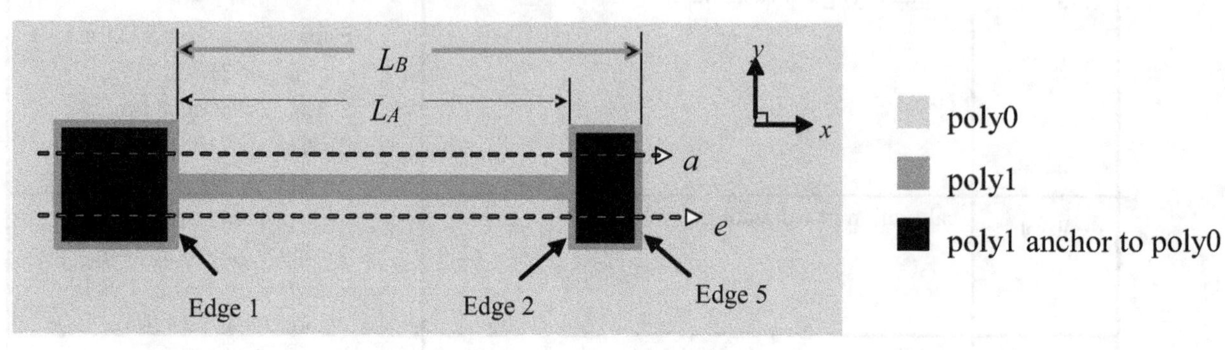

Figure L7. One of the poly1 fixed-fixed beam test structures depicted in Fig. L6.

For the *reproducibility* data, a round robin test chip was passed from laboratory to laboratory with eight laboratories participating. Each participant was asked to obtain both L_A and L_B from five poly1 fixed-fixed beams in an array of test structures that had either a 0° orientation or a 90° orientation. Following the 2002 version of ASTM standard test method E 2244 [45] for in-plane length measurements, the raw, uncalibrated measurements were recorded on Data Analysis Sheet A (similar to the existing Data Analysis Sheet L.1 [13]) for measurements of L_A. Data Analysis Sheet B (similar to the existing Data Analysis Sheet L.2) was used for measurements of L_B. (In this section, for simplicity, we will refer to the existing Data Analysis Sheets L.1 and L.2.)

Tables L4 and L5 present the *repeatability* and *reproducibility* results, respectively, for the in-plane length measurements with transitional edges facing each other, such as Edges 1 and 2 in Fig. L7, and Tables L6 and L7 present similar results for in-plane length measurements with transitional edges facing the same direction, such as Edges 1 and 5 in Fig L7. In these tables, *n* is the number of measurements and is listed first followed by the microscope magnifications (*mag*) and L_{ave}, which is the average of the *repeatability* or *reproducibility* measurement results. For the *repeatability* measurements only, $\sigma_{repeat(samp)}'$, is given next, which is the standard deviation of the *repeatability* in-plane length measurements. Then, in all four tables, the ±2 σ_L limits are given, where σ_L is the standard deviation of the in-plane length measurements, followed by the average of the *repeatability* or *reproducibility* standard uncertainty values (i.e., u_{Lave} and/or u_{cLave}). The last row in each table specifies values for ΔL, where $\Delta L = L_{ave} - L_{des}$.

Table L4. *Repeatability and Uncertainties of NIST Measurement Results for In-Plane Length Measurements When the Transitional Edges Face Each Other*

Design Length (L_{des})	25 µm	80 µm	200 µm	500 µm	1000 µm
n	48	48	48	48	48
mag	80×	40×	20×	10×	5×
L_{ave}	24.37 µm	79.76 µm	199.10 µm	495.0 µm	995.5 µm
$\sigma_{repeat(samp)}{}'$	0.10 µm	0.086 µm	0.15 µm	0.80 µm	2.5 µm
$\pm 2\sigma_L$ limits	±0.20 µm (±0.81 %)	±0.17 µm (±0.22 %)	±0.30 µm (±0.15 %)	±1.6 µm (±0.32 %)	±5.0 µm (±0.50 %)
u_{Lave}[a]	0.23 µm (0.95 %)	1.0 µm (1.3 %)	0.95 µm (0.48 %)	1.7 µm (0.33 %)	3.2 µm (0.32 %)
u_{cL1ave}[b]	0.33 µm (1.3 %)	1.1 µm (1.4 %)	1.1 µm (0.54 %)	1.9 µm (0.39 %)	3.6 µm (0.36 %)
ΔL	−0.63 µm	−0.24 µm	−0.90 µm	−5.0 µm	−4.5 µm

[a] Where u_{Lave} is the sum of the u_L values divided by n

[b] The average of the u_{cL1} values as determined using Eq. (L18)

In particular, note in **Tables L5 and L7** the magnifications used for the specified values of the design length (L_{des}). Not all laboratories had the same magnifications available; therefore, each laboratory was instructed to use the highest magnification available for the given measurement. The underlined value is the magnification used for the *repeatability* measurements.

Table L5 includes the in-plane length results from 650 µm long fixed-fixed beams that were used for residual strain measurements. Similar 650 µm length measurements are not included in Table L4 since *repeatability* measurements at a magnification of 5× are already available with the 1000 µm long measurements.

Table L5. *Reproducibility and Uncertainty of Round Robin Measurement Results for In-Plane Length Measurements When the Transitional Edges Face Each Other*

Design Length (L_{des})	25 µm	80 µm	200 µm	500 µm	650 µm	1000 µm
n	7	7	6	6[a]	6	6[a]
mag	100×, <u>80×</u>,[b] 50×, 39×, 20×, 10×, w[c]	50×, <u>40×</u>, 25×, 25×, 10×, 10×, w	25×, 20.4×, <u>20×</u>, 10×, 10×, 10×	10.2×, <u>10×</u>, 10×, 10×, 5×, 5×	25×, 7.8×, <u>5×</u>, 5×, 5×, w	<u>5×</u>, 5×, 5×, 5×, 5×, 5×
L_{ave}	24.91 µm	79.70 µm	200.61 µm	497.8 µm	651.4 µm	999.8 µm
$\pm 2\sigma_L$ limits	±4.3 µm (±17 %)	±5.3 µm (±6.6 %)	±4.1 µm (±2.0 %)	±3.9 µm (±0.78 %)	±5.3 µm (±0.81 %)	±4.9 µm (±0.49 %)
u_{Lave}[d]	0.60 µm (2.4 %)	0.71 µm (0.89 %)	0.86 µm (0.43 %)	1.5 µm (0.30 %)	1.6 µm (0.25 %)	2.5 µm (0.25 %)
ΔL	−0.09 µm	−0.30 µm	0.61 µm	−2.2 µm	1.4 µm	−0.2 µm

[a] Three of these measurements were taken from the same instrument by two different operators.

[b] Underlined values correspond to the magnifications used for the *repeatability* measurements.

[c] The symbol "w" stands for "unknown." The magnification was not reported by the round robin participant.

[d] Where u_{Lave} is the sum of the u_L values divided by n

Table L6. *Repeatability and Uncertainties of NIST Measurement Results for In-Plane Length Measurements When the Transitional Edges Face the Same Direction*

Design Length (L_{des})	60 µm	115 µm	235 µm	535 µm	1035 µm
n	48	48	48	48	48
mag	80×	40×	20×	10×	5×
L_{ave}	59.56 µm	115.96 µm	234.67 µm	532.2 µm	1035.0 µm
$\sigma_{repeat(samp)}$	0.13 µm	0.19 µm	0.23 µm	0.25 µm	0.61 µm
±$2\sigma_L$ limits	±0.27 µm (±0.45 %)	±0.39 µm (±0.33 %)	±0.47 µm (±0.20 %)	±0.50 µm (±0.094 %)	±1.2 µm (±0.12 %)
u_{Lave} [a]	0.067 µm (0.11 %)	0.14 µm (0.12 %)	0.26 µm (0.11 %)	0.54 µm (0.10 %)	1.1 µm (0.10 %)
u_{cL1ave} [b]	0.54 µm (0.91 %)	0.55 µm (0.47 %)	0.64 µm (0.27 %)	1.1 µm (0.21 %)	2.0 µm (0.20 %)
ΔL	−0.44 µm	0.96 µm	−0.33 µm	−2.8 µm	0.0 µm

[a] Where u_{Lave} is the sum of the u_L values divided by n

[b] The average of the u_{cL1} values as determined using Eq. (L18)

Table L7. *Reproducibility and Uncertainty of Round Robin Measurement Results for In-Plane Length Measurements When the Transitional Edges Face the Same Direction*

Design Length (L_{des})	60 µm	115 µm	235 µm	535 µm	1035 µm
n	6 [a]	6	6 [a]	6 [b]	6 [b]
mag	80×, 80×, 50×, 39×, 20×, 10×	40×, 25×, 25×, 25×, 10×, 10×	20.4×, 20×, 20×, 10×, 10×, 5×	10.2×, 10×, 10×, 10×, 5×, 5×	5.9×, 5×, 5×, 5×, 5×, 5×
L_{ave}	59.68 µm	115.34 µm	235.79 µm	533.8 µm	1035.1 µm
±$2\sigma_L$ limits	±1.2 µm (±2.1 %)	±5.0 µm (±4.4 %)	±4.1 µm (±1.7 %)	±5.6 µm (±1.0 %)	±5.0 µm (±0.48 %)
u_{Lave} [c]	0.22 µm (0.37 %)	0.32 µm (0.27 %)	0.43 µm (0.18 %)	0.67 µm (0.13 %)	1.1 µm (0.10 %)
ΔL	−0.32 µm	0.34 µm	0.79 µm	−1.2 µm	0.1 µm

[a] Two of these measurements were taken from the same instrument by different operators.

[b] Three of these measurements were taken from the same instrument by two different operators.

[c] Where u_{Lave} is the sum of the u_L values divided by n

The test method for in-plane length measurements emphasizes two values, the in-plane length measurement, L, and the combined standard uncertainty [19-21], u_{cL}, of that measurement. Figure L8 illustrates the offsets between the measured length results and the designed values for both the NIST measurements and the round robin averages for L. This figure consists of four plots of ΔL versus L_{des} where $\Delta L = L_{ave} - L_{des}$. These plots are for the *repeatability* and *reproducibility* measurements obtained from Data Analysis Sheet L.1 and Data Analysis Sheet L.2. The average ΔL value (ΔL_{ave}) for each plot is given. There do not seem to be obvious systematic offsets in these data; however, the *repeatability* data from Data Analysis Sheet L.1 show the highest $|\Delta L_{ave}|$ value with all the measurements of L_{ave} being less than L_{des} (as given in Table L4). Therefore, there may be a bias towards measuring lower values of L when using Data Analysis Sheet L.1 in this laboratory, and the degree of the resulting bias varies with magnification. Even though these data were calibrated, it should be emphasized that calibration of the microscope magnification is considered mandatory for in-plane length measurements. The interferometric software may treat data at transitional edges differently (for example, if neighboring data points are averaged together) which could result in a bias to an in-plane length measurement when the transitional edges face each other. With the microscope magnifications calibrated in the x-and y-directions, one is able to determine an accurate value for L_{offset} for that magnification of that interferometric microscope when using similar in-plane length calculations on similar test structures. Therefore, when measuring in-plane lengths that face each other in the manner specified for the round robin, the ΔL values in Table L4 can be equated with $-L_{offset}$ for similar test structures using the interferometric microscope used at NIST.

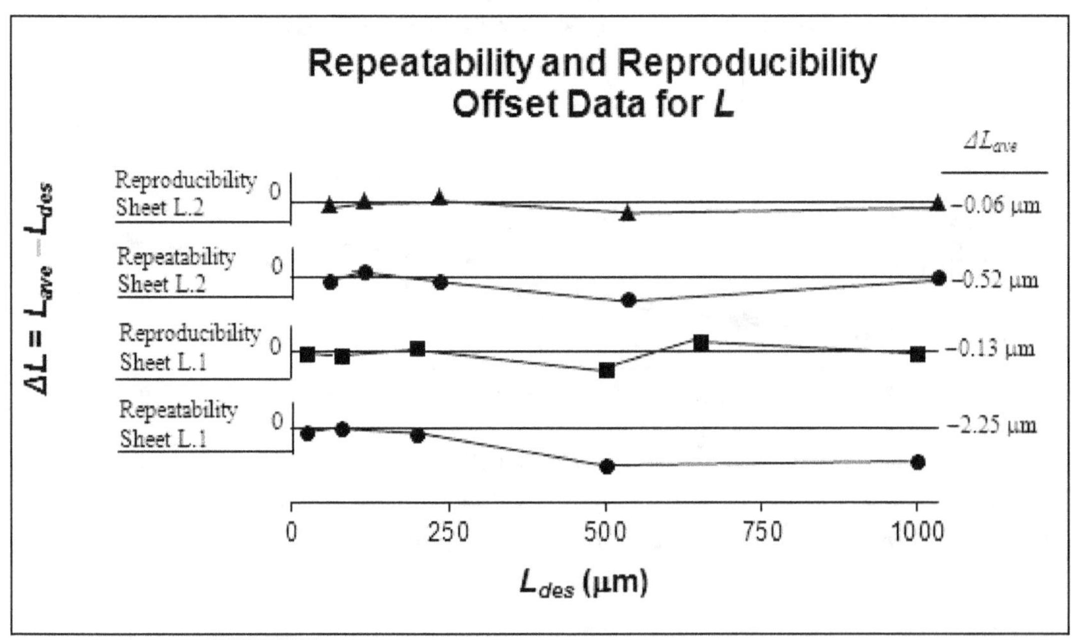

Figure L8. Repeatability and reproducibility offset data for L[59].

Now, consider the trends associated with u_{Lave} as given in Fig. L9 and Fig. L10 for Data Analysis Sheet L.1 and Data Analysis Sheet L.2, respectively. In both of these figures, u_{Lave} increases with increasing length primarily due to the increase in the pixel-to-pixel spacings associated with the lower-powered objectives that are used for the longer length measurements. Also, these figures show that the *repeatability* and *reproducibility* measurements are somewhat comparable. For Data Analysis Sheet L.1, u_L is given by Eq. (L13) with L_{minL} and L_{maxL} given in Eqs. (L19) and (L20), respectively, and if the data points determining L_{minL} and L_{maxL} in these equations are chosen in the manner specified in the standard test method [45], the comparable results in Fig. L9 are expected. For Data Analysis Sheet L.2, u_L is given by Eq. (L23) where x_{res} is the uncalibrated resolution of the interferometric microscope in the x-direction. Therefore, u_L can be calculated before the measurement is even taken since this equation does not rely upon any data points. Therefore, the *repeatability* and *reproducibility* measurements in Fig. L10 should be comparable.

Probing deeper into this, Fig. L10 (in combination with the magnifications specified in Table L7) indicates that the interferometric microscope used at NIST for the *repeatability* measurements has a comparable value for x_{res} for the highest L_{des} (=1035 µm) measurement taken at the lowest (5×) magnification to those used by the other laboratories that participated in this round robin. However, this laboratory benefited by having five different magnifications with which to take measurements. Therefore, for the smaller values of L_{des} given in Table L7, NIST apparently used higher-powered objectives than those used by most of the other laboratories and was able to achieve lower values for u_{Lave}. This could imply that the effective technical lifetime of an interferometric microscope can be extended by purchasing multiple objectives.

Comparing u_{Lave} in Figs. L9 and L10, the values for u_{Lave} from Data Analysis Sheet L.2 are considerably less than those from Data Analysis Sheet L.1. This implies that more precise in-plane length measurements are possible when the transitional edges face the same direction.

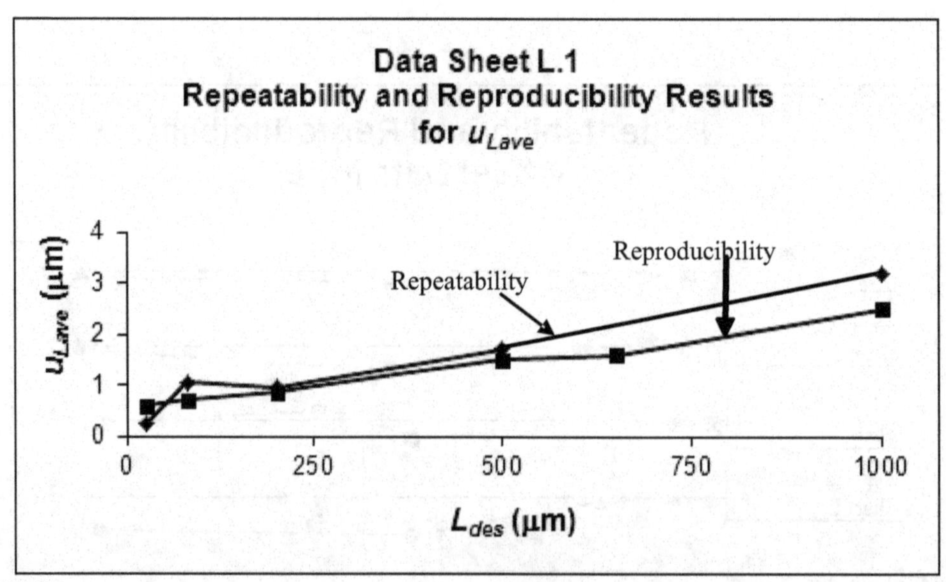

Figure L9. Comparing repeatability and reproducibility results for u_{Lave} in Data Analysis Sheet L.1.

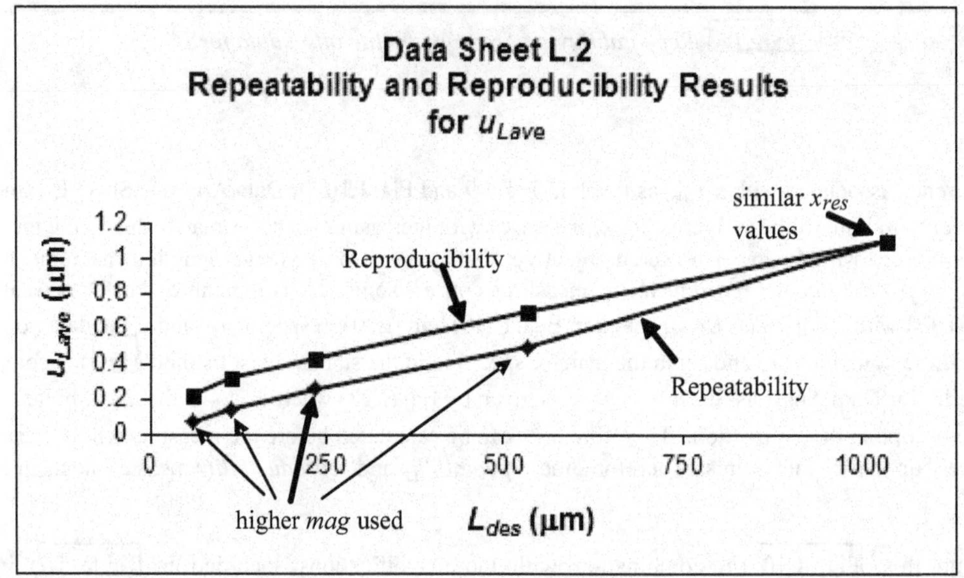

Figure L10. Comparing repeatability and reproducibility results for u_{Lave} in Data Analysis Sheet L.2.

The precision and bias obtained from the round robin data can be stated as follows:

Precision: The precision data for L are given in Tables L4 through L7. In particular, the ± 2 σ_L limits indicate that the *reproducibility* results (e.g., ± 2.0 % for Data Analysis Sheet L.1 in Table L5 for L_{des}=200 μm) are much poorer than the corresponding *repeatability* results (i.e., ± 0.15 % in Table L4). This might be due to resolution limits if one is measuring the smaller features with low power objectives, errors in the calibration factors of the different instruments and different objectives, and different persons taking the measurements and analyzing the data.

Bias: The data in Fig. L8 suggest no significant offsets associated with the length data, except for a tendency for *repeatability* Data Analysis Sheet L.1 length data to be less than L_{des} for this laboratory for all magnifications. The degree of the resulting bias is different for each magnification.

6.6 Using the MEMS 5-in-1 to Verify In-Plane Length Measurements

To compare your in-house length measurements with NIST measurements, you will need to fill out Data Analysis Sheet L.0. (This data analysis sheet is accessible via the URL specified in the reference [13], a reproduction of which is given in Appendix 5). After calibrating the instrument, locating the test structure, taking the measurements, and performing the calculations, the data on the completed form can be compared with the data on the SRM Certificate and the completed data analysis sheet supplied with the MEMS 5-in-1. Details of the procedure are given below.

Calibrate the instrument: Calibrate the instrument as given in Sec. 6.2. Obtain the inputs for Table 1 in Data Analysis Sheet L.0.

Locate the in-plane length test structure: In the fifth grouping of test structures, shown in Figs. L1(a and b), on the MEMS 5-in-1 chips shown in Fig. 1 and Fig. 2 for SRM 2494 and SRM 2495, respectively, in-plane length measurements are made. In-plane length test structures are provided for this purpose, such as those shown in Fig. L2(a) or Fig. L4 for SRM 2494 and in Fig. L3(a) for a poly1 test structure on SRM 2495. Data Analysis Sheet L.0 requires measurements from one in-plane length test structure. The specific test structure to be measured can be deduced from the data entered on the NIST-supplied Data Analysis Sheet L.0 that accompanies the SRM.

For the in-plane length grouping of test structures on SRM 2494, as shown in Fig. L1(a), the target test structure and applicable measurement can be found as follows:
1. The input *design length* (i.e., input #5 on Data Analysis Sheet L.0, a reproduction of which is given in Appendix 5) specifies the design length of the in-plane length test structure. The length of the in-plane length test structure (in micrometers) is given at the top of each column of test structures in Fig. L1(a); therefore *design length* can be used to locate the column in which the target test structure resides. Design lengths for the in-plane length test structures are given in Table L1.
2. The input *which* (i.e., input #6) specifies which in-plane length test structure in the column to measure (i.e., the "first," "second," "third," etc.). Since there are three instances of each test structure, the radio button corresponding to "first," "second," or "third" is used to identify the target test structure.
3. The input *type* (i.e., input #4) specifies the type of in-plane length measurement (L_{oo}, L_{ii}, or L_{io}) as shown in Fig. L4. For SRM 2494 measurements, an inside edge-to-inside edge length measurement as given by L_{ii} in Fig. L4 is requested.

For the in-plane length grouping of test structures for SRM 2495, as shown in Fig. L1(b), the target test structure and applicable measurement can be found as follows:
1. The input *material* (i.e., input #3 on Data Analysis Sheet L.0, a reproduction of which is given in Appendix 5) specifies the composition of the in-plane length test structure, which should be either "poly1" or "poly2" since there are four arrays of poly1 [or P1 as given in Fig. L1(b)] structures and two arrays of poly2 (or P2) structures.
2. The input *type* (i.e., input #4) specifies the type of in-plane length measurement (L_{oo}, L_{ii}, or L_{io}) as shown in Fig. L4. For SRM 2495 measurements, an inside edge-to-inside edge length measurement (L_{ii}) is requested. Since an array consists of all fixed-fixed beams or all cantilevers, we will assume this measurement is between the anchor lips of a fixed-fixed beam, as given by L in Fig. L3(a); therefore, the target test structure is within a fixed-fixed beam array. There are two poly1 fixed-fixed beam arrays in the in-plane length grouping of test structures and they are located on the left side of this grouping, as can be seen in Fig. L1(b). There is also one poly2 fixed-fixed beam array located in the middle of the bottom row of arrays.
3. The input *orientation* (i.e., input #8) is used to locate the target array. There are two orientations (a 0° orientation and a 90° orientation) of poly1 fixed-fixed beam arrays. An array with a 0° orientation has the length of the beam parallel to the x-axis of the chip, the axes of which are shown in Fig. 2 and Fig. L1(b). Therefore, the poly1 array of fixed-fixed beams with a 0° orientation is the bottom left array and the poly array of fixed-fixed beams with a 90° orientation is the top left array. There is one poly2 fixed-fixed beam array with a 0° orientation located in the middle of the bottom row of this grouping of test structures.
4. The input *design length* (i.e., input #5) specifies the design length of the fixed-fixed beam. The design length of the fixed-fixed beam (in micrometers) is given next to the second of three fixed-fixed beams of the same length, as can barely be seen in Fig. L1(b). Therefore, *design length* can be used to locate a set of three possible target test structures. Design lengths for the fixed-fixed beam test structures are given in Table L1.
5. The input *which* (i.e., input #6) is used to locate which iteration of the test structure is the target test structure, where "first" corresponds to the topmost or leftmost test structure in the array of the same material that has the specified length.

6. As stated previously, *type* (i.e., input #4) specifies that an inside edge-to-inside edge length measurement, L_{ii}, is requested. This measurement is taken between the anchor lips of the fixed-fixed beam, as given by L in Fig. L3(a).

Take the measurements: Following the steps in ASTM standard test method E 2244 [5] for in-plane length measurements, the chip is oriented ^{60} under the interferometric microscope as shown in Fig. L2(a) or Fig. L3(a) and one 3-D data set is obtained using the highest magnification objective that is available and feasible. The data are leveled and zeroed, and four 2-D data traces (a', a, e, and e') are obtained. From each of the four data traces, the raw, uncalibrated measurements for $x1_{uppert}$ and $x2_{uppert}$ are obtained along with $n1_t$ and $n2_t$ (as specified in Sec. 6.3) and recorded in Data Analysis Sheet L.0. The uncalibrated values for ya' and ye' are also recorded in Data Analysis Sheet L.0.

Perform the calculations: Enter the data into Data Analysis Sheet L.0 as follows:
1. Press one of the "Reset this form" buttons. (One of these buttons is located near the top of the data analysis sheet and the other is located near the middle of the data analysis sheet.)
2. Supply the inputs to Table 1 and Table 2.
3. Press one of the "Calculate and Verify" buttons to obtain the results for the in-plane length test structure. (One of these buttons is located near the top of the data analysis sheet and the other is located near the middle of the data analysis sheet.)
4. Verify the data by checking to see that all the pertinent boxes in the verification section at the bottom of the data analysis sheet say "ok". If one or more of the boxes say "wait," address the issue, if necessary, by modifying the inputs and recalculating.
5. Print out the completed data analysis sheet to compare both the inputs and outputs with those on the NIST-supplied data analysis sheet.

Compare the measurements: The MEMS 5-in-1 is accompanied by a Certificate. This Certificate specifies an in-plane length value, L, and the expanded uncertainty, U_L, (with $k=2$) intending to approximate a 95 % level of confidence. It is your responsibility to determine an appropriate criterion for acceptance, such as given below:

$$D_L = \left| L_{(customer)} - L \right| \le \sqrt{U^2_{L(customer)} - U^2_L} , \qquad (L27)$$

where D_L is the absolute value of the difference between your in-plane length value, $L_{(customer)}$, and the in-plane length value on the SRM Certificate, L, and where $U_{L(customer)}$ is your expanded uncertainty value and U_L is the expanded uncertainty on the SRM Certificate. If your measured value for in-plane length (as obtained in the newly filled out Data Analysis Sheet L.0) satisfies your criterion for acceptance and there are no pertinent "wait" statements at the bottom of your Data Analysis Sheet L.0, you can consider yourself to be appropriately measuring in-plane length according to the ASTM E 2244 in-plane length standard test method [5] according to your criterion for acceptance.

Any questions concerning the measurements, analysis, or comparison can be directed to mems-support@nist.gov.

---

^{60} This orientation assumes that the pixel-to-pixel spacing in the *x*-direction of the interferometric microscope is smaller than or equal to the pixel-to-pixel spacing in the *y*-direction.

7 Residual Stress and Stress Gradient

Residual stress is defined as the remaining forces per unit area within the structural layer of interest after the original causes(s) during fabrication have been removed yet before the constraint of the sacrificial layer (or substrate) is removed (in whole or in part). (Residual) stress gradient is defined as a through-thickness variation (of the residual stress) in the structural layer of interest before it is released. These measurements are an aid in the design and fabrication of MEMS devices [28-29] and ICs [27].

In this section, Sec. 7.1 provides the equations for residual stress and stress gradient. The uncertainty analysis is presented in Sec. 7.2. Following this, Sec. 7.3 describes how to use the MEMS 5-in-1 to verify residual stress and stress gradient measurements.

7.1 Residual Stress and Stress Gradient Equations

Equations for residual stress and stress gradient are presented in this section, given the Young's modulus value, E, obtained from Data Analysis Sheet YM.3 (as specified in Sec. 2) [1,13].

Residual stress: To calculate the residual stress, σ_r, of a thin film layer, the residual strain of the thin film layer, ε_r, must also be known. This value of residual strain and its combined standard uncertainty value, $u_{c\varepsilon r3}$, are found (as specified in Sec. 3) from measurements of a fixed-fixed beam test structure comprised of that layer using Data Analysis Sheet RS.3 [13]. Then, the residual stress is calculated using the following equation:

$$\sigma_r = E\varepsilon_r \ . \tag{X1}$$

Stress gradient: To calculate the stress gradient, σ_g, of a thin film layer, the strain gradient of the thin film layer, s_g, must also be known. This value of strain gradient and its combined standard uncertainty value, u_{csg3}, are found (as specified in Sec. 4) from measurements of a cantilever test structure comprised of that layer using Data Analysis Sheet SG.3 [13]. Then, the stress gradient is calculated using the following equation:

$$\sigma_g = E s_g \ . \tag{X2}$$

7.2 Residual Stress and Stress Gradient Uncertainty Analysis

In this section, two sets of combined standard uncertainty equations are presented for residual stress and stress gradient. The first set of combined standard uncertainty equations are used for the MEMS 5-in-1. The propagation of uncertainty technique [19-21] (a brief overview of which is given in Appendix 8) is used, which results in relative uncertainties that can be of more value to the user than absolute uncertainties. For example, relative uncertainties can be used to determine what parameters residual stress and stress gradient are most sensitive to and how accurate the parameters must be to assure a pre-determined accuracy. (Relative uncertainties are obtained in Sec. 2.4.1 in the uncertainty analysis for Young's modulus.) The second set of combined standard uncertainty equations presented for residual stress and stress gradient are earlier equations [31,32] which use a technique which adds absolute uncertainties in quadrature.

7.2.1 Residual Stress and Stress Gradient Uncertainty Analyses for the MEMS 5-in-1

This section presents the combined standard uncertainty equations for residual stress and stress gradient that are used for the MEMS 5-in-1.

Residual stress: The combined standard uncertainty equation for residual stress is found by applying the propagation of uncertainty technique for parameters that multiply (as presented in Appendix 8) to Eq. (X1). The one sigma uncertainty of the value of the residual stress, $\sigma_{\sigma r}$ can be written as follows:

$$\sigma_{\sigma r} = |\sigma_r| \sqrt{\left(\frac{\sigma_E}{E}\right)^2 + \left(\frac{\sigma_{\varepsilon r}}{\varepsilon_r}\right)^2} \ , \tag{X3}$$

where σ_E and $\sigma_{\varepsilon r}$ are the standard deviations of Young's modulus and residual strain, respectively. Rewriting the above equation in terms of combined standard uncertainties (by equating $\sigma_{\sigma r}$ with $u_{c\sigma r3}$, σ_E with u_{cE3}, and $\sigma_{\varepsilon r}$ with $u_{c\varepsilon r3}$) results in the following equation:

$$u_{c\sigma r3} = |\sigma_r| \sqrt{\left(\frac{u_{cE3}}{E}\right)^2 + \left(\frac{u_{c\varepsilon r3}}{\varepsilon_r}\right)^2} \ , \tag{X4}$$

where a number following the subscript "σr" in "$u_{c\sigma r}$" and "E" in "u_{cE}" indicates the Young's modulus data analysis sheet that is used to obtain these combined standard uncertainty values. Therefore, both $u_{c\sigma r3}$ and u_{cE3} imply that Data Analysis Sheet YM.3 is used to obtain these values. However, the number "3" following the subscript "εr" in "$u_{c\varepsilon r}$" indicates that Data Analysis Sheet RS.3 is used to obtain that value. In determining the combined standard uncertainty, a Type B evaluation [19-21] (i.e., one that uses means other than the statistical Type A analysis) is used for each source of uncertainty.

The expanded uncertainty for residual stress, $U_{\sigma r}$, is calculated using the following equation:

$$U_{\sigma r} = k u_{c\sigma r3} = 2 u_{c\sigma r3} \ , \tag{X5}$$

where the k value of 2 approximates a 95 % level of confidence.

Reporting residual stress results: Since it can be assumed that the estimated values of the uncertainty components have an approximate Gaussian distribution with approximate combined standard uncertainty $u_{c\sigma r3}$, the residual stress is believed to lie in the interval $\sigma_r \pm u_{c\sigma r3}$ (expansion factor $k=1$) representing a level of confidence of approximately 68 %.

Stress gradient: The combined standard uncertainty equation for stress gradient is found by applying the propagation of uncertainty technique for parameters that multiply (as presented in Appendix 8) to Eq. (X2). The one sigma uncertainty of the value of the stress gradient, $\sigma_{\sigma g}$ can be written as follows:

$$\sigma_{\sigma g} = \sigma_g \sqrt{\left(\frac{\sigma_E}{E}\right)^2 + \left(\frac{\sigma_{sg}}{s_g}\right)^2} \ , \tag{X6}$$

where σ_E and σ_{sg} are the standard deviations of Young's modulus and strain gradient, respectively. Rewriting the above equation in terms of combined standard uncertainties (by equating $\sigma_{\sigma g}$ with $u_{c\sigma g3}$, σ_E with u_{cE3}, and σ_{sg} with u_{csg3}) results in the following equation:

$$u_{c\sigma g3} = \sigma_g \sqrt{\left(\frac{u_{cE3}}{E}\right)^2 + \left(\frac{u_{csg3}}{s_g}\right)^2} \ , \tag{X7}$$

where a number following the subscript "σg" in "$u_{c\sigma g}$" and "E" in "u_{cE}" indicates the Young's modulus data analysis sheet that is used to obtain these combined standard uncertainty values. Therefore, both $u_{c\sigma g3}$ and u_{cE3} imply that Data Analysis Sheet YM.3 is used to obtain the se values. However, the number "3" following the subscript "sg" in "u_{csg}" indicates that Data Analysis Sheet SG.3 is used to obtain that value. In determining the combined standard uncertainty, a Type B evaluation [19-21] (i.e., one that uses means other than the statistical Type A analysis) is used for each source of uncertainty.

The expanded uncertainty for stress gradient, $U_{\sigma g}$, is calculated using the following equation:

$$U_{\sigma_g} = k u_{c\sigma_{g3}} = 2 u_{c\sigma_{g3}} \quad , \tag{X8}$$

where the k value of 2 approximates a 95 % level of confidence.

Reporting stress gradient results: Since it can be assumed that the estimated values of the uncertainty components have an approximate Gaussian distribution with approximate combined standard uncertainty $u_{c\sigma_{g3}}$, the (residual) stress gradient is believed to lie in the interval $\sigma_g \pm u_{c\sigma_{g3}}$ (expansion factor $k=1$) representing a level of confidence of approximately 68 %.

7.2.2 Previous Residual Stress and Stress Gradient Uncertainty Analyses

This section presents previous combined standard uncertainty equations for residual stress and stress gradient.

Residual Stress: A combined standard uncertainty equation that was used previously for residual stress in Data Analysis Sheet YM.2 [13] is as follows:

$$u_{c\sigma r 2} = \sqrt{u_{E(\sigma r)}^2 + u_{\varepsilon r(\sigma r)}^2} \quad , \tag{X9}$$

where $u_{E(\sigma r)}$ is due to the measurement uncertainty of E and where $u_{\varepsilon r(\sigma r)}$ is due to the measurement uncertainty of ε_r. A number following the subscript "σr" in "$u_{c\sigma r}$" indicates the Young's modulus data analysis sheet that is used to obtain the combined standard uncertainty value. Therefore, $u_{c\sigma r 2}$ implies that Data Analysis Sheet YM.2 is used. In determining the combined standard uncertainty, a Type B evaluation [19-21] (i.e., one that uses means other than the statistical Type A analysis) is used for each source of uncertainty. Table X1 gives the equation for both uncertainty components.

The first uncertainty component in Eq. (X9) and listed in Table X1 is $u_{E(\sigma r)}$, which is determined from the minimum and maximum residual stress values (namely, σ_{rmin} and σ_{rmax}, respectively). Assuming a Gaussian distribution, $u_{E(\sigma r)}$ is calculated using the equation given in Table X1 where u_{cE2} is the combined standard uncertainty of the Young's modulus measurement as given in Sec. 2.4.2 using Data Analysis Sheet YM.2.

The second uncertainty component in Eq. (X9) and listed in Table X1 is $u_{\varepsilon r(\sigma r)}$, which is determined from the minimum and maximum residual stress values (namely, σ_{rmin} and σ_{rmax}, respectively). Assuming a Gaussian distribution, $u_{\varepsilon r(\sigma r)}$ is calculated using the equation in Table X1 where $u_{c\varepsilon r3}$ is the combined standard uncertainty of the residual strain measurement as given in Sec 3.4 using Data Analysis Sheet RS.3.

Eq. (X9) can be shown to be equivalent to Eq. (X4) if u_{cE2} is equated with u_{cE3} and if the values for Young's modulus are the same (in other words if $f_{correction}$ is set equal to 0 Hz in Data Analysis Sheet YM.3).

The combined standard uncertainty equation for $u_{c\sigma r1}$ (obtained using Data Analysis Sheet YM.1 [31,32]) is similar to Eq. (X9), however the calculation of the first component, $u_{E(\sigma r)}$, uses u_{cE1} instead of u_{cE2}.

Table X1 Residual Stress Uncertainty Equations [10]

	source of uncertainty	G or U[a] / A or B[b]	equation
1. $u_{E(\sigma r)}$	Young's modulus	G / B	$u_{E(\sigma r)} = u_{cE2} \lvert \varepsilon_r \rvert$
2. $u_{\varepsilon r(\sigma r)}$	residual strain	G / B	$u_{\varepsilon r(\sigma r)} = u_{c\varepsilon r3} E$

[a] "G" indicates a Gaussian distribution and "U" indicates a uniform distribution

[b] Type A or Type B analysis

Stress gradient: A combined standard uncertainty equation that was used previously for stress gradient in Data Analysis Sheet YM.2 [13] is as follows:

$$u_{c\sigma_g 2} = \sqrt{u_{E(\sigma_g)}^2 + u_{sg(\sigma_g)}^2} \quad , \tag{X10}$$

where $u_{E(\sigma_g)}$ is due to the measurement uncertainty of E, and where $u_{sg(\sigma_g)}$ is due to the measurement uncertainty of s_g. A number following the subscript "σ_g" in "$u_{c\sigma_g}$" indicates the Young's modulus data analysis sheet that is used to obtain the combined standard uncertainty value. Therefore, $u_{c\sigma_g 2}$ implies that Data Analysis Sheet YM.2 is used. In determining the combined standard uncertainty, a Type B evaluation [19-21] (i.e., one that uses means other than the statistical Type A analysis) is used for each source of uncertainty. Table X2 gives the equation for both uncertainty components.

The first uncertainty component in Eq. (X10) and listed in Table X2 is $u_{E(\sigma_g)}$, which is determined from the minimum and maximum stress gradient values (namely, σ_{gmin} and σ_{gmax}, respectively). Assuming a Gaussian distribution, $u_{E(\sigma_g)}$ is calculated using the equation given in Table X2 where u_{cE2} is the combined standard uncertainty of the Young's modulus measurement as given in Sec. 2.4.2 using Data Analysis Sheet YM.2.

The second uncertainty component in Eq. (X10) and listed in Table X2 is $u_{sg(\sigma_g)}$, which is determined from the minimum and maximum stress gradient values (namely, σ_{gmin} and σ_{gmax}, respectively). Assuming a Gaussian distribution, $u_{sg(\sigma_g)}$ is calculated using the equation given in Table X2 where u_{csg3} is the combined standard uncertainty of the strain gradient measurement as given in Sec. 4.4 using Data Analysis Sheet SG.3.

Eq. (X10) can be shown to be equivalent to Eq. (X7) if u_{cE2} is equated with u_{cE3} and if the values for Young's modulus are the same (in other words if $f_{correction}$ is set equal to 0 Hz in Data Analysis Sheet YM.3).

The combined standard uncertainty equation for $u_{c\sigma_g 1}$ (obtained using Data Analysis Sheet YM.1 [31,32]) is similar to Eq. (X10), however the calculation of the first component, $u_{E(\sigma_g)}$, uses u_{cE1} instead of u_{cE2}.

Table X2. *(Residual) Stress Gradient Uncertainty Equations* [10]

	source of uncertainty	G or U[a] / A or B[b]	equation
1. $u_{E(\sigma_g)}$	Young's modulus	G / B	$u_{E(\sigma_g)} = u_{cE2}S_g$
2. $u_{sg(\sigma_g)}$	strain gradient	G / B	$u_{sg(\sigma_g)} = u_{csg3}E$

[a] "G" indicates a Gaussian distribution and "U" indicates a uniform distribution

[b] Type A or Type B analysis

7.3 Using the MEMS 5-in-1 to Verify Residual Stress and Stress Gradient Measurements

To compare your residual stress and stress gradient measurements with NIST measurements, you will need to provide the optional inputs to Table 4 (in addition to Table 1 and Table 2) in Data Analysis Sheet YM.3. (This data analysis sheet is accessible via the URL specified in the reference [13], a reproduction of which is given in Appendix 1.) This data analysis sheet is used for Young's modulus measurements in Sec. 2. Two inputs to Table 4 (namely, ε_r and $u_{c\varepsilon r3}$) come from the outputs of Data Analysis Sheet RS.3 (see Sec. 3) and the other two inputs to Table 4 (namely, s_g and u_{csg3}) come from the outputs of Data Analysis Sheet SG.3 (see Sec. 4).

Therefore, as specified in Sec. 2.6, to perform the calculations, enter the data into Data Analysis Sheet YM.3 as follows:
1. Press the "Reset this form" button located near the middle of the data analysis sheet. (One of these buttons is located near the top of the data analysis sheet and the other is located near the middle of the data analysis sheet.)

2. Fill out Table 1, Table 2, and Table 4.
3. Press the "Calculate and Verify" button to obtain the results for the cantilever. (One of these buttons is located near the top of the data analysis sheet and the other is located near the middle of the data analysis sheet.)
4. Verify the data by checking to see that all the pertinent boxes in the verification section at the bottom of the data analysis sheet say "ok". If one or more of the boxes say "wait," address the issue, if necessary, by modifying the inputs and recalculating.
5. Print out the completed data analysis sheet to compare both the inputs and outputs with those on the NIST-supplied data analysis sheet.

Compare the measurements: The MEMS 5-in-1 is accompanied by a Certificate. This Certificate specifies an effective residual stress value, σ_r, an effective stress gradient value, σ_g, and their corresponding expanded uncertainty values, namely U_{σ_r} and U_{σ_g}, respectively, (with $k=2$ in both cases) intending to approximate a 95 % level of confidence. It is your responsibility to determine an appropriate criterion for acceptance, such as given below:

$$D_{\sigma_r} = \left| \sigma_{r(customer)} - \sigma_r \right| \leq \sqrt{U_{\sigma_r(customer)}^2 - U_{\sigma_r}^2} \,, \tag{X11}$$

$$D_{\sigma_g} = \left| \sigma_{g(customer)} - \sigma_g \right| \leq \sqrt{U_{\sigma_g(customer)}^2 - U_{\sigma_g}^2} \,, \tag{X12}$$

where D_{σ_r} is the absolute value of the difference between your residual stress value, $\sigma_{r(customer)}$, and the residual stress value on the SRM Certificate, σ_r, and where $U_{\sigma_r(customer)}$ is your expanded uncertainty value and U_{σ_r} is the expanded uncertainty on the SRM Certificate. Similarly, D_{σ_g} is the absolute value of the difference between your stress gradient value, $\sigma_{g(customer)}$, and the stress gradient value on the SRM Certificate, σ_g, and where $U_{\sigma_g(customer)}$ is your expanded uncertainty value and U_{σ_g} is the expanded uncertainty on the SRM Certificate. If your measured values for residual stress and stress gradient (as obtained in the newly filled out Data Analysis Sheet YM.3) satisfies your criterion for acceptance and there are no pertinent "wait" statements at the bottom of your Data Analysis Sheet YM.3, you can consider yourself to be appropriately measuring residual stress and stress gradient according to the SEMI MS4 Young's modulus standard test method [1] according to your criterion for acceptance.

An effective residual stress and an effective stress gradient are reported since an effective Young's modulus value is used in the parametric calculations (due to non-idealities associated with the geometry and/or composition of the cantilevers as specified in Sec. 2.6). In addition, effective values of residual strain (as specified in Sec. 3.6) and strain gradient (as specified in Sec. 4.6) may also be used in the calculation of residual stress and stress gradient, respectively. When you use SEMI standard test method MS4 [1], ASTM standard test method E 2245 [2] and ASTM standard test method E 2246 [3] with your own test structures, you must be cognizant of the geometry and composition of your test structures because these test methods assume an ideal geometry and composition, implying that you would be obtaining "effective" values if the geometry and/or composition of your test structures deviate from the ideal.

Any questions concerning the measurements, analysis, or comparison can be directed to mems-support@nist.gov.

8   Thickness

Thickness is defined as the height in the *z*-direction of one or more designated thin-film layers. Step height test structures can be used to obtain inputs to thickness calculations. Thickness measurements are an aid in the design and fabrication of MEMS devices [28-29] and ICs [27].

In this section, the NIST-developed thickness test structures on SRM 2494 and SRM 2495, as shown in Fig. 1 and Fig. 2 in the Introduction, respectively, are given in Sec. 8.1. Then, Sec. 8.2 discusses the calibration procedure for thickness measurements, Sec. 8.3 discusses the use of Data Analysis Sheet T.1 to obtain the composite oxide beam thickness for SRM 2494 chips, and Sec. 8.4 discusses the use of Data Analysis Sheet T.3 to obtain the poly1 or poly2 thickness for SRM 2495 chips. Following this, Sec. 8.5 describes how to use the MEMS 5-in-1 to verify thickness measurements.

8.1 Thickness Test Structures

Thickness measurements for SRM 2494, as depicted in Fig. 1, are taken in the fourth grouping of test structures, as shown in Fig. T1(a). For SRM 2495, depicted in Fig. 2, thickness measurements are taken from the thickness test structures, as shown in Fig. T1(b), located in the sixth grouping.

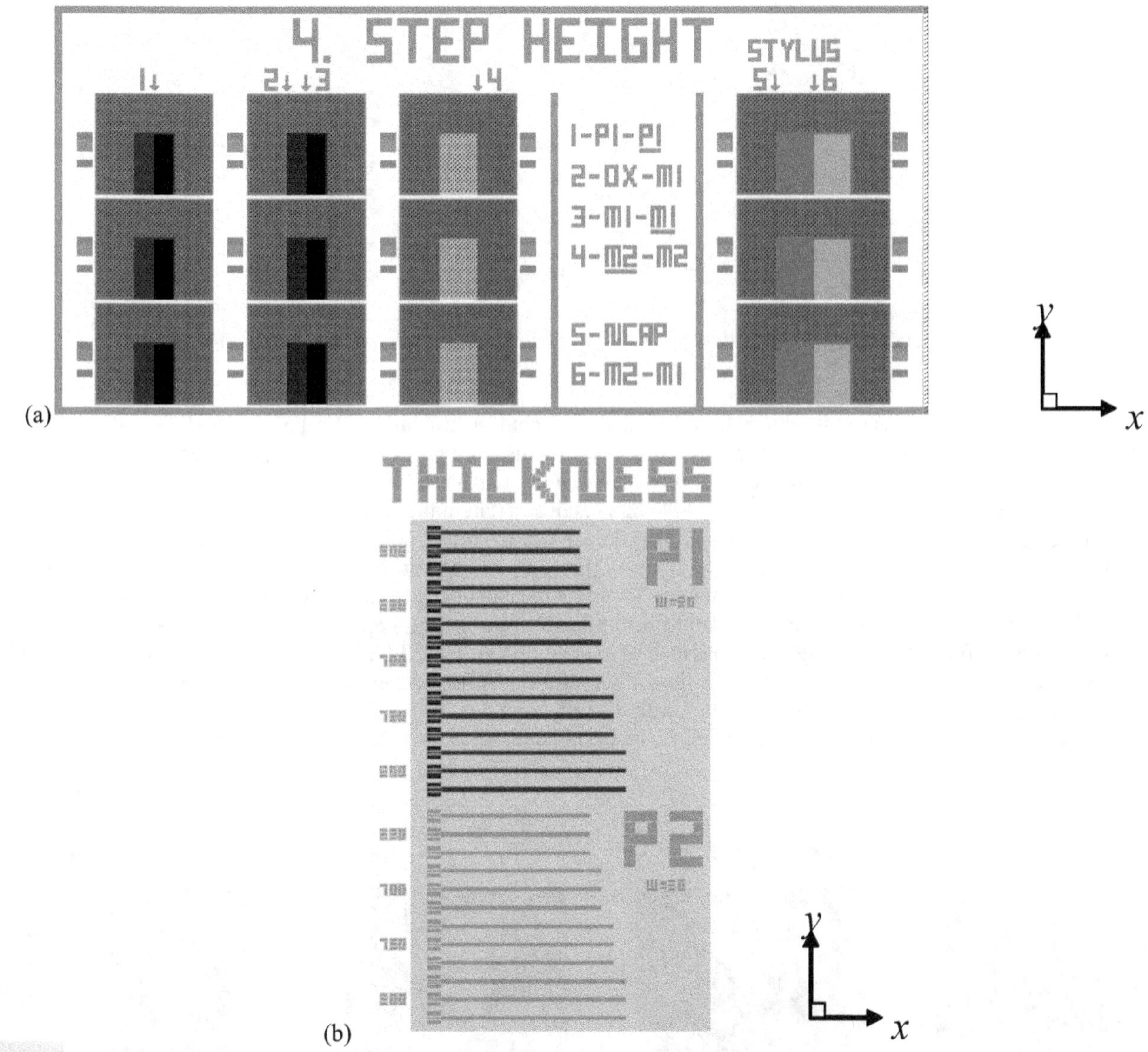

Figure T1. The test structures used for thickness measurements on (a) SRM 2494, fabricated on a multi-user 1.5 µm CMOS process [8] followed by a bulk-micromachining etch, as depicted in Fig. 1, and (b) SRM 2495, fabricated using a polysilicon multi-user surface-micromachining MEMS process [9] with a backside etch, as depicted in Fig. 2.

For SRM 2494: The thickness of the SiO$_2$ beam, t_{SiO2}, is obtained for SRM 2494 thickness measurements. This is the thickness of the composite oxide beam used in Sec. 2 through Sec. 4 for Young's modulus, residual strain, and strain gradient measurements, respectively, however, the oxide beam thickness is only a required input for Young's modulus and residual strain calculations in Data Analysis Sheet YM.3 and Data Analysis Sheet RS.3, respectively [13]. As shown in Fig. T2, four oxide thicknesses (t_1, t_2, t_3, and t_4) sum together to obtain t_{SiO2}. Fig. T2 also includes a more descriptive nomenclature for these thicknesses as defined and used in [6].

Before the post-processing of the SRM 2494 chips, bulk silicon is directly beneath the bottommost oxide layer of thickness t_1 in Fig. T2. During the post-processing XeF$_2$ etch (as specified in Sec. 1.4.1), any exposed silicon beside the designed cantilevers and fixed-fixed beams as well as beneath these beams is etched away. Therefore, the bottommost oxide layer in the designed cantilevers and fixed-fixed beams is of thickness t_1, as shown in Fig. T2. Also before the post-processing, a silicon nitride cap is on top of the topmost oxide layer of thickness t_4, in Fig. T2. During the post-processing CF$_4$+O$_2$ etch (as specified in Sec. 1.4.1), this nitride cap is removed such that the topmost oxide layer is of thickness t_4. Therefore, the oxide beam thickness is comprised of the four oxide thicknesses given in Fig. T2. Consult [6] for additional specifics associated with the process and these four SiO$_2$ layers.

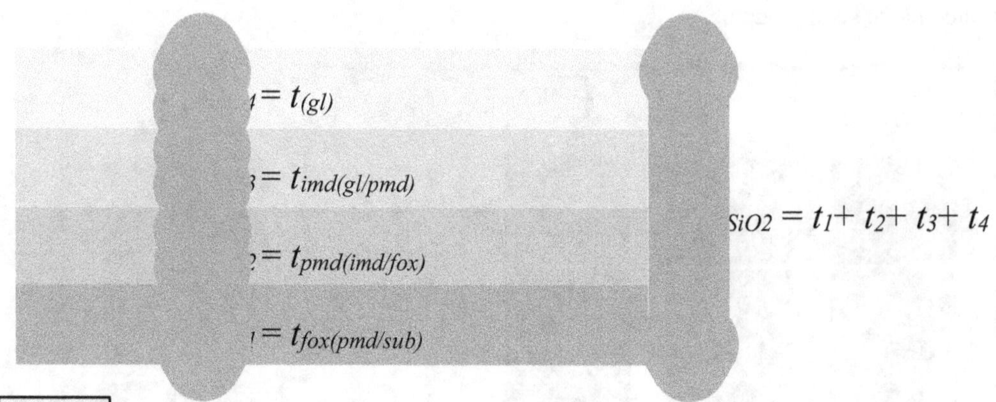

Figure T2. _The four SiO$_2$ thicknesses that comprise the composite oxide beam thickness_

There are four distinct thickness test structures (also called step height test structures in Sec. 5) on SRM 2494 shown in Fig. T1(a) (with three occurrences of each structure). The four test structures are given in Fig. T3, from which six step height measurements are obtained. These six measurements can be used in calculations to determine the thickness of the composite oxide beams for the determination of Young's modulus and residual strain in the first and second groupings of test structures (as specified in Secs. 2 and 3, respectively). The arrow(s) at the top of each test structure locate(s) the step(s) to be measured. As seen in this figure, one measurement is made on the first and third step height test structures and two measurements are made on the second and fourth step height test structures in order to obtain the composite beam oxide thickness. The fourth test structure (associated with the fifth and sixth arrows) does not have a reflective top surface for each platform and as such is intended to be used with a stylus profilometer (or comparable instrument) as specified in Table T1. (If the stylus makes its initial contact with the sample surface on top of the oxide between the third and fourth test structures, indentations in the sample surface are not expected.) Table T1 also includes details associated with the test structures. Cross sections for the test structures shown in Fig. T3 are given in Fig. T4 through Fig. T7 respectively. The design dimensions for these test structures are given in Table SH1 (in Sec. 5).

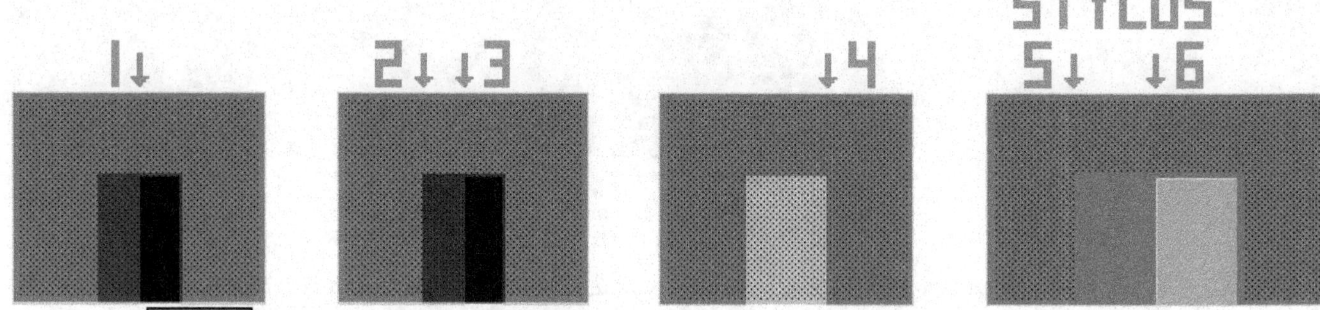

Figure T3. _For SRM 2494, four step height test structures used to obtain step height measurements._
Measurements from the test structures shown above can be used to determine the thickness of the composite oxide beams.

Measurement #[a]	Test Structure[b]	Step[c]	Cross section	Measuring Instrument
1	1st	$step1_{AB}$	see Fig. T4	optical interferometer or comparable instrument
2	2nd	$step2_{rA}$	see Fig. T5	optical interferometer or comparable instrument
3	2nd	$step1_{EF}$	see Fig. T5	optical interferometer or comparable instrument
4	3rd	$step1_{GH}$	see Fig. T6	optical interferometer or comparable instrument
5	4th	$step3_{AB}(n)^{-}$ [d]	see Fig. T7	stylus profilometer or comparable instrument
6	4th	$step3_{BC}(0)$ [e]	see Fig. T7	stylus profilometer or comparable instrument

[a] As given by the arrows in Fig. T3.

[b] Designates one of the four test structures depicted in Fig. T3.

[c] The names of these steps match the names of similar steps in the Certification Plus grouping of test structures on this chip as given in [6].

[d] The "(n)" indicates this measurement is taken after the chip is post processed using n cycles of a XeF$_2$ etch [6]. The trailing "$^{-}$" indicates that the nitride cap has been removed.

[e] The "(0)" indicates this measurement is taken before the chip is post processed, implying that the nitride cap is still present [6].

The first step height test structure (shown in Fig. T4) is a metal2 (m2)-over-poly1 (p1) step going from active area (aa) to field oxide (fox) as can be seen in the cross section given in Fig. T4. The name of this step ($step1_{AB}$) and the other steps in this grouping of step height test structures are such that they match the names of similar steps for the thickness test structures (given in the Certification Plus grouping of test structures on this chip) from which the thicknesses of all the layers in the process can be obtained using the electro-physical technique [6]. Consult the reference [6] for more details.

The reference platform around three of the four sides of this first step height test structure (and the other test structures in Fig. T3) consists of the deposited oxides sandwiched between active area and metal2.

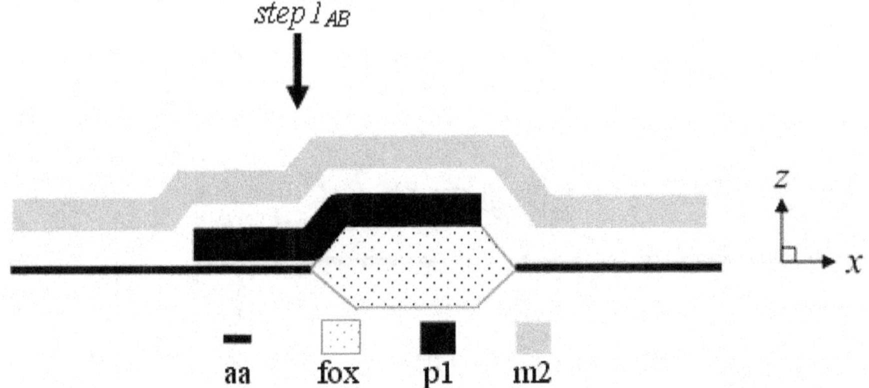

Figure T4. The cross section of the first step height test structure shown in Fig. T3.

The cross section of the second step height test structure (shown in Fig. T3) from which $step2_{rA}$ and $step1_{EF}$ are obtained is given in Fig. T5. The cross section of the third step height test structure from which $step1_{GH}$ is obtained is given in Fig. T6. And, the cross section of the fourth step height test structure from which $step3_{AB}(n)^{-}$ is obtained using a stylus profilometer (or comparable instrument) is given in Fig. T7. This same test structure is used at NIST to obtain $step3_{BC}(0)$ before the chip is post processed.

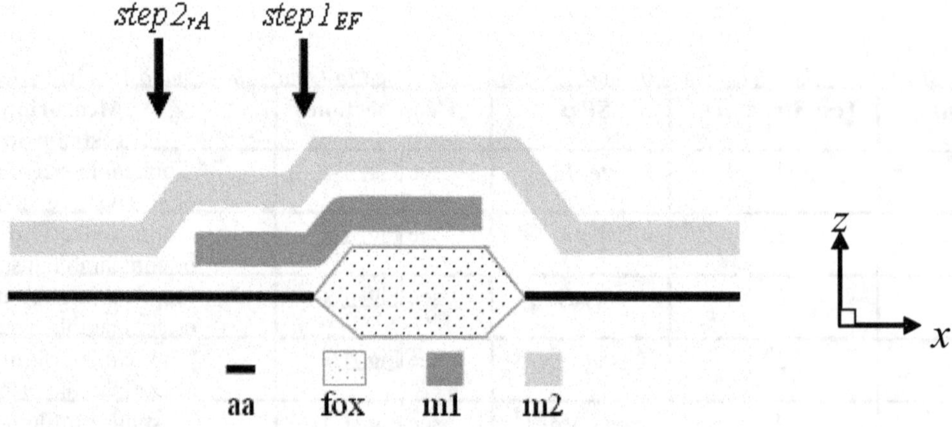

Figure T5. *The cross section of the second step height test structure shown in* Fig. T3.

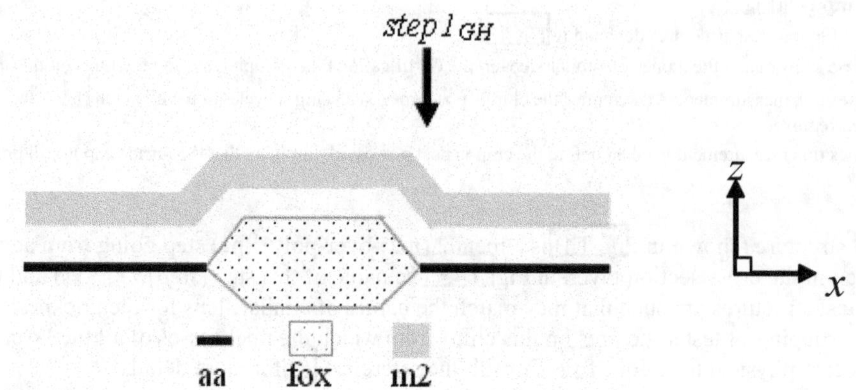

Figure T6. *The cross section of the third step height test structure shown in* Fig. T3.

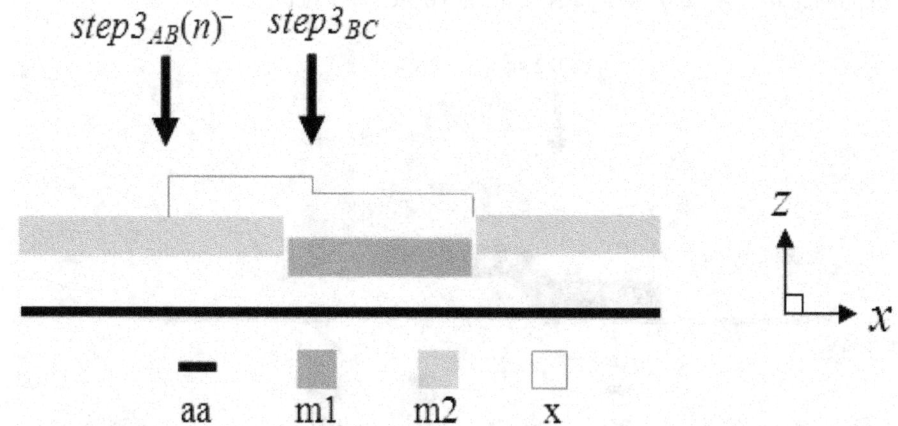

Figure T7. *The cross section of the fourth step height test structure shown in* Fig. T3.
*(The label "x" refers to the presence of an oxide and/or nitride layer,
depending upon when the measurement is taken.)*

For SRM 2495: The poly1 and poly2 thickness test structures are shown in Fig. T1(b) with one of the poly1 cantilevers given in Fig. T8(a). The poly2 cantilevers have a similar design. To obtain the thickness of this poly1 cantilever (or a poly2 cantilever) using the optomechanical technique [7], stiction is required. Stiction is defined as the adhesion between the portion of a structural layer that is intended to be freestanding and its underlying layer. As can be seen in Figs. T8(b and c), the cantilever beam in Fig. T8(a) is adhered to the top of the underlying layer. (The dimension J in Fig. T8(b) is depicted in Fig. T9 as the positive vertical distance between the bottom of the suspended structural layer and the top of the underlying layer, which takes into consideration the roughness of each surface, any residue present between the adhering elements, and a tilting component in the *y*-direction.)

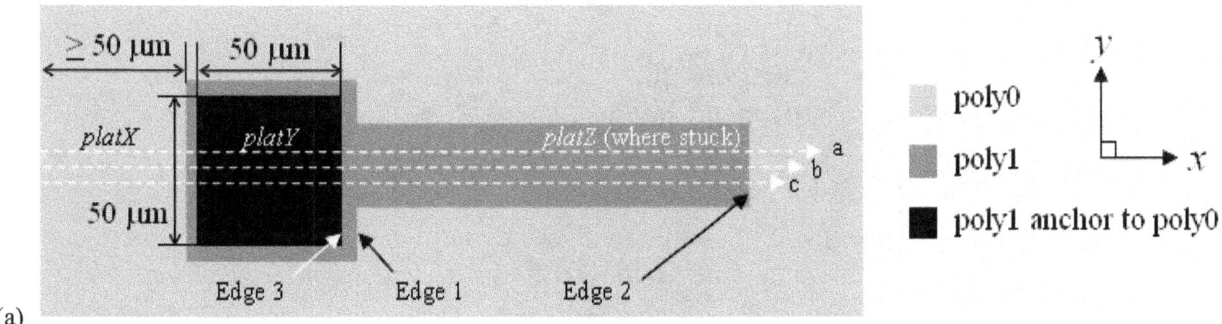

(a)

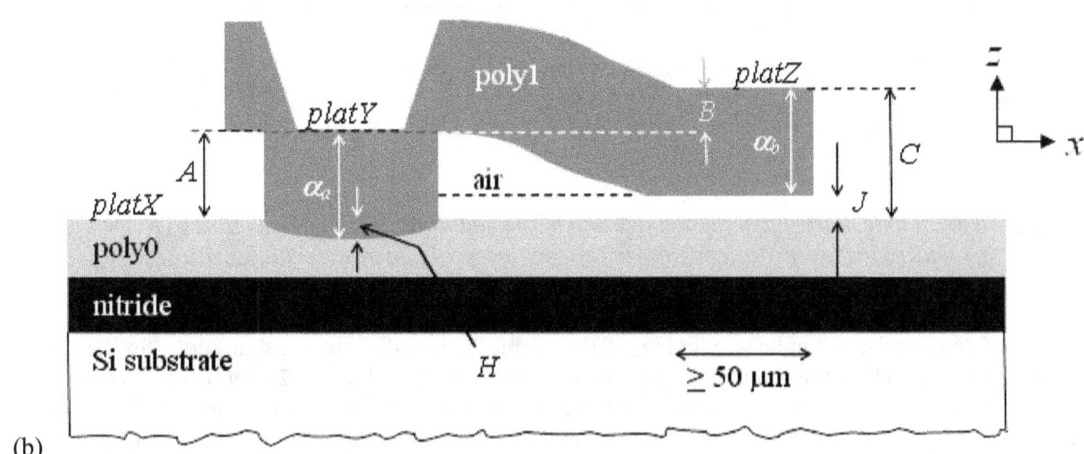

(b)

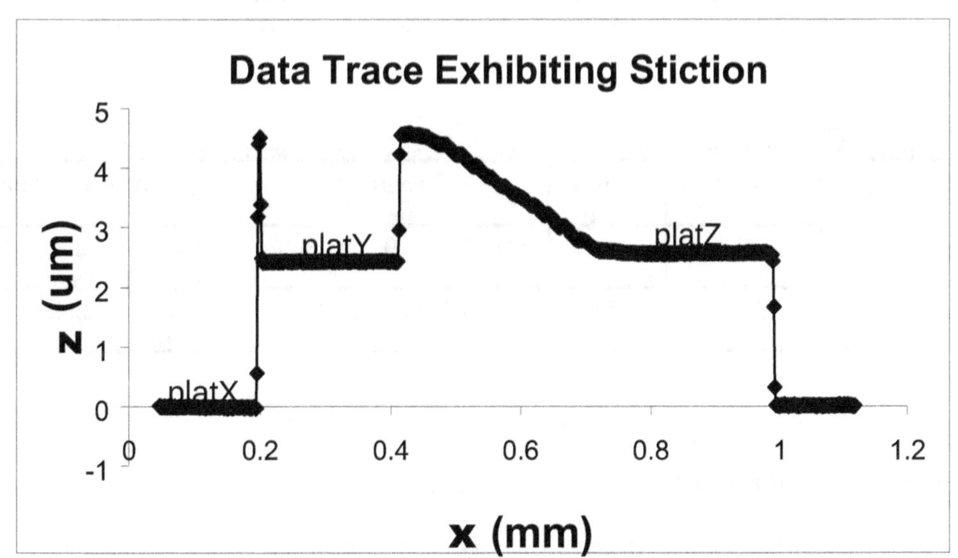

(c)

Figure T8. For a poly1 cantilever shown in Fig. T1(b), (a) a design rendition,
(b) a cross section showing the cantilever adhered to the top of the underlying layer, and
(c) a 2-D data trace [such as Trace a, b, or c in (a)] taken along the length of this cantilever.[61]

---

[61] *Copyright, ASTM International, 100 Barr Harbor Drive, West Conshohocken, PA 19428, USA. Reproduced via permissions with ASTM International.*

138

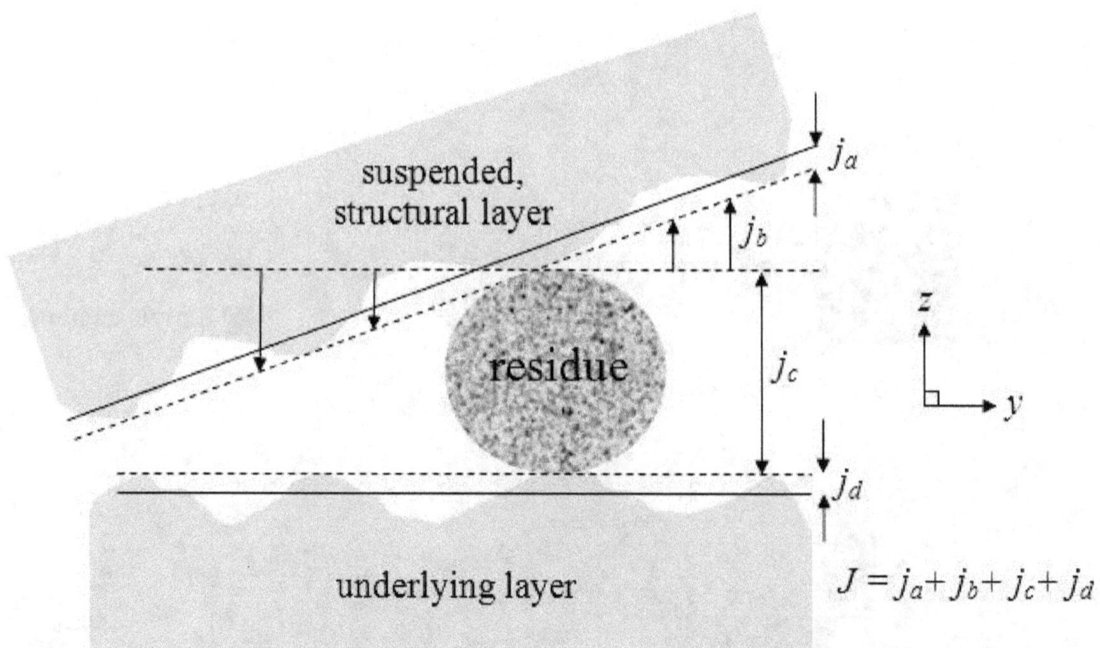

Figure T9. A schematic illustration, along the width of the cantilever where it has adhered to the top of the underlying layer, depicting the component parts of dimension J in Fig. T8(b).

There are two arrays of cantilevers, as can be seen in Fig. T1(b). Table T2 specifies the configurations for these cantilevers. The top array consists of 15 poly1 cantilevers and the bottom array consists of twelve poly2 cantilevers. As can barely be seen in this figure, the design length for a cantilever is given next to the second of three iterations of the same length cantilever. Both the poly1 and the poly2 arrays are designed with poly0 as the underlying layer. This is the layer to which a cantilever exhibiting stiction adheres. The poly0 layer is designed to encompass an entire array of cantilevers and, to aid in the thickness measurements, it extends at least 50 μm beyond each anchor on the opposite side of the anchor than the side from which the cantilever extends [as shown in Fig. T8(a) for one cantilever]. As also shown in Fig. T8(a), the anchor is designed to be 50 μm by 50 μm in order to obtain a sufficient amount of data in this region.

Table T2. Cantilever Configurations for Thickness Measurements on SRM 2495.

Structural Layer	Orientation	Width (μm)	Length (μm)	Quantity of Beams
poly1	$0°$	20	600, 650, 700, 750, 800	three of each length (or 15 beams)
poly2	$0°$	20	650, 700, 750, 800	three of each length (or 12 beams)

8.2 Calibration Procedures for Thickness Measurements

For SRM 2494: On SRM 2494, thickness measurements are taken on step height test structures. Therefore, see Sec. 5.2 in the step height section for calibration in the z-direction.

For SRM 2495: On SRM 2495, thickness measurements are taken on cantilevers that have adhered to the top of the underlying layer. In particular, a step height measurement is taken from the top of the poly0 layer to the top of the anchor, preferably using a stylus profilometer (or comparable instrument). Then, using an optical interferometer (or comparable instrument), a step height measurement is taken from the top of the anchor to the top of the portion of the cantilever that has adhered to the top of the underlying layer. To calibrate both instruments in the z-direction, see Sec. 5.2 for step height measurements. The NIST-supplied data will use an optical interferometer for both measurements to maintain the integrity of the SRM since indentations can be created when the stylus makes contact with the sample surface.

8.3 Using Data Analysis Sheet T.1

Data Analysis Sheet T.1 uses electrical and physical techniques [6] to calculate the thickness of the composite oxide beams on SRM 2494. The electrical techniques use capacitance values for the dielectric layers and sheet resistance and resistivity values for the metal2 layer. The physical technique is the step height procedure discussed in Sec. 5. Let us look at each of the four tables given in Data Analysis Sheet T.1 (a reproduction of which is given in Appendix 6) one at a time.

For Table 1 in Appendix 6, the first five step heights [namely, $step1_{AB}$, $step2_{rA}$, $step1_{EF}$, $step1_{GH}$, and $step3_{AB}(n)^-$] in Fig. T4 through Fig. T7 and their combined standard uncertainty values, u_{cSH}, are requested from the four step height test structures given in Fig. T3. These five values are obtained from Data Analysis Sheet SH.1.a as discussed in Sec. 5. The sixth requested step height in Table 1 of Appendix 6 is $step3_{BC}(0)$. The NIST-supplied values for $step3_{BC}(0)$ and its combined standard uncertainty on Data Analysis Sheet T.1, which accompanies SRM 2494, should be used for these values since $step3_{BC}(0)$ is a measurement taken before the post processing. Also requested in Table 1 is a residual uncertainty component, u_{res}, for each step height. This component is a place holder to describe additional sources of uncertainty that may become apparent during the measurement. It can be set equal to zero (i.e., $u_{res}=0$) such that the combined standard uncertainty value for each step height, u_{cstep}, is the value obtained from Data Analysis Sheet SH.1.a. In other words:

$$u_{cstep} = \sqrt{u_{cSH1a}^2 + u_{res}^2} = u_{cSH1a} \ . \tag{T1}$$

A number (or a number and a letter) following the subscript "SH" in "u_{cSH}" indicates the data analysis sheet that is used to obtain the combined standard uncertainty value. Therefore, u_{cSH1a} implies that Data Analysis Sheet SH.1.a is used.

At the top of Table 2 in Appendix 6, σ_ε is requested where σ_ε is the estimated standard deviation of the value stated for the permittivity of SiO_2, ε_{SiO2} (where $\varepsilon_{SiO2}=34.5$ aF/μm). Then, in Table 2, select values of capacitance per unit area (C_a) and their standard deviations (σ_{Ca}), as specified in the first column of Table T3, are obtained from the semiconductor fabrication service [8]. [The NIST-supplied data on Data Analysis Sheet T.1 that accompanies SRM 2494 should be used for these entries.] The corresponding thicknesses (t), as specified in the second column of Table T3 are calculated using the following equation:

$$t = \varepsilon_{SiO2} / C_a \ . \tag{T2}$$

The combined standard uncertainty values for these thicknesses (u_{ctCa}) are obtained by applying the propagation of uncertainty technique (as presented in Appendix 8) to Eq. (T2) for uncorrelated parameters in a multiplicative relationship. The one sigma uncertainty of the value of the thickness, σ_{tCa} (which can be equated here with the combined standard uncertainty for the thickness, u_{ctCa}) can be written as follows:

$$u_{ctCa} = \sigma_{tCa} = t\sqrt{\left(\frac{\sigma_\varepsilon}{\varepsilon_{SiO2}}\right)^2 + \left(\frac{\sigma_{Ca}}{C_a}\right)^2} \ , \tag{T3}$$

where each uncertainty component is obtained using a Type B analysis. If a residual component is added, the equation can be written as follows:[62]

$$u_{ctCa} = \sigma_{tCa} = t\sqrt{\left(\frac{\sigma_\varepsilon}{\varepsilon_{SiO2}}\right)^2 + \left(\frac{\sigma_{Ca}}{C_a}\right)^2 + \left(\frac{\sigma_{resCa}}{r_{res}}\right)^2} \ , \tag{T4}$$

where (σ_{resCa}/r_{res}) represents the relative uncertainty due to a residual component (which is typically assumed to be zero).

---

[62] This uncertainty equation is different than the one presented in reference [6].

Table T3. For SRM 2494, the Inputs Requested for Table 2 of Data Analysis Sheet T.1

Obtain C_a and σ_{Ca} for the following:	For a calculation of the corresponding thickness[a] using $t = \varepsilon_{SiO2}/C_a$
p1-to-substrate (sub)	$t_{fox(p1/sub)elec}$
p1-to-aan[b]	$t_{thin(p1/aan)elec}$
m1-to-sub	$[t_{fox\ m1(pmd/sub)} + t_{pmd(m1/fox)}]elec$
m1-to-aan	$t_{pmd(m1/aan)elec}$
m2-to-sub	$[t_{fox\ m2(pmd/sub)} + t_{pmd(imd/fox)} + t_{imd(m2/pmd)}]elec$
m2-to-aan	$[t_{pmd(imd/aan)} + t_{imd(m2/pmd)}]elec$

[a] Consult the reference [6] for a detailed description of the nomenclature used.

[b] Poly1-to-active area with an n-implant

For Table 3 in Appendix 6, the metal2 (m2) thickness, $t_{(m2)elec}$, is calculated using the following equation:

$$t_{(m2)elec} = \frac{\rho}{R_s} \quad , \tag{T5}$$

where R_s (the sheet resistance) and ρ (the resistivity) are obtained from the semiconductor fabrication service [8]. [The NIST-supplied data on Data Analysis Sheet T.1 that accompanies SRM 2494 should be used for these entries.] The combined standard uncertainty value for this thickness (u_{ctRs}) is obtained by applying the propagation of uncertainty technique (as presented in Appendix 8) to Eq. (T5) for uncorrelated parameters in a multiplicative relationship. The one sigma uncertainty of the value of the thickness, σ_{tRs} (which can be equated here with the combined standard uncertainty for the thickness, u_{ctRs}) can be written as follows:

$$u_{ctRs} = \sigma_{tRs} = t\sqrt{\left(\frac{\sigma_\rho}{\rho}\right)^2 + \left(\frac{\sigma_{Rs}}{R_s}\right)^2} \quad , \tag{T6}$$

where σ_{Rs} (the standard deviation of the sheet resistance) and σ_ρ (the standard deviation of the resistivity) are obtained from the semiconductor fabrication service [8]. [The NIST-supplied data on Data Analysis Sheet T.1 that accompanies SRM 2494 should be used for these entries.] Each uncertainty component is obtained using a Type B analysis. If a residual component is added, the equation can be written as follows:[63]

$$u_{ctRs} = \sigma_{tRs} = t\sqrt{\left(\frac{\sigma_\rho}{\rho}\right)^2 + \left(\frac{\sigma_{Rs}}{R_s}\right)^2 + \left(\frac{\sigma_{resRs}}{r_{res}}\right)^2} \quad , \tag{T7}$$

where (σ_{resRs}/r_{res}) represents the relative uncertainty due to a residual component (which is typically assumed to be zero).

To discuss Table 4 in Appendix 6, let us turn our attention to Table T4, which includes the composite oxide beam thickness, t_{SiO2}, and the four SiO_2 thicknesses it is comprised of [namely, $t_{fox(pmd/sub)}$, $t_{pmd(imd/fox)}$, $t_{imd(gl/pmd)}$, and $t_{(gl)}$,[64] as given by t_1, t_2, t_3, and t_4 as shown in Fig. T2]. Note in Table T4 that t_1 can be calculated four different ways as given by t_{1a}, t_{1b}, t_{1c}, and t_{1d}. This table also includes the equations [6] used to obtain these thicknesses. These equations have been rewritten in this table

---

[63] This uncertainty equation is different than the one presented in reference [6].

[64] Consult the reference [6] for a detailed description of the nomenclature used.

using a simplified notation (capital letters) for the thicknesses or step heights as given in Table T5. It is assumed that these thicknesses or step heights are uncorrelated and that the standard deviations are known. As shown in Table T4, the thickness of the oxide beam, t_{SiO2}, is given by Z, in the equation that follows:

$$Z = t_{SiO2} = t_1 + t_2 + t_3 + t_4 = X + Y \quad , \tag{T8}$$

where

$$X = t_3 + t_4 - t_{imd(m2/m1)elec} = B + C - D + E \;, \tag{T9}$$

and where Y can be calculated in one of the following four possible ways:

$$Y_1 = t_{1d} + t_2 + t_{imd(m2/m1)elec} = M \;, \tag{T10}$$

$$Y_2 = t_{1c} + t_2 + t_{imd(m2/m1)\;elec} = A + K - L \;, \tag{T11}$$

$$Y_3 = t_{1b} + t_2 + t_{imd(m2/m1)\;elec} = A - J + O \;, \tag{T12}$$

or

$$Y_4 = t_{1a} + t_2 + t_{imd(m2/m1)\;elec} = A + F + O \;, \tag{T13}$$

where

$$O = G - H + \frac{HI}{I - G} \;. \tag{T14}$$

Table T4. For SRM 2494, the Four SiO_2 Oxide Thicknesses That Compose the Oxide Beam

	Thickness[a]	Equation[a]
1a	$t_{1a} = t_{fox(pmd/sub)}$	$t_{1a} = step1_{EF} + t_{1plus} = F + t_{1plus}$ $t_{1plus} = t_{fox(p1/sub)elec} - step1_{AB} + t_{1plus3} = G - H + t_{1plus3}$ $t_{1plus3} = [step1_{AB}/(t_{thin(p1/aan)elec} - t_{fox(p1/sub)elec})]t_{thin(p1/aan)elec} = HI/(I - G)$ so $t_{1a} = F + G - H + HI/(I - G)$
1b	$t_{1b} = t_{fox(pmd/sub)}$	$t_{1b} = -step1_{GH} + t_{1plus} = -J + t_{1plus}$ so $t_{1b} = -J + G - H + HI/(I - G)$
1c	$t_{1c} = t_{fox(pmd/sub)}$	$t_{1c} = [t_{fox,m1(pmd/sub)} + t_{pmd(m1/fox)}]_{elec} - t_{pmd(m1/aan)elec} = K - L$
1d	$t_{1d} = t_{fox(pmd/sub)}$	$t_{1d} = [t_{fox,m2(pmd/sub)} + t_{pmd(imd/fox)} + t_{imd(m2/pmd)}]_{elec} - [t_{pmd(imd/aan)} + t_{imd(m2/pmd)}]_{elec}$ $= M - A$
1	t_1	$t_1 =$ the thickness (i.e., t_{1a}, t_{1b}, t_{1c}, or t_{1d}) with the smallest value for u_c
2	$t_2 = t_{pmd(imd/fox)}$	$t_2 = [t_{pmd(imd/aan)} + t_{imd(m2/pmd)}]_{elec} - t_{imd(m2/m1)elec} = A - t_{imd(m2/m1)elec}$

3	$t_3 = t_{imd(gl/pmd)}$	$t_3 = t_{(m2)elec} + step3_{BC}(0) - step2_{rA} + t_{imd(m2/m1)elec} = B + C - D + t_{imd(m2/m1)elec}$
4	$t_4 = t_{(gl)}$	$t_4 = step3_{AB}(n)^- = E$
	t_{SiO2}	$t_{SiO2} = t_1 + t_2 + t_3 + t_4 = Z$

[a] Consult the reference [6] as needed.

<table T5 title>

Table T5. Simplified Notation for Thicknesses or Step Heights and Their Standard Deviations

	Thickness or Step Height[a]	Simplified Notation	Standard Deviation
1	$[t_{pmd(imd/aan)} + t_{imd(m2/pmd)}]_{elec}$	A	σ_A
2	$t_{(m2)elec}$	B	σ_B
3	$step3_{BC}(0)$	C	σ_C
4	$step2_{rA}$	D	σ_D
5	$step3_{AB}(n)^-$	E	σ_E
6	$step1_{EF}$	F	σ_F
7	$t_{fox(p1/sub)elec}$	G	σ_G
8	$step1_{AB}$	H	σ_H
9	$t_{thin(p1/aan)elec}$	I	σ_I
10	$step1_{GH}$	J	σ_J
11	$[t_{fox,m1(pmd/sub)} + t_{pmd(m1/fox)}]_{elec}$	K	σ_K
12	$t_{pmd(m1/aan)elec}$	L	σ_L
13	$[t_{fox,m2(pmd/sub)} + t_{pmd(imd/fox)} + t_{imd(m2/pmd)}]_{elec}$	M	σ_M

[a] Consult the reference [6] as needed.

Table T6 presents these equations for O, X, Y_1, Y_2, Y_3, Y_4, and the four possible calculations of Z (namely, Z_1, Z_2, Z_3, and Z_4) along with the equations to calculate the standard deviation values. (Table 4 in Appendix 6 provides the calculation results.) For all of the standard deviation calculations (except for σ_O), the propagation of uncertainty technique [19-21] (a brief overview of which is given in Appendix 8) is used for uncorrelated parameters in an additive relationship.

	Thickness or Step Height	Standard deviation
1	$O = G - H + \dfrac{HI}{I-G}$	$\sigma_O = \sqrt{\left[1 + \dfrac{HI}{(I-G)^2}\right]^2 \sigma_G^2 + \dfrac{G^2}{(I-G)^2}\sigma_H^2 + \dfrac{(GH)^2}{(I-G)^4}\sigma_I^2}$
2	$X = B + C - D + E$	$\sigma_X = \sqrt{\sigma_B^2 + \sigma_C^2 + \sigma_D^2 + \sigma_E^2}$
3	$Y_1 = M$	$\sigma_{Y1} = \sigma_M$
4	$Y_2 = A + K - L$	$\sigma_{Y2} = \sqrt{\sigma_A^2 + \sigma_K^2 + \sigma_L^2}$
5	$Y_3 = A - J + O$	$\sigma_{Y3} = \sqrt{\sigma_A^2 + \sigma_J^2 + \sigma_O^2}$
6	$Y_4 = A + F + O$	$\sigma_{Y4} = \sqrt{\sigma_A^2 + \sigma_F^2 + \sigma_O^2}$
7	$Z_1 = X + Y_1$	$\sigma_{Z1} = \sqrt{\sigma_X^2 + \sigma_{Y1}^2}$ $\sigma_{Z1} = \sqrt{\sigma_B^2 + \sigma_C^2 + \sigma_D^2 + \sigma_E^2 + \sigma_M^2}$
8	$Z_2 = X + Y_2$	$\sigma_{Z2} = \sqrt{\sigma_X^2 + \sigma_{Y2}^2}$ $\sigma_{Z2} = \sqrt{\sigma_A^2 + \sigma_B^2 + \sigma_C^2 + \sigma_D^2 + \sigma_E^2 + \sigma_K^2 + \sigma_L^2}$
9	$Z_3 = X + Y_3$	$\sigma_{Z3} = \sqrt{\sigma_X^2 + \sigma_{Y3}^2}$ $\sigma_{Z3} = \sqrt{\sigma_A^2 + \sigma_B^2 + \sigma_C^2 + \sigma_D^2 + \sigma_E^2 + \sigma_J^2 + \sigma_O^2}$
10	$Z_4 = X + Y_4$	$\sigma_{Z4} = \sqrt{\sigma_X^2 + \sigma_{Y4}^2}$ $\sigma_{Z4} = \sqrt{\sigma_A^2 + \sigma_B^2 + \sigma_C^2 + \sigma_D^2 + \sigma_E^2 + \sigma_F^2 + \sigma_O^2}$

The equation for O [see Eq. (T14)] contains what we are assuming are uncorrelated input parameters in a relationship that is not solely additive or multiplicative. Therefore, as given in Appendix 8, the partial derivatives are found such that σ_O can be calculated as follows:

$$\sigma_O = \sqrt{\left[\frac{\partial O}{\partial G}\right]^2 \sigma_G^2 + \left[\frac{\partial O}{\partial H}\right]^2 \sigma_H^2 + \left[\frac{\partial O}{\partial I}\right]^2 \sigma_I^2} \quad , \tag{T15}$$

where

$$\frac{\partial O}{\partial G} = 1 + \frac{HI}{(I-G)^2} \, , \tag{T16}$$

$$\frac{\partial O}{\partial H} = \frac{G}{I-G} \, , \tag{T17}$$

and

144

$$\frac{\partial O}{\partial I} = -\frac{GH}{(I-G)^2}, \tag{T18}$$

such that

$$\sigma_O = \sqrt{\left[1+\frac{HI}{(I-G)^2}\right]^2 \sigma_G^2 + \frac{G^2}{(I-G)^2}\sigma_H^2 + \frac{(GH)^2}{(I-G)^4}\sigma_I^2} \; . \tag{T19}$$

After equating σ_{Z1}, σ_{Z2}, σ_{Z3}, and σ_{Z4} in Table T6 with u_{Z1}, u_{Z2}, u_{Z3}, and u_{Z4}, respectively, the smallest of the combined standard uncertainty values u_{Z1}, u_{Z2}, u_{Z3}, and u_{Z4} is called u_{cSiO2}. The corresponding Z value (namely, Z_1, Z_2, Z_3, or Z_4, respectively) is chosen to represent t_{SiO2}.

For SRM 2494 that uses Data Analysis Sheet T.1, the expanded uncertainty is U_{SiO2}, as calculated using the following equation:

$$U_{SiO2} = ku_{cSiO2} = 2u_{cSiO2}, \tag{T20}$$

where the k value of 2 approximates a 95 % level of confidence.

Reporting results [19-21]. Since it can be assumed that the estimated values of the uncertainty components are either approximately uniformly or Gaussianly distributed with approximate combined standard uncertainty u_{cSiO2}, the composite oxide beam thickness is believed to lie in the interval $t_{SiO2} \pm u_{cSiO2}$ (coverage factor k=1) representing a level of confidence of approximately 68 %.

8.4 Using Data Analysis Sheet T.3

Data Analysis Sheet T.3 uses the optomechanical technique [7] to calculate the poly1 or poly2 thickness on SRM 2495 using a cantilever test structure [such as shown in Fig. T8(a) for poly1] that has adhered to the top of the underlying poly0 layer, as can be seen in Figs. T8(b and c). ASTM standard test method E 2246 [3] can be used to determine if the cantilever is exhibiting stiction. Once assured that the cantilever is exhibiting stiction, two step height measurements are made using SEMI standard test method MS2 [4]; therefore, Sec. 5 can be consulted, as appropriate, for measurement and calculation details. For the first step height measurement, data from three 2-D data traces are used to calculate A, as shown in Fig. T8(b), and for the second step height measurement, data from three 2-D data traces are used to calculate B, also shown in Fig. T8(b). This technique recommends a stylus measurement for the measurement of A, and an optical measurement for the measurement of B for a lower combined standard uncertainty value than if an optical instrument were used for both measurements. The NIST-supplied data will use an optical interferometer for both measurements to maintain the integrity of the SRM since indentations can be created when the stylus makes contact with the sample surface.

Let us look at each of the five tables (i.e., Tables 1, 2, 3a, 3b, and 3c) given in Data Analysis Sheet T.3 (a reproduction of which is given in Appendix 7) one at a time. In these tables, the trailing N subscript to specific parameters is A when referring to the measurement of A and B when referring to the measurement of B.

For Table 1 in Appendix 7, the first two inputs given below are used to specify the environmental conditions:
1. The input $temp_N$ (i.e., input #1) specifies the temperature during the measurement.
2. The input $relative\ humidity_N$ (i.e., input #2) specifies the relative humidity during the measurement.

Referring to Fig. T1(b), the next five inputs given below are used to locate the target test structure on SRM 2495:
1. The input mat (i.e., input #3) is used to identify if the target test structure is in the upper poly1 array [that is given the designation P1 in Fig. T1(b)] or in the lower poly2 array (that is given the designation P2).
2. The input $test\ structure$ (i.e., input #4) specifies the test structure to be measured. The upper and lower arrays in Fig. T1(b) consist of only cantilevers. Therefore, the radio button for cantilever should be selected.
3. The input $design\ length$ (i.e., input #5) specifies the design length of the target test structure. As can barely be seen in Fig. T1(b), the design length is specified to the left of the anchor of the second of three identically designed cantilevers. The possible poly1 and poly2 design lengths are given in Table T2.

4. The input *which* (i.e., input #6) specifies which cantilever of the given length is the target test structure. Since there are three instances of each cantilever, the radio button corresponding to "first," "second," or "third" is used to locate the target test structure, where "first" corresponds to the topmost cantilever in the array that has the given length.

5. The input *orient* (i.e., input #7) specifies the orientation of the target test structure. The orientation is 0° for all of the cantilevers in Fig. T1(b).

The next three inputs as given below are for general information and reminders to the user:

1. The input mag_N (i.e., input #8) specifies the magnification used for the measurement of A or B, as appropriate.

2. The input $align_N$ (i.e., input #9) specifies whether or not the data obtained have been aligned properly [4]. The purpose of this input is to remind the user to align the test structure, with respect to the optics of the instrument, before taking a measurement.

3. The input $level_N$ (i.e., input #10) specifies whether or not the data have been leveled. The purpose of this input is to remind the user to level the data before recording a measurement.

The next nine inputs, in Table 1 of Appendix 7, are associated with the calibration in the z-direction of the instrument(s) used to measure A and B. (A stylus profilometer is recommended for the measurement of A, and an optical interferometer is recommended for the measurement of B; however, the NIST-supplied data use an optical interferometer for both measurements to maintain the integrity of the SRM since indentations can be created when the stylus makes contact with the sample surface.) Consult Sec 5.2 for details associated with these inputs which follow:

1. The input $cert_N$ (i.e., input #11),

2. The input σ_{certN} (i.e., input #12),

3. The input σ_{6aveN} (i.e., input #13),

4. The input $\bar{z}_{6aveN}$ (i.e., input #14),

5. The input σ_{6sameN} (i.e., input #15),

6. The input $\bar{z}_{6sameN}$ (i.e., input #16),

7. The input z_{driftN} (i.e., input #17),

8. The input cal_{zN} (i.e., input #18), and

9. The input z_{linN} (i.e., input #19),

where once again N is A when associated with the measurement of A and B when associated with the measurement of B.

The remaining eight inputs in Table 1 of Appendix 7 are associated with the processing of SRM 2495. (The NIST-supplied data on Data Analysis Sheet T.3 that accompanies SRM 2495 can be used for the first five inputs.) A description of the eight inputs[65] is given in Data Analysis Sheet T.3 [13], a reproduction of which is given in Appendix 7, and repeated below:

1. The input $\sigma_{repeat(samp)N}$ (i.e., input #20) specifies the relative step height *repeatability* standard deviation as obtained from step height test structures fabricated in a process similar to that used to fabricate the sample.

2. The input H (i.e., input #21) specifies the anchor etch depth, as shown in Fig. T8(b).

3. The input ΔH (i.e., input #22) specifies the range of the anchor etch depth (as provided by the processing facility).

4. The input J_{est} (i.e., input #23) is an estimated value for the dimension J (if known) [7], as shown in Fig. T8(b). If it is not known, zero should be inputted.

5. The input u_{cJest} (i.e., input #24) is an estimated value for the combined standard uncertainty of J_{est} (if J_{est} is known and inputted); otherwise zero should be inputted.

6. The input s_{roughX} (i.e., input #25) is the uncalibrated surface roughness of *platX*, shown in Fig. T8(a), calculated as the smallest of all the measured values obtained for s_{platXt}, as discussed below. (Consult the Definition of Symbols Section, if needed.) However, if the surfaces of *platX*, *platY*, and *platZ* all have identical compositions, then it is calculated as the smallest of all the values obtained for s_{platXt}, $s_{platYt1}$, $s_{platYt2}$, and s_{platZt} in which case $s_{roughX} = s_{roughY} = s_{roughZ}$.

7. The input s_{roughY} (i.e., input #26) is the uncalibrated surface roughness of *platY*, shown in Fig. T8(a), calculated as the smallest of all the measured values obtained for $s_{platYt1}$ and $s_{platYt2}$, as discussed below. (Consult the Definition of Symbols Section, if needed.) However, if the surfaces of *platX*, *platY*, and *platZ* all have identical

---

[65] Consult the Definition of Symbols section, if needed.

compositions, then it is calculated as the smallest of all the values obtained for s_{platXt}, $s_{platYt1}$, $s_{platYt2}$, and s_{platZt} in which case $s_{roughX}=s_{roughY}=s_{roughZ}$.

8. The input s_{roughZ} (i.e., input #27) is the uncalibrated surface roughness of *platZ*, shown in Fig. T8(a), calculated as the smallest of all the measured values obtained for s_{platZt}, as discussed below. (Consult the Definition of Symbols Section, if needed.) However, if the surfaces of *platX*, *platY*, and *platZ* all have identical compositions, then it is calculated as the smallest of all the values obtained for s_{platXt}, $s_{platYt1}$, $s_{platYt2}$, and s_{platZt} in which case $s_{roughX}=s_{roughY}=s_{roughZ}$.

For Table 2 in Appendix 7, uncalibrated inputs are requested from *platX*, *platY*, and *platZ*, as shown in Fig. T8(a) using SEMI standard test method MS2 [4]. In particular, both the platform height and standard deviation values are requested from Trace a, Trace b, and Trace c, where the data are leveled and zeroed with respect to the top of the underlying poly0 layer. Therefore, for the measurements of *A* taken with a stylus profilometer, the uncalibrated inputs *platXa*, *platXb*, *platXc*, s_{platXa}, s_{platXb}, and s_{platXc} are requested from *platX* and the uncalibrated inputs *platYa1*, *platYb1*, *platYc1*, $s_{platYa1}$, $s_{platYb1}$, and $s_{platYc1}$ are requested from *platY*. For the measurements of *B* taken with an optical interferometer, the uncalibrated inputs *platYa2*, *platYb2*, *platYc2*, $s_{platYa2}$, $s_{platYb2}$, and $s_{platYc2}$ are requested from *platY* and the uncalibrated inputs *platZa*, *platZb*, *platZc*, s_{platZa}, s_{platZb}, and s_{platZc} are requested from *platZ*. If an interferometric microscope is used for the measurements of *A* and *B*, the *platYt* and s_{platYt} data for both measurements can be the same (e.g., *platYa1*=*platYa2* and $s_{platYa1}=s_{platYa2}$) if the measurements of *platXt*, *platYt*, *platZt*, and the corresponding standard deviations are obtained from the same data trace.

For Table 3a in Appendix 7, calibrated values for A_a, A_b, A_c, B_a, B_b, and B_c are calculated using the following equations:[66]

$$A_t = (platYt1 - platXt)cal_{zA} \text{, and} \tag{T21}$$

$$B_t = (platZt - platYt2)cal_{zB}, \tag{T22}$$

where *t* is the data trace (i.e., Trace a, Trace b, or Trace c) being considered. Also, the following standard deviations are calculated:

$$s_{platXave} = cal_{zA}\frac{\left(s_{platXa} + s_{platXb} + s_{platXc}\right)}{3}, \tag{T23}$$

$$s_{platY1ave} = cal_{zA}\frac{\left(s_{platYa1} + s_{platYb1} + s_{platYc1}\right)}{3}, \tag{T24}$$

$$s_{platY2ave} = cal_{zB}\frac{\left(s_{platYa2} + s_{platYb2} + s_{platYc2}\right)}{3}, \text{ and} \tag{T25}$$

$$s_{platZave} = cal_{zB}\frac{\left(s_{platZa} + s_{platZb} + s_{platZc}\right)}{3}. \tag{T26}$$

For Table 3b in Appendix 7, *A*, *B*, and their combined standard uncertainty values (namely, u_{cSHA} and u_{cSHB}, respectively) are calculated using the following equations[67] (where *N* is *A* when referring to the measurement of *A*, and *B* when referring to the measurement of *B*):

$$N = \frac{(N_a + N_b + N_c)}{3}, \text{ and} \tag{T27}$$

$$u_{cSHN} = \sqrt{u_{LstepN}^2 + u_{WstepN}^2 + u_{certN}^2 + u_{calN}^2 + u_{repeat(shs)N}^2 + u_{driftN}^2 + u_{linearN}^2 + u_{repeat(samp)N}^2}, \tag{T28}$$

---

[66] These equations are similar to those found in Sec. 5 for step height measurements.
[67] These equations are similar to those found in Sec. 5 for step height measurements.

where $\quad u_{LstepA} = \sqrt{s_{platXave}^2 - (cal_{zA}s_{roughX})^2 + s_{platY1ave}^2 - (cal_{zA}s_{roughY})^2}$, $\qquad$ (T29)

$$u_{LstepB} = \sqrt{s_{platY2ave}^2 - (cal_{zB}s_{roughY})^2 + s_{platZave}^2 - (cal_{zB}s_{roughZ})^2} \; , \qquad \text{(T30)}$$

$$u_{WstepN} = \sigma_{WstepN} = STDEV(N_a, N_b, N_c) \; , \qquad \text{(T31)}$$

$$u_{certN} = \left| \frac{\sigma_{certN} N}{cert_N} \right| \; , \qquad \text{(T32)}$$

$$u_{calN} = \left| \frac{\sigma_{6aveN} N}{\bar{z}_{6aveN}} \right| \; , \qquad \text{(T33)}$$

$$u_{repeat(shs)N} = \left| \frac{\sigma_{6sameN} N}{\bar{z}_{6sameN}} \right| \; , \qquad \text{(T34)}$$

$$u_{driftN} = \left| \frac{(z_{driftN} cal_{zN})N}{2\sqrt{3} cert_N} \right| \; , \qquad \text{(T35)}$$

$$u_{linearN} = \left| \frac{z_{linN} N}{\sqrt{3}} \right| \; , \text{ and} \qquad \text{(T36)}$$

$$u_{repeat(samp)N} = \sigma_{repeat(samp)N} |N| \; . \qquad \text{(T37)}$$

Then, in Table 3c of Appendix 7, the following calculations are made [7]

$$C = A + B \text{ and } u_{cC} = \sqrt{u_{cSHA}^2 + u_{cSHB}^2} \; , \qquad \text{(T38)}$$

$$J = B - H \text{ and } u_{cJ} = \sqrt{u_{cSHB}^2 + u_{cH}^2} \text{ where } u_{cH} = \Delta H / 6 \; , \qquad \text{(T39)}$$

$$\alpha_a = A + H \text{ and } u_{c\alpha_a} = \sqrt{u_{cSHA}^2 + u_{cH}^2} \; , \text{ and} \qquad \text{(T40)}$$

$$\alpha_b = C - J_{est} \text{ and } u_{c\alpha_b} = \sqrt{u_{cC}^2 + u_{cJest}^2} \; . \qquad \text{(T41)}$$

The thickness of the suspended poly1 or poly2 layer, α, is taken to be the value specified for α_a or α_b (whichever has the smaller combined standard uncertainty value, $u_{c\alpha}$) unless $J_{est} = 0$ in which case $\alpha = \alpha_a$. Also, each of the standard uncertainty components is obtained using a Type B analysis, except for u_{WstepN}, u_{calN}, $u_{repeat(shs)N}$, and $u_{repeat(samp)N}$, which use a Type A analysis.

For SRM 2495 using Data Analysis Sheet T.3, the expanded uncertainty is $U\alpha$, as calculated using the following equation:

$$U_\alpha = ku_{c\alpha} = 2u_{c\alpha} \quad , \tag{T42}$$

where the k value of 2 approximates a 95 % level of confidence.

Reporting results [19-21]: Since it can be assumed that the estimated values of the uncertainty components are either approximately uniformly or Gaussianly distributed with approximate combined standard uncertainty $u_{c\alpha}$, the thickness is believed to lie in the interval $\alpha \pm u_{c\alpha}$ (expansion factor $k=1$) representing a level of confidence of approximately 68 %.

8.5 Using the MEMS 5-in-1 to Verify Thickness Measurements

To compare your thickness measurements with NIST measurements, you will need to fill out Data Analysis Sheet T.1 or T.3 when using SRM 2494 or SRM 2495, respectively. These data analysis sheets are accessible via the URL specified in the reference [13] and reproductions of them are given in Appendix 6 and Appendix 7, respectively. After calibrating the instrument, locating the test structure, taking the measurements, and performing the calculations, the data on your completed form can be compared with the data on the SRM Certificate and the completed data analysis sheet supplied with the MEMS 5-in-1. Details of the procedure are given below.

Calibrate the instrument: Calibrate the instrument(s) as given in Sec. 8.2. For Data Analysis Sheet T.1, consult Sec. 5.6 for specifics associated with Data Analysis Sheet SH.1.a. For Data Analysis Sheet T.3, see Sec. 8.4 for the inputs requested for both the measurement of A and the measurement of B.

Locate the target test structure: For SRM 2494, as shown in Fig. 1, thickness measurements are taken on the step height test structures given in the fourth grouping of test structures, which is shown in Fig. T1(a). For SRM 2495, as shown in Fig. 2, thickness measurements are taken on the cantilever test structures, as shown in Fig. T1(b), that are located within the Certification Plus grouping of test structures. The specific test structure to be measured (on SRM 2494 or SRM 2495) can be deduced from the data entered on the NIST-supplied Data Analysis Sheet T.1 or T.3, respectively, which accompanies the SRM.

For SRM 2494, the target test structure in Fig. T1(a) can be found as follows:
1. Five data analysis sheets for step height measurements are used to obtain the first five step height inputs [namely, $step1_{AB}$, $step2_{rA}$, $step1_{EF}$, $step1_{GH}$, and $step3_{AB}(n)^-$] for Data Analysis Sheet T.1. To locate the specific test structure that was measured for the first measurement (namely, $step1_{AB}$), consult the specific Data Analysis Sheet SH.1.a that was used to obtain that measurement for the SRM Certificate. See Sec. 5.6 for specifics. For the measurements of $step2_{rA}$, $step1_{EF}$, $step1_{GH}$, and $step3_{AB}(n)^-$, any one of the three instances of the applicable test structure can be measured. For the sixth step height input, namely $step3_{BC}(0)$, the NIST-supplied value is used since this is a measurement taken before the post-processing.

For SRM 2495, see Sec. 8.4 for the inputs to Table 1 of Appendix 7 that are used to locate the target test structure in Fig. T1(b).

Take the measurements: For SRM 2494, the first five step heights [namely, $step1_{AB}$, $step2_{rA}$, $step1_{EF}$, $step1_{GH}$, and $step3_{AB}(n)^-$] are measured from the four step height test structures in Fig. T3 and inputted in Data Analysis Sheet T.1. Recall that $step3_{AB}(n)^-$ is taken with a stylus profilometer (or comparable instrument). Then, the NIST-supplied value for the sixth step height [namely, $step3_{BC}(0)$] is inputted since this is a measurement taken before the chip is etched. Also, NIST-supplied data can be used for the Table 2 and Table 3 inputs. Consult Sec. 8.1 and Sec. 5.6 for the step height measurements taken for Data Analysis Sheet T.1.

For SRM 2495, using SEMI standard test method MS2 [4], a stylus profilometer is recommended to measure A, as shown in Fig. T8(b) using three 2-D data traces, as shown in Fig. T8(a); however the stylus measurement should not go beyond the anchor and on to the cantilever because it could damage the test structure. The data are leveled and zeroed with respect to the top of the underlying poly0 layer. For Data Analysis Sheet T.3, uncalibrated measurements of $platXa$, $platXb$, $platXc$, s_{platXa}, s_{platXb}, and s_{platXc} are requested from $platX$ and for $platY$ uncalibrated measurements of $platYa1$, $platYb1$, $platYc1$, $s_{platYa1}$, $s_{platYb1}$, and $s_{platYc1}$ are requested.

Then, using SEMI standard test method MS2 [4], an optical interferometer is used to measure B, as shown in Fig. T8(b), using the highest magnification objective that is available and feasible. The data are once again leveled and zeroed with respect to the top of the underlying poly0 layer, and Trace a, Trace b, and Trace c, as shown in Fig. T8(a) are obtained. For Data Analysis Sheet T.3, uncalibrated measurements of $platYa2$, $platYb2$, $platYc2$, $s_{platYa2}$, $s_{platYb2}$, and $s_{platYc2}$ are requested from $platY$ and for $platZ$ uncalibrated measurements of $platZa$, $platZb$, $platZc$, s_{platZa}, s_{platZb}, and s_{platZc} are requested.

The NIST-supplied data on Data Analysis Sheet T.3 were obtained using an optical interferometer for the measurements of both A and B to maintain the integrity of the SRM since indentations can be created when the stylus makes contact with the sample surface.

Data Analysis Sheet T.3 also requests the following process specific data: $\sigma_{repeat(samp)N}$, H, ΔH, J_{est}, u_{cJest}, s_{roughX}, s_{roughY}, and s_{roughZ}, of which the first five values can be obtained from the NIST-supplied Data Analysis Sheet T.3 that accompanies SRM 2495.

Perform the calculations: Enter the data into Data Analysis Sheet T.1 or T.3 as follows:
1. Press the "Reset this form" button located near the top and/or middle of the data analysis sheet.
2. Supply the appropriate inputs to Table 1, Table 2, and Table 3 for Data Analysis Sheet T.1 and the inputs to Table 1 and Table 2 for Data Analysis Sheet T.3.
3. Press the "Calculate and Verify" button (also located near the top and/or middle of the data analysis sheet) to obtain the results.
4. Verify the data by checking to see that all the pertinent boxes in the verification section at the bottom of the data analysis sheet say "ok". If one or more of the boxes say "wait," address the issue, if necessary, by modifying the inputs and recalculating.
5. Print out the completed data analysis sheet to compare both the inputs and outputs with those on the NIST-supplied data analysis sheet.

Compare the measurements: The MEMS 5-in-1 is accompanied by a Certificate. This Certificate specifies a thickness value (t_{SiO2} for SRM 2494 and α for SRM 2495) and the expanded uncertainty (U_{SiO2} for SRM 2494 and $U\alpha$ for SRM 2495) with $k=2$ intending to approximate a 95 % level of confidence. It is your responsibility to determine an appropriate criterion for acceptance, such as given below:

$$D_{SiO2} = \left| t_{SiO2(customer)} - t_{SiO2} \right| \leq \sqrt{U_{SiO2(customer)}^2 - U_{SiO2}^2} \, , \tag{T43}$$

$$D_\alpha = \left| \alpha_{(customer)} - \alpha \right| \leq \sqrt{U_{\alpha(customer)}^2 - U_\alpha^2} \, , \tag{T44}$$

where D_{SiO2} is the absolute value of the difference between your composite oxide beam thickness value, $t_{SiO2(customer)}$, and the thickness value on the SRM Certificate, t_{SiO2}, and where $U_{SiO2(customer)}$ is your expanded uncertainty value and U_{SiO2} is the expanded uncertainty on the SRM Certificate. Similarly, D_α is the absolute value of the difference between your poly1 (or poly2) thickness value, $\alpha_{(customer)}$, and the thickness value on the SRM Certificate, α, and where $U_{\alpha(customer)}$ is your expanded uncertainty value and U_α is the expanded uncertainty on the SRM Certificate. If your measured thickness value (as obtained in the newly filled out Data Analysis Sheet T.1 or T.3, respectively) satisfies your criterion for acceptance and there are no pertinent "wait" statements at the bottom of your Data Analysis Sheet T.1 or T.3, you can consider yourself to be appropriately measuring the composite oxide beam thickness or the poly1 (or poly2) thickness, respectively, according to the SEMI MS2 standard test method [4] according to your criterion for acceptance.

Any questions concerning the measurements, analysis, or comparison can be directed to mems-support@nist.gov.

9 Summary

The MEMS 5-in-1 is a standard reference device sold as a NIST Standard Reference Material (SRM). The purpose of the MEMS 5-in-1 is to allow users to compare their in-house measurements with NIST measurements using five ASTM and SEMI documentary standard test methods, thereby validating their use of the standard test methods. The five standard test methods [1-5] are for measuring Young's modulus, residua l strain, strain gradient, step height, and in-plane length, respectively. Additional measurements for comparison include residual stress, stress gradient, and beam thickness. (The calculations for residual stress and stress gradient are provided in the Young's modulus standard test method and the beam thickness ca lculations rely upon step height measurements.) Therefore, eight measurements can be compared using five standard test methods. The calculations are performed on-line using the data analysis sheets (reproductions of which are given in Appendices 1 through 7) accessible via the NIST MEMS Calculator Website [13].

In summary, we present the following:
- A *knowledge base* that has been developed for the user to take measurements on the MEMS 5-in-1 and verify them with NIST measurements.

Therefore, the *proven-in skill set* obtained from the proper use of the MEMS 5-in-1:
- Enables the user to take similar measurements on similarly (or differently) processed test structures. (When a material property is extracted from measurements taken on a test structure, the user must have an understanding of the geometry and composition of the test structure in order to obtain meaningful results.   In other words, an "effective" value may be obtained as opposed to a "true" value if there are non -idealities associated with the geometry and/or composition of the test structure.)
- Enables measurements to be meaningfully compared between laboratories on similarly (or differently) processed test structures where differences in community measurements has been tightened due to the use of a generally accepted standard test method with the SRM used as a tool to verify the proper use of the applicable standard test method. (Once again, when a material property is extracted from measurements taken on a test structure, the user must have an understanding of the geometry and composition of the test structure in order to determine if a "true" or "effective" valu e has been obtained.)

From an *applications* point of view:
- These measurements can be used in the design and fabrication of MEMS devices and ICs. For example, high residual stress values can result in failure mechanisms in ICs such as electromigration, stress migration, and delamination. So, residual stress measurements can be used to improve the yield in CMOS fabrication processes.
- Young's modulus is used in the calculation of residual stress. The article entitled "Young's Modulus Measurements in Standard IC CMOS Processes using MEMS Test Structures" [27] stretches the imagination   by obtaining Young's modulus for each layer in a CMOS process using an optimization program.

Therefore, the MEMS 5-in-1 (as described in the remaining paragraphs) can be considered a stepping stone that provides the groundwork for all sorts of comparisons and applications.

There are two SRMs available for purchase (SRM 2494 and SRM 2495). SRM 2494 is the MEMS 5-in-1 fabricated on a multi-user 1.5 μm CMOS process [8] followed by a bulk-micromachining etch, as shown in Fig. 1. For this SRM, the material properties of the composite oxide layer are measured. SRM 2495 is the MEMS 5-in-1 fabricated using a polysilicon multi-user surface-micromachining MEMS process [9] with a backside etch, as shown in Fig. 2. For this SRM, the material properties of the first or second polysilicon layer are measured.

Each MEMS 5-in-1 is accompanied by a Certificate, data analysis sheets, and this NIST Special Publication, SP 260. For a current example of the SRM 2494 and 2495 Certificates (each of which typically includes the eight NIST measurements for comparison) see the **Data and Information Files** link on https://www-s.nist.gov/srmors/view_detail.cfm?srm=8096 and https://www-s.nist.gov/srmors/view_detail.cfm?srm=8097, respectively. The data analysis sheets that accompany the MEMS 5-in-1 include the raw data used at NIST to obtain the measurements on the Certificates. In Sec. 2.6, 3.6, 4.6, 5.6, 6.6, 7.3, or 8.5 of this SP 260 for the parameter of interest, the user is instructed to follow the procedures in the applicable ASTM or SEMI standard test method in the taking of the measurements on the same test structures that NIST measurements were taken. The user's measurements can then be compared with the NIST measurements supplied on the SRM Certificates (and the data analysis sheets) to facilitate the validation of the use of the documentary standard test methods.

This SP 260 provides overall use and background information for the MEMS 5-in-1 (SRM 2494 and SRM 2495). Sec. 1 of this SP 260 provides details associated with the following:
- The instruments:

- ○ Specifications for the optical vibrometer, stroboscopic interferometer, or comparable instrument used for Young's modulus measurements and a validation procedure for frequency measurements, and
  - ○ Specifications for the optical interferometer or comparable instrument used for residual strain, strain gradient, step height, and in-plane length measurements and a validation procedure for height measurements.
- The contents of the MEMS 5-in-1:
  - ○ For SRM 2494, and
  - ○ For SRM 2495.
- The classification of the SRM 2494 chips,
- The post processing of the SRM 2494 and SRM 2495 chips,
- The pre-package inspection (including the classification of the SRM 2495 chips),
- The packaging of the MEMS 5-in-1,
- NIST measurements taken on the MEMS 5-in-1,
- The SRM Certificates,
- Traceability,
- Material available for the MEMS 5-in-1,
- Storage and handling,
- Measurement conditions and procedures for the customer,
- Stability tests, and
- Length of Certification.

Then, Sec. 2 through Sec. 6 discuss the test structures, the calibration and measurement procedures, the uncertainty analysis, the round robin results, and how the user can use the MEMS 5-in-1 to verify their in-house measurements for the following standard test methods, associated parameters, and data analysis sheets:

- SEMI standard test method MS4 for Young's modulus measurements using Data Analysis Sheet YM.3 (a reproduction of Data Analysis Sheet YM.3 is given in Appendix 1),
- ASTM standard test method E 2245 for residual strain measurements using Data Analysis Sheet RS.3 (a reproduction of Data Analysis Sheet RS.3 is given in Appendix 2),
- ASTM standard test method E 2246 for strain gradient measurements using Data Analysis Sheet SG.3 (a reproduction of Data Analysis Sheet SG.3 is given in Appendix 3),
- SEMI standard test method MS2 for step height measurements using Data Analysis Sheet SH.1.a (a reproduction of Data Analysis Sheet SH.1.a is given in Appendix 4), and
- ASTM standard test method E 2244 for in-plane length measurements using Data Analysis Sheet L.0 (a reproduction of Data Analysis Sheet L.0 is given in Appendix 5).

Section 7 of this SP 260 provides the user with details concerning residual stress and stress gradient calculations. These calculations can be performed in Data Analysis Sheet YM.3. And finally, Sec. 8 provides the user with details concerning thickness calculations. For SRM 2494, the electro-physical technique [6] as presented in this section is used with Data Analysis Sheet T.1 (a reproduction of which is given in Appendix 6). For SRM 2495, the optomechanical technique [7] as presented in this section is used with Data Analysis Sheet T.3 (a reproduction of which is given in Appendix 7).

The NIST SRM Program Office [18] can be contacted (http://ts.nist.gov/measurementservices/referencematerials/index.cfm) to purchase a MEMS 5-in-1 and accompanying material.

10 Acknowledgements

Acknowledgments go to   Richard A. Allen and   Craig D. McGray in the Test Structures Subgroup   within the MEMS Measurement Science and Services Project at NIST and to Jason J. Gorman in the Engineering Laboratory. The SEMI North American MEMS Standards Committee and the ASTM Subcommittee E08.05 on Cyclic Deformation and Fatigue Crack Formation are also acknowledged.

For supporting this work, acknowledgments go to the Enabling Devices and ICs Group, the Semiconductor Electronics Division, and the Physical Measurement Laboratory as well as the former Office of Microelectronics Programs and the former Electronics and Electrical Engineering Laboratory. MOSIS and MEMSCAP are thanked for the industry cost-share and engineering support received for the SRM 2494 and SRM 2495 chips, respectively.

Stefan D. Leigh (Statistical Engineering Division) is acknowledged for his diligent teaching with respect to the ISO GUM rules concerning combined standard uncertainties, expanded uncertainties, and for guidance with respect to general statistical practices. Andrew Rukhin (Statistical Engineering Division) is thanked for his behind the scenes statistical wizardry. Young-Man Choi (Intelligent Systems Division) is thanked for his FEM simulations of resonating cantilevers for   Young's modulus measurements.

For reviewing this paper, the following people from NIST are acknowledged: Stefan D. Leigh, Richard A. Allen, W. Robert Thurber, and Erik M. Secula.

References

[1]   SEMI MS4, "Test Method for Young's Modulus Measurements of Thin, Reflecting Films Based on the Frequency of Beams in Resonance," *as to be submitted for ballot as of the writing of this SP 260.*

[2]   ASTM E08, "E 2245 Standard Test Method for Residual Strain Measurements of Thin, Reflecting Films Using an Optical Interferometer," *as to be submitted for ballot as of the writing of this SP 260.*

[3]   ASTM E08, "E 2246 Standard Test Method for Strain Gra   dient Measurements of Thin, Reflecting Films Using an Optical Interferometer," *as to be submitted for ballot as of the writing of this SP 260.*

[4]   SEMI MS2, "Test Method for Step Height Measurements of Thin Films," *as to be submitted for ballot as of the writing of this SP 260.*

[5]   ASTM E08, "E 2244 Standard Test Method for In   -Plane Length Measurements of Thin, Reflecting Films Using an Optical Interferometer," *as to be submitted for ballot as of the writing of this SP 260.*

[6]   J. C. Marshall and P. T. Vernier, "Electro   -physical technique for post-fabrication measurements of CMOS process layer thicknesses," *NIST J. Res.*, Vol. 112, No. 5, p. 223-256, 2007.

[7]   J. C. Marshall, "New Optomechanical Technique for Measuring Layer Thickness in MEMS Processes,"   *Journal of Microelectromechanical Systems*, Vol. 10, No. 1, pp. 153-157, March 2001.

[8]   The URL for the MOSIS Website is http://www.mosis.com   "MOSIS provides access to fabrication of prototype and low-volume production quantities of integrated circuits…by combining designs from many customers onto multi   -project wafers…decreasing the cost of each design." The 1.5 μm On Semiconductor (formerly AMIS) CMOS process was used for SRM 2494.

[9]   The SRM 2495 chips were fabricated at MEMSCAP using MUMPs-Plus! (PolyMUMPs with a backside etch). The URL for the MEMSCAP Website is http://www.memscap.com.

[10]   J. Marshall, R. A. Allen, C. D. McGray, and J. Geist, "MEMS Young's Modulus and Step Height Measurements wit h Round Robin Results," *NIST J. Res.*, Vol. 115, No. 5, p. 303-342, 2010.

[11]   Marshall, J. C., Scace, R. I., and Baylies, W. A., "MEMS Length and Strain Round Robin Results with Uncertainty Analysis," *NISTIR 7291*, National Institute of Standards and Technology, January 2006.

[12]   Marshall, J. C., Secula, E. M., and Huang, J., "Ro   und Robin for Standardization of MEMS Length and Strain Measurements,"   SEMI Technology Symposium: Innovations in Semiconductor Manufacturing (STS: ISM), SEMICON West 2004, San Francisco, CA, July 12-14, 2004.

[13]   The data analysis sheets, design files, and details associated with the MEMS 5-in-1 can be found at the MEMS Calculator Web Site (Standard Reference Database 166) accessible via the National Institute of Standards and Technology (NIST) Data Gateway (http://srdata.nist.gov/gateway/) with the keyword "MEMS Calculator."

[14]   Marshall, J. C., "MEMS Length and Strain Measurements Using an Optical Interferometer,"   *NISTIR 6779*, National Institute of Standards and Technology, August 2001.

[15]   May, W.E., Parris, R.M., Beck, C.M., Fassett, J.D., Greenberg, R.R., Guenther, F.R., Kramer, G.W., Wise, S.A., Gills, T.E., Colbert, J.C., Gettings, R.J., and MacDonald, B.R., "Definitions of Terms and Modes Used at NIST for Value Assignment of Reference Materials for Chemical Measurements," NIST SP 260-136 (2000) available at http://www.nist.gov/srm/publications.cfm.

[16]   Taylor, B. N. and Thompson, A., "The International System of Units (SI),"   *NIST Special Publication 330, 2008 Edition*, National Institute of Standards and Technology, March 2008.

[17]   Thompson, A. and Taylor, B. N., "Guide for the Use of the International System of Units (SI),"   *NIST Special Publication 811, 2008 Edition*, National Institute of Standards and Technology, March 2008.

[18]   Contact the NIST SRM Program Office to obtain a MEMS 5-in-1, its certificate, and accompanying data analysis sheets by visiting http://ts.nist.gov/measurementservices/referencematerials/index.cfm. This   office can also be contacted to initiate the recalibration of a MEMS 5-in-1.

[19]   Taylor, B. N. and Kuyatt, C. E., "Guidelines for Evaluating and Expressing the Uncertainty of NIST Measurement Results," *NIST Technical Note 1297*, National Institute of Standards and Technology, September 1994.

[20]   EUROCHEM / CITAC Guide CG 4, "Quantifying Uncertainty in Analytical Measurement," Second Edition, QUAM: 2000.1.

[21]   International Organization for Standardization, "Guide to the Expression of Uncertainty in Measurement," ANSI /NCSL Z540-2-1997, 1997.

[22]   W. Weaver, S. P. Timoshenko, and D. H. Young, Vibration Problems in Engineering, 5 [th] edition, John Wiley & Sons, 1990.

[23]   L. Kiesewetter, J.-M. Zhang, D. Houdeau, and A. Steckenborn, Determination of Young's moduli of micromechanical thin films using the resonance method, Sensors and Actuators A, Vol. 35, pp. 153-159, 1992.

[24]   Draft International Standard ISO 25178-604, *Geometrical product specifications (GPS)   — Surface texture: Areal — Part 604: Nominal characteristics of non-contact (coherence scanning interferometry) instruments*, (International Organization for Standardisation, Geneva, 2010).

[25]   F. Chang *et al* ., "Gas-phase silicon micromachining with xenon difluoride," in   *Proc. SPIE Symp. Micromach. Microfab.*, Austin, TX, 1995, pp. 117-128.

[26]   Gregory T. A. Kovacs, Micromachined Transducers Sourcebook, McGraw-Hill, New York, NY, p. 312, 1998.

[27]   J. C. Marshall, D. L. Herman, P. T. Vernier, D. L. DeVoe, and M. Gaitan, "Young's Modulus Measurements in Standard IC CMOS Processes using MEMS Test Structures,"   *IEEE Electron Device Letters*, Vol. 28, No. 11, p. 960-963, 2007.

[28]   An Assessment of the US Measurement System: Addressing Measurement Barriers to Accelerate Innovation, Appendix B, "Case Study –   Measurement Needs Technology at Issue: Micro-/Nano-   Technology", (NIST Special Publication 1048, page 309, 2006).

[29]   J. V. Clark, "Electro Micro-Metrology," Ph.D. dissertation, University of California at Berkeley, Dec. 2005.

[30]   MEMS and Nanotechnology Clearinghouse, MEMS and Nanotechnology Exchange, http://www.memsnet.org/material, (accessed April 12, 2010).

[31]   SEMI MS4-1109, "Test Method for Young's Modulus Measurements of Thin, Reflecting Films Based on the Frequency of Beams in Resonance," November 2009 (visit http://www.semi.org for ordering information).

[32]   SEMI MS4-1107, "Test Method for Young's Modulus Measurements of Thin, Reflecting Films Based on the Frequency of Beams in Resonance," November 2007 (visit http://www.semi.org for ordering information).

[33]   T. B. Gabrielson, Mechanical-Thermal Noise in Micromachined Acoustic and Vibration Sensors, IEEE Transactions on Electron Devices, Vol. 40, No. 5. pp. 903-909, May 1993.

[34]   P. I. Oden, Gravimetric sensing of metallic deposits using an end-loaded microfabricated beam structure, Sensors and Actuators B, Vol. 53, pp. 191-196, 1998.

[35]   E. M. Lawrence, C. Rembe, MEMS Characterization using New Hybrid Laser Doppler Vibrometer / Strobe Video System, SPIE Proceedings Vol. 5343, Reliability, Testing, and Characterization of MEMS/MOEMS III, pp. 45-54, January 2003.

[36]   ASTM E08, "E 2245 Standard Test Method for Residual Strain Measurements of Thin, Reflecting Films Using an Optical Interferometer," ASTM E 2245−05, *Annual Book of ASTM Standards*, Vol. 03.01, 2006. (Visit http://www.astm.org for ordering information.)

[37]   ASTM E08, "E 2245 Standard Test Method for Residual Strain Measurements of Thin, Reflecting Films Using an Optical Interferometer," ASTM E 2245−02, *Annual Book of ASTM Standards*, Vol. 03.01, 2003. (Visit http://www.astm.org for ordering information.)

[38]   ASTM E08, "E 2246 Standard Test Method for Strain Gradient Measurements of Thin, Reflecting Films Using an Optical Interferometer," ASTM E 2246−05, *Annual Book of ASTM Standards*, Vol. 03.01, 2006. (Visit http://www.astm.org for ordering information.)

[39]   ASTM E08, "E 2246 Standard Test Method for Strain Gradient Measurements of Thin, Reflecting Films Using an Optical Interferometer," ASTM E 2246−02, *Annual Book of ASTM Standards*, Vol. 03.01, 2003. (Visit http://www.astm.org for ordering information.)

[40]   SEMI MS2-1109, "Test Method for Step Height Measurements of Thin Films," November 2009, (visit http://www.semi.org for ordering information).

[41]   SEMI MS2-0307, "Test Method for Step Height Measurements of Thin, Reflecting Films Using an Optical Interferometer," March 2007, (visit http://www.semi.org for ordering information).

[42]   T. V. Vorburger, J. F. Song, T. B. Renegar, and A. Zheng, "NIST Surface Roughness and Step Height Calibrations: Measurement Conditions and Sources of Uncertainty," January 23, 2008. (Visit http://www.nist.gov/mel/ped/smm/upload/nistsurfcalib.pdf.)

[43]   ASTM E42, "E 2530 Standard Practice for Calibrating the *Z*-Magnification of an Atomic Force Microscope at Subnanometer Displacement Levels Using Si *(111)* Monatomic Steps," *Annual Book of ASTM Standards*, Vol. 03.06, 2006. (Visit http://www.astm.org for ordering information.)

[44]   ASTM E08, "E 2244 Standard Test Method for In-Plane Length Measurements of Thin, Reflecting Films Using an Optical Interferometer," ASTM E 2244−05, *Annual Book of ASTM Standards*, Vol. 03.01, 2006. (Visit http://www.astm.org for ordering information.)

[45]   ASTM E08, " E 2244 Standard Test Method for In-Plane Length Measurements of Thin, Reflecting Films Using an Optical Interferometer," ASTM E 2244‒02, *Annual Book of ASTM Standards*, Vol. 03.01, 2003.   (Visit http://www.astm.org for ordering information.)

Data Analysis Sheet YM.3

Data analysis sheet for determining the Young's modulus value of a thin film layer for use with the MEMS 5-in-1 SRMs

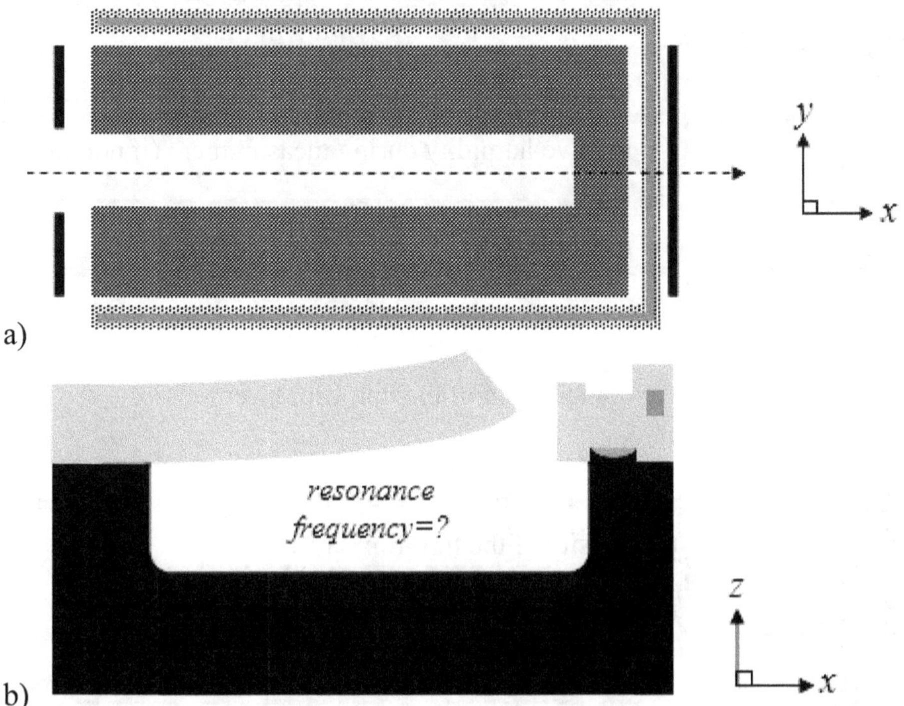

Figure YM.3.1. For CMOS cantilever a) a design rendition and b) a cross section

To obtain the following measurements, consult SEMI standard test method MS4 entitled "Test Method for Young's Modulus Measurements of Thin, Reflecting Films Based on the Frequency of Beams in Resonance." (Updates to this standard will be balloted.)

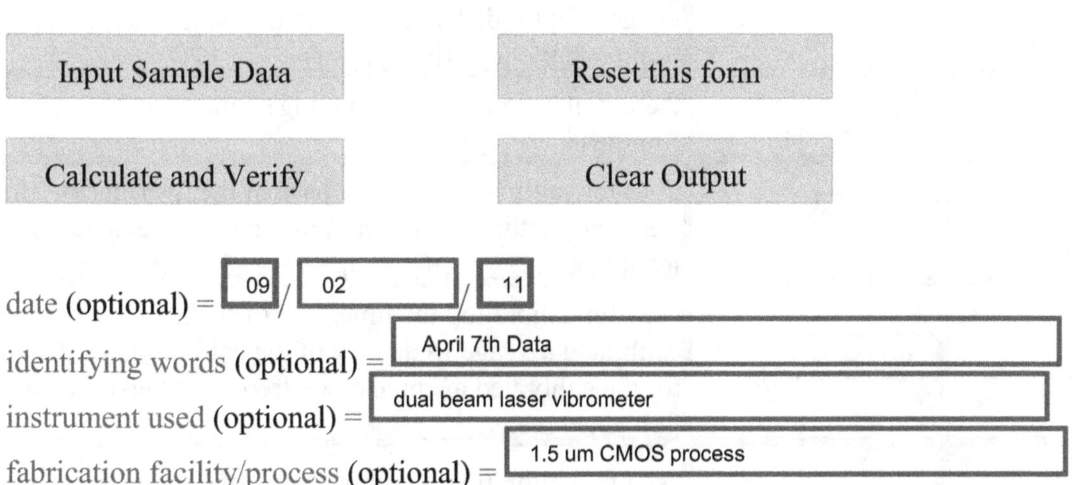

test chip name (optional) = | SRM 2494 Monitor 6106.1 |

test chip number (optional) = | s/n 0001 |

root filename (optional) = | |

comments (optional) = | |

Table 1 - Preliminary INPUTS			Description
1	$temp=$	21.1 $^\circ$C	the temperature during measurement (should be held constant)
2	*relative humidity=*	46 %	the relative humidity during measurement (if not known, enter -1)
3	$mag=$	20 $\times$	the magnification
4	$mat=$	poly1 $\bigcirc$ poly2 $\bigcirc$ SiO$_2$ $\odot$ other $\bigcirc$	the composition of the thin film layer
5*	$\rho=$	2.2 g/cm^3	the density of the thin film layer
6	$\sigma_\rho=$	0.05 g/cm^3	the one sigma uncertainty of the value of ρ
7*	$\mu=$	1.84 $\times 10^{-5}$ Ns/m^2	the viscosity of the ambient surrounding the cantilever
8*	$W=$	28.0 μm	the suspended beam width
9*	$t=$	2.5846 μm	the thickness of the thin film layer (as found using Data Sheet T.1 or Data Sheet T.3)
10	$\sigma_{thick}=$	0.1088 μm	the one sigma uncertainty of the value of t (as found using Data Sheet T.1 or Data Sheet T.3)
11	$d_{gap}=$	30.0 μm	the gap depth (distance between the bottom of the suspended beam and the underlying layer)
12*	$E_{init}=$	70 GPa	the initial estimate for the Young's modulus value of the thin film layer
13	$f_{instrument}=$	102.400000 MHz	used for calibrating the time base of the instrument: the frequency setting for the calibration measurements (or the manufacturer's specification for the clock frequency)
14	$f_{meter}=$	102.399437 MHz	used for calibrating the time base of the instrument: the calibrated average frequency of the calibration measurements (or the calibrated average clock frequency) taken with a frequency meter
15	$\sigma_{meter}=$	1 Hz	used for calibrating the time base of the instrument: the

| 16 | $u_{certf} =$ | 0 | Hz | used for calibrating the time base of the instrument: the certified uncertainty of the frequency measurements as specified on the frequency meter's certificate |
| | | | | standard deviation of the frequency measurements taken with the frequency meter |

* The five starred entries in this table are required inputs for the calculations in the Preliminary Estimates Table.

Table 2 - Cantilever INPUTS				Description
17	$name=$	SRM cantilever		the cantilever name (optional)
18	$orient=$	0° ⦿ 90° ○ 180° ○ 270° ○ other ○		the orientation of the cantilever
19*	$L_{can}=$	300	μm	the suspended cantilever length
20	$whichcan=$	first ⦿ second ○ third ○ fourth ○ fifth ○ sixth ○ other ○		indicates which cantilever on the test chip, where "first" corresponds to the topmost cantilever in the column or array that has the specified length?
21	$\sigma_L=$	0.2	μm	the one sigma uncertainty of the value of L_{can}
22	$f_{resol}=$	1.25	Hz	the uncalibrated frequency resolution for the given set of measurement conditions
23	$f_{meas1}=$	24.6425	kHz	the first uncalibrated, damped resonance frequency measurement (or the first uncalibrated, undamped resonance frequency measurement, for example, if the measurements were performed in a vacuum)
24	$f_{meas2}=$	24.6275	kHz	the second uncalibrated, damped resonance frequency measurement (or the second uncalibrated, undamped resonance frequency

				Description
				measurement, for example, if the measurements were performed in a vacuum)
25	$f_{meas3}=$	24.64	kHz	the third uncalibrated, damped resonance frequency measurement (or the third uncalibrated, undamped resonance frequency measurement, for example, if the measurements were performed in a vacuum)
26	$f_{correction}=$	0	kHz	the correction term for the cantilever's resonance frequency
27	$\sigma_{support}=$	0	kHz	the uncertainty in the cantilever's resonance frequency due to a non-ideal support (or attachment conditions)
28	$\sigma_{cantilever}=$	0	kHz	the uncertainty in the cantilever's resonance frequency due to geometry and/or composition deviations from the ideal

* The starred entry in this table is a required input for the calculations in the Preliminary Estimates Table.

Table 3 - Fixed-Fixed Beam INPUTS (if cantilever not available)			Description
29	$name2=$		the fixed-fixed beam name (optional)
30	$orient2=$	0° 90° ● other	the orientation of the fixed-fixed beam
31*	$L_{ffb}=$	μm	the suspended fixed-fixed beam length
32	$whichffb=$	first second third fourth fifth ● other	indicates which fixed-fixed beam on the test chip, where "first" corresponds to the topmost fixed-fixed beam in the column or array that has the specified length?
33	$f_{ffb}=$	kHz	the average uncalibrated resonance frequency of the fixed-fixed beam

* The starred entry in this table is a required input for the calculations in the Preliminary Estimates Table.

Table 4 - Optional INPUTS			Description
For residual stress calculations:			**Description**
34	$\varepsilon_r=$	-2656.0 $\times 10^{-6}$	the residual strain of the thin film layer (as found using ASTM E 2245 and Data Sheet RS.3 for compressive residual strain)
35	$u_{c\varepsilon r}=$	100.32 $\times 10^{-6}$	the combined standard uncertainty value for residual strain (as found using Data Sheet RS.3 for compressive residual strain)
For stress gradient calculations:			
36	$s_g=$	890.54 m^{-1}	the strain gradient of the thin film layer (as found using ASTM E 2246 and Data Sheet SG.3)
37	$u_{csg}=$	55.56 m^{-1}	the combined standard uncertainty value for strain gradient (as found using Data Sheet SG.3)

Input Sample Data	Reset this form
Calculate Estimates	Clear Output

Table 5 - Preliminary ESTIMATES*			Description
38	$f_{caninit}=$	26.2 kHz	$= \text{SQRT}[E_{init}\,t^2 / (38.330\,\rho\,L_{can}^4)]$ (the estimated resonance frequency of the cantilever)
39	$f_{ffbinithi}=$	Infinity kHz	$= \text{SQRT}[E_{init}\,t^2 / (0.946\,\rho\,L_{ffb}^4)]$ (the estimated upper bound for the resonance frequency of the fixed-fixed beam)
40	$f_{ffbinitlo}=$	Infinity kHz	$= \text{SQRT}[E_{init}\,t^2 / (4.864\,\rho\,L_{ffb}^4)]$ (the estimated lower bound for the resonance frequency of the fixed-fixed beam)
41	$Q=$	58.4	$= W\,t^2\,\text{SQRT}(\rho\,E_{init}) / (24\,\mu\,L_{can}^2)$ (the estimated Q-factor)
42	$p_{diff}=$	0.0037 %	$=\{1-\text{SQRT}[1-1 / (4\,Q^2)]\}\times100\ \%$ should be $< 2\ \%$ (the estimated percent difference between the damped and undamped resonance frequency of the cantilever)

* The seven starred inputs in the first three tables are required for the calculations in this table.

Calculate and Verify	Clear Output

OUTPUTS:

Table 6 - Frequency calculations:			Description
43	$cal_f=$	0.99999450	$= f_{meter} / f_{instrument}$ (the calibration factor for a frequency measurement)
44	$f_{measave}=$	24.6365 kHz	$= \text{AVE } [f_{meas1}, f_{meas2}, f_{meas3}] cal_f$ (the average calibrated damped resonance frequency of the cantilever, $f_{dampedave}$, or the average calibrated undamped resonance frequency of the cantilever if, for example, the measurements were performed in a vacuum)
45	$f_{undamped1}=$	24.6433 kHz	$= f_{damped1} / \text{SQRT}[1-1/(4Q^2)]$ where $f_{damped1}=f_{meas1}(cal_f)$ (the first calibrated undamped resonance frequency calculated from the cantilever's first damped resonance frequency measurement, if applicable)
46	$f_{undamped2}=$	24.6283 kHz	$= f_{damped2} / \text{SQRT}[1-1/(4Q^2)]$ where $f_{damped2}=f_{meas2}(cal_f)$ (the second calibrated undamped resonance frequency calculated from the cantilever's second damped resonance frequency measurement, if applicable)
47	$f_{undamped3}=$	24.6408 kHz	$= f_{damped3} / \text{SQRT}[1-1/(4Q^2)]$ where $f_{damped3}=f_{meas3}(cal_f)$ (the third calibrated undamped resonance frequency calculated from the cantilever's third damped resonance frequency measurement, if applicable)
48	$f_{undampedave}=$	24.6374 kHz	$= \text{AVE } [f_{undamped1}, f_{undamped2}, f_{undamped3}]$ (the average calibrated undamped resonance frequency of the cantilever assuming $f_{meas1}, f_{meas2},$ and f_{meas3} from the second table are damped resonance frequencies)
49	$\sigma_{fundamped}=$	0.0080 kHz	$= \text{STDEV } (f_{undamped1}, f_{undamped2}, f_{undamped3})$ (the one sigma uncertainty of the value of $f_{undampedave}$ assuming $f_{meas1}, f_{meas2},$ and f_{meas3} from the second table are damped resonance frequencies)
50	$f_{can}=$	24.6374 kHz	$= f_{undampedave} + f_{correction}$ (the modified resonance frequency of the cantilever for use if $f_{meas1}, f_{meas2},$ and f_{meas3} from the second table are damped resonance frequencies)
51	$f_{measavenew}=$	24.6365 kHz	$= f_{measave} + f_{correction}$ (the modified resonance frequency of the cantilever for use if $f_{meas1}, f_{meas2},$ and f_{meas3} from the second table are undamped resonance frequencies)

1. Young's modulus calculation (as obtained from the cantilever assuming clamped-free boundary conditions):

a. $E = 38.330\, \rho\, f_{can}^{2}\, L_{can}^{4}\, /\, t^{2} = \boxed{62.1}$ GPa
(Use this value if f_{meas1}, f_{meas2}, and f_{meas3} in the second table are ~~damped~~ resonance frequencies.)

b. $E = 38.330\, \rho\, f_{measavenew}^{2}\, L_{can}^{4}\, /\, t^{2} = \boxed{62.1}$ GPa
(Use this value if f_{meas1}, f_{meas2}, and f_{meas3} in the second table are ~~undamped~~ resonance frequencies.)

c. $u_{cE} = \sigma_{E} = E\, \mathrm{SQRT}[(\sigma_{\rho}/\rho)^{2} + 4(\sigma_{fcan}/f_{can})^{2} + 16(\sigma_{L}/L_{can})^{2} + 4(\sigma_{thick}/t)^{2}] = \boxed{5.42}$ GPa *

or $\sigma_{E} / E = \boxed{0.087}$ *

where $\sigma_{fcan}/f_{can} = \mathrm{SQRT}[(\sigma_{fundamped}/f_{can})^{2} + (\sigma_{fresol}/f_{can})^{2} + (\sigma_{freqcal}/f_{can})^{2} + (\sigma_{support}/f_{can})^{2} + (\sigma_{cantilever}/f_{can})^{2}],$

$\sigma_{fresol} = f_{resol}\, cal_{f}\, /\, [2\mathrm{SQRT}(3)],$

and $\sigma_{freqcal} = f_{undampedave}\, [\mathrm{SQRT}(\sigma_{meter}^{2} + u_{certf}^{2})\, /\, f_{meter}]$

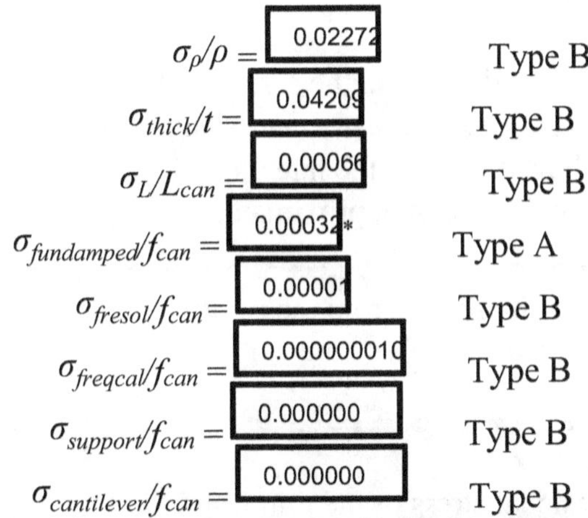

* assumes f_{meas1}, f_{meas2}, and f_{meas3} in the second table are damped resonance frequencies

$U_{E} = 2u_{cE} = \boxed{10.83}$ GPa (expanded uncertainty)

$3u_{cE} = \boxed{16.25}$ GPa

d. $E - U_{E} = \boxed{51.24}$ GPa (a lower bound for E)

$E + U_{E} = \boxed{72.90}$ GPa (an upper bound for E)
(assuming f_{meas1}, f_{meas2}, and f_{meas3} in the second table are damped resonance frequencies)

e. Report the results as follows: Since it can be assumed that the estimated values of the uncertainty components are either approximately uniformly or

Gaussianly distributed with approximate combined standard uncertainty u_{cE}, the Young's modulus value is believed to lie in the interval $E \pm u_{cE}$ (expansion factor k=1) representing a level of confidence of approximately 68 %.

2. Young's modulus calculation (as obtained from a fixed-fixed beam...~~not recommended~~):

a. $E_{simple} = 4.864 \, \rho \, (f_{ffb} \, cal_f)^2 \, L_{ffb}^4 / t^2 = \boxed{0.0}$ GPa

    (as obtained from the fixed-fixed beam assuming simply-supported boundary conditions for both supports)

b. $E_{clamped} = 0.946 \, \rho \, (f_{ffb} \, cal_f)^2 \, L_{ffb}^4 / t^2 = \boxed{0.0}$ GPa

    (as obtained from the fixed-fixed beam assuming clamped-clamped boundary conditions)

c. $E = (E_{simple} + E_{clamped}) / 2 = \boxed{0.0}$ GPa   (use this value, if must)

d. $u_E = (E_{simple} - E_{clamped}) / 6 = \boxed{0.0}$ GPa   (as obtained from a Type B analysis)

e. Report the results as follows: Since it can be assumed that the estimated value of the standard uncertainty, u_E, is approximately Gaussianly distributed, the Young's modulus value is believed to lie in the interval $E \pm u_E$ (expansion factor k=1) representing a level of confidence of approximately 68 %.

Table 7 - Optional OUTPUTS (using E and u_{cE} from the cantilever and assuming f_{meas1}, f_{meas2}, and f_{meas3} in the second table are damped resonance frequencies)

	For residual stress:		Description
52	$\sigma_r =$	-164.84 MPa	$= E \, \varepsilon_r$ (the residual stress of the thin film layer)
53	$u_{cor} =$	15.673 MPa	$= \|\sigma_r\| \, \mathrm{SQRT}[(\sigma_E / E)^2 + (\sigma_{\varepsilon r} / \varepsilon_r)^2]$ (the combined standard uncertainty value for residual stress where $\sigma_{\varepsilon r}$ is equated with $u_{c\varepsilon r}$)
54	$\sigma_{\sigma r} / \|\sigma_r\| =$	0.095	where $\sigma_{\sigma r}$ is equated with $u_{c\sigma r}$
55	$\sigma_E / E =$	0.087	as obtained from this data sheet
56	$\sigma_{\varepsilon r} / \|\varepsilon_r\| =$	0.038	where $\sigma_{\varepsilon r}$ is equated with $u_{c\varepsilon r}$ and where ε_r and $u_{c\varepsilon r}$ were obtained from Data Sheet RS.3
57	$2u_{cor} =$	31.345 MPa	$= U_{\sigma r}$ the expanded uncertainty for residual stress
58	$3u_{cor} =$	47.018 MPa	three times the combined standard uncertainty for residual stress

#		Value		Description
59	$\sigma_r - U_{\sigma r} =$	-196.19	MPa	a lower bound for σ_r
60	$\sigma_r + U_{\sigma r} =$	-133.50	MPa	an upper bound for σ_r
	For stress gradient:			
61	$\sigma_g =$	55271.8	GPa/m	$= E\,s_g$ (the stress gradient of the thin film layer)
62	$u_{c\sigma g} =$	5928.43	GPa/m	$= \sigma_g\,\mathrm{SQRT}[(\sigma_E / E)^2 + (\sigma_{sg} / s_g)^2]$ (the combined standard uncertainty value for stress gradient where σ_{sg} is equated with u_{csg})
63	$\sigma_{\sigma g} / \sigma_g =$	0.107		where $\sigma_{\sigma g}$ is equated with $u_{c\sigma g}$
64	$\sigma_E / E =$	0.087		as obtained from this data sheet
65	$\sigma_{sg} / s_g =$	0.062		where σ_{sg} is equated with u_{csg} and where s_g and u_{csg} were obtained from Data Sheet SG.3
66	$2u_{c\sigma g} =$	11856.8	GPa/m	$= U_{\sigma g}$ the expanded uncertainty for stress gradient
67	$3u_{c\sigma g} =$	17785.2	GPa/m	three times the combined standard uncertainty for stress gradient
68	$\sigma_g - U_{\sigma g} =$	43415.0	GPa/m	a lower bound for σ_g
69	$\sigma_g + U_{\sigma g} =$	67128.7	GPa/m	an upper bound for σ_g

Modify the input data, given the information supplied in any flagged statement below, if applicable, then recalculate:

#		Statement
1.	ok	Please provide inputs to Tables 1 and 2 for calculations using data from a cantilever.
2.	ok	The value for *temp* should be between 19.4 $^\circ$C and 21.6 $^\circ$C, inclusive.
3.	ok	The value for *relative humidity* (if known) should be between 0 % and 60 %, inclusive.
4.	wait	Please provide inputs to Table 3, ρ, W, t, and E_{init} for calculations using data from a fixed-fixed beam, if applicable.
5.	ok	The value for *mag* should be greater than or equal to 20×.
6.	ok	The value for ρ should be between 1.00 g/cm^3 and 5.00 g/cm^3.
7.	ok	The value for σ_ρ should be between 0.0 g/cm^3 and 0.10 g/cm^3.
8.	ok	The value for μ should be between $0.70{\times}10^{-5}$ Ns/m^2 and $3.0{\times}10^{-5}$ Ns/m^2.
9.	ok	The value for W should be greater than t and less than L_{can}.
10.	wait	The value for W should be greater than t and less than L_{ffb}, if inputted.

11.	ok	The value for t should be between 0.000 µm and 10.000 µm.
12.	ok	The value for σ_{thick} should be between 0.0 µm and 0.3 µm.
13.	ok	Squeeze film damping expected for the cantilever since $d_{gap} < W / 3$.
14.	ok	The value for E_{init} should be between 10 GPa and 300 GPa.
15.	ok	The values for σ_{meter} and u_{certf} should be between 0.0 Hz and 25.0 Hz, inclusive.
16.	ok	The value for L_{can} should be between 0 µm and 1000 µm.
17.	ok	The value for σ_L should be between 0.0 µm and 2.0 µm.
18.	ok	The value for f_{resol} should be between 0 Hz and 50 Hz.
19.	ok	The values for f_{meas1}, f_{meas2}, and f_{meas3} should be between 5.00 kHz and 300.0 kHz.
20.	ok	The value for $f_{correction}$ should be between -10 kHz and 10 kHz, inclusive.
21.	ok	The values for $\sigma_{support}$ and $\sigma_{cantilever}$ should be between 0 kHz and 10 kHz, inclusive.
22.	wait	If inputted, the value for L_{ffb} should be between 0 µm and 1000 µm.
23.	wait	If inputted, the value for f_{ffb} should be between 5.0 kHz and 1200 kHz.
24.	ok	If inputted, the value for ε_r should be between -4500×10^{-6} and 4500×10^{-6} and not equal to 0.0.
25.	ok	If inputted, the value for u_{cer} should be between 0.0 and 300.0×10^{-6}.
26.	ok	If inputted, the value for s_g should be between 0.0 m^{-1} and 1500.0 m^{-1}.
27.	ok	If inputted, the value for u_{csg} should be between 0.0 m^{-1} and 100.0 m^{-1}.
28.	ok	The values for f_{meas1}, f_{meas2}, and f_{meas3} are not within 20 kHz of $f_{caninit}$.
29.	wait	If inputted, the value for f_{ffb} should be between $f_{ffbinitlo}$ and $f_{ffbinithi}$.
30.	ok	The value for p_{diff} should be between 0 % and 2 %.
31.	ok	The value for cal_f should be between 0.9990 and 1.0010.
32.	ok	The value for $\sigma_{fundamped}$ should be between 0.0 kHz and 0.5 kHz, inclusive.
33.	ok	The value of E obtained from the cantilever should be within 20 GPa of E_{init}.
34.	ok	The value of u_{cE} obtained from the cantilever should be between 0 GPa and 10 GPa.
35.	wait	If applicable, the value of E obtained from the fixed-fixed beam should be within 30 GPa of E_{init}.

<table>
<tr><td>36.</td><td>wait</td><td>If applicable, the value of u_E obtained from the fixed-fixed beam should be between 0 GPa and 20 GPa.</td></tr>
</table>

Return to Main MEMS Calculator Page.

Email questions or comments to mems-support@nist.gov.

NIST is an agency of the U.S. Commerce Department.
The Semiconductor and Dimensional Metrology Division is within the Physical Measurement Laboratory.
The MEMS Measurement Science and Services Project is within the Microelectronics Device Integration Group.

Date created: 6/5/2006
Last updated: 9/2/2011

Data Analysis Sheet RS.3

Data analysis sheet for residual strain measurements for use with the MEMS 5-in-1 SRMs

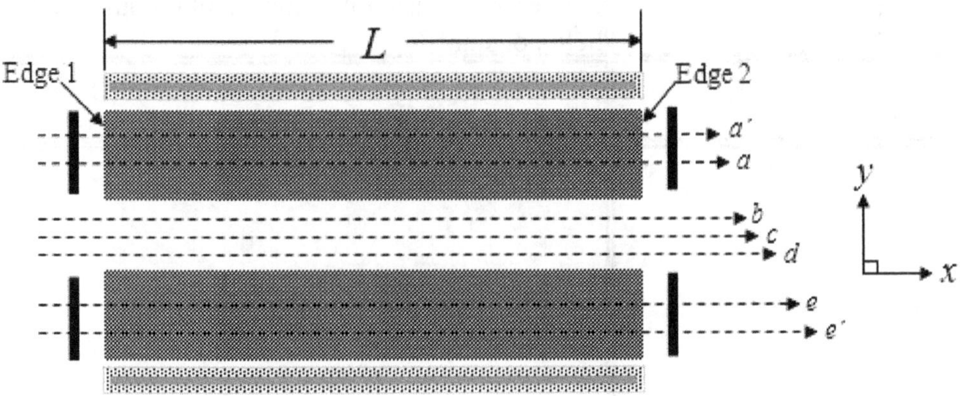

Figure RS.3.1. Top view of a fixed-fixed beam used to measure residual strain.

To obtain the following measurements, consult ASTM standard test method E 2245 entitled "Standard Test Method for Residual Strain Measurements of Thin, Reflecting Films Using an Optical Interferometer." (Updates to this standard will be balloted.)

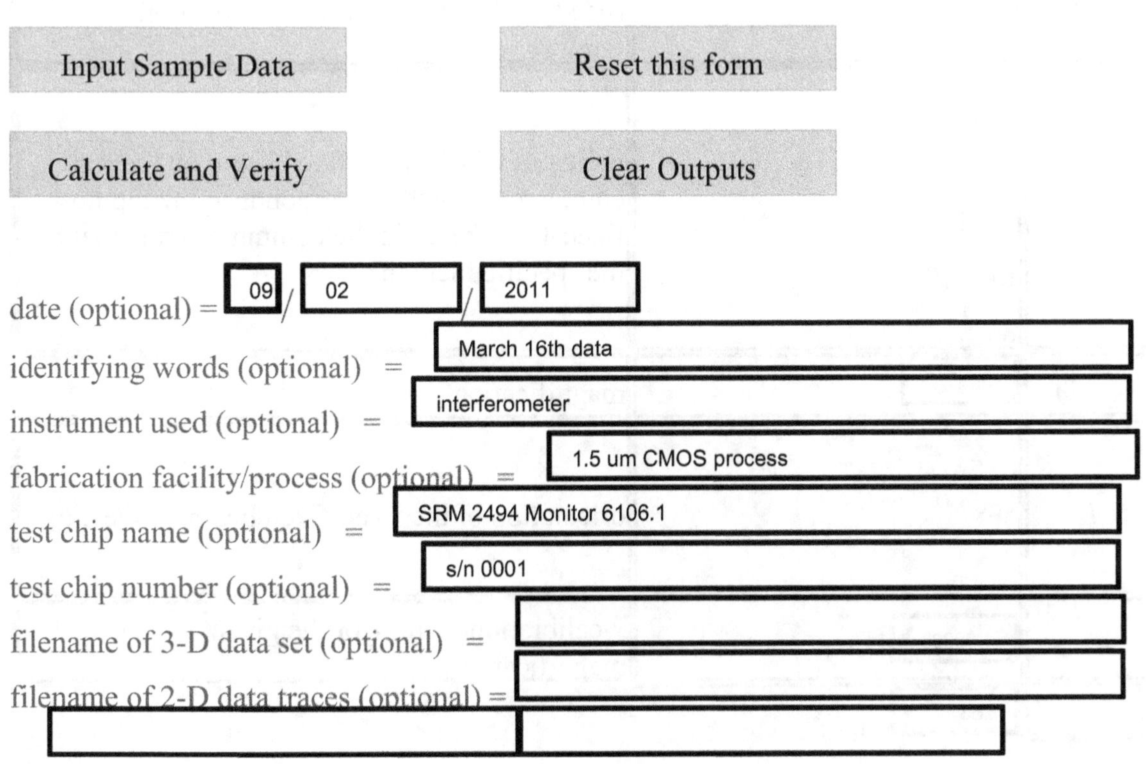

Input Sample Data	Reset this form
Calculate and Verify	Clear Outputs

date (optional) = 09 / 02 / 2011

identifying words (optional) = March 16th data

instrument used (optional) = interferometer

fabrication facility/process (optional) = 1.5 um CMOS process

test chip name (optional) = SRM 2494 Monitor 6106.1

test chip number (optional) = s/n 0001

filename of 3-D data set (optional) =

filename of 2-D data traces (optional) =

comments (optional) =

Table 1 - Preliminary ESTIMATES			Description
1	$temp =$	21.1 °C	temperature during measurement (should be held constant)
2	$relative\ humidity =$	51 %	relative humidity during measurement (if not known, enter -1)
3	$material =$	○ Poly1 ○ Poly2 ○ stacked Poly1 and Poly2 ◉ SiO_2 ○ SiC-2 ○ SiC-3 ○ other	material
4	$t =$	2.5846 µm	beam thickness
5	$design\ length =$	200 µm	design length
6	$design\ width =$	40 µm	design width (needed for test structure identification purposes only)
7	$which\ beam?$	◉ first ○ second ○ third ○ fourth ○ other	indicates which fixed-fixed beam on the test chip, where "first" corresponds to the topmost fixed-fixed beam in the column or array that has the specified length?
8	$magnification =$	25 ×	magnification
9	$orientation =$	◉ 0° ○ 90° ○ other	orientation of the fixed-fixed beam on the chip
10	$cal_x =$	1.00293	x-calibration factor (for the given magnification)
11	$ruler_x =$	253 µm	